The Feathery Tribe

Robert Ridgway, ca. 1875.
Smithsonian Institution, Division of Birds.

The Feathery Tribe

Robert Ridgway and the
Modern Study of Birds

DANIEL LEWIS

Yale

UNIVERSITY PRESS

NEW HAVEN & LONDON

Published with assistance from the foundation established in memory of
Amasa Stone Mather of the Class of 1907, Yale College.

Yale University Press books may be purchased in quantity for
educational, business, or promotional use. For information,
please e-mail sales.press@yale.edu (US office) or
sales@yaleup.co.uk (UK office).

Designed by James J. Johnson
Set in Fairfield type by Tseng Information Systems, Inc.
Printed in the United States of America.

Library of Congress Cataloging-in-Publication Data
Lewis, Daniel, 1959–
The feathery tribe : Robert Ridgway and the modern study of birds / Daniel Lewis.
p.   cm.
Includes bibliographical references and index.
ISBN 978-0-300-17552-3 (hardback)
1. Ridgway, Robert, 1850–1929. 2. Ornithologists — United States —
Biography. 3. Ornithology. I. Title.
QL31.R47L49 2012
598.092 — dc23
[B]   2011050618

A catalogue record for this book is available from the British Library.

This paper meets the requirements of ANSI/NISO Z39.48–1992 (Permanence of Paper).

10 9 8 7 6 5 4 3 2 1

For Pamela

Half angel
Half bird

Contents

Preface

"There is . . . a considerable proportion of the community at large who are strongly prejudiced against the 'scientists,' both as a body and as individuals," complained Smithsonian curator of birds Robert Ridgway in 1906 to his close friend and counterpart William Brewster of the Harvard Museum of Comparative Zoology. It was true that scientists—and specifically, the changes they were making to scientific practices at the end of the nineteenth century—were baffling and annoying to members of the public who paid any attention to their work. Ornithologists were changing many long-established and much-loved names of birds and talking about them, not in terms of their habits or life histories, as they had done for centuries, but in terms of confusing notions like systematics, intergradation of species, and nomenclatural reform—things that seemed to have more to do with being inside a building studying birds than with being out in nature looking at them. "I am a field, not *Closet* ornithologist," scoffed Kansas naturalist Nathaniel Goss in 1883, writing to Joel Allen, president of the month-old American Ornithologists' Union (AOU), which had as a primary goal the untangling of issues of systematics and nomenclature.

This book is about what it meant to be a professional studying birds in the last quarter of the nineteenth century, how a professional class emerged, what it looked like, what roles amateurs played, and how these changes led to the science of ornithology as we practice it today. Part of the challenge in uncovering these meanings is defining just what constitutes the "professional" and in describing and understanding its counterpart, the "amateur"—and in fact, determining whether those are even meaningful categories of analysis at all. They prove to be extremely useful, and not least because the participants usually tried to define themselves in those terms.

The mix of attitudes about status was complex. For instance, some amateurs, in contrast to Goss's view, were willing to accept that there was something inferior in their status, or at least were perfectly happy with it. Writing to Allen, one Canadian amateur remarked, "Please do not consider it necessary to apologize for using the term 'amateur' to me. I am designedly and confessedly an amateur and am ready to share in whatever obloquy the name carries." This self-described amateur, Montague Chamberlain, was like many others in his lack of desire to be something he felt he was not. Some amateurs groveled while writing to what they considered to be their betters. "[Your book] is not meant for ignoramuses like me but for experts," wrote "D.R." from Malvern Wells in England to ornithologist Henry Eliot Howard on the publication of his two-volume work on British warblers. British correspondents, perhaps owing to centuries-old notions of class, were often quite submissive. "I hope this letter will not be a great nuisance to you, if so and if you are, as I know you are, very busy put it in your W.P.B. [waste paper basket]!" one woman wrote to a prominent British bird author, practically begging him to trash her letter.

Entangling matters, amateurs were not simply have-nots in science. Many had considerable wealth, both inherited and earned, in contrast to the relatively poorly paid professionals. This allowed some amateurs the full-time luxury of collecting and studying birds. Some, like Goss, were very proud of their status and disdained what they described as "technical" ornithology. Others, like the ones described above, were deeply insecure, despite their often-formidable knowledge of birds and their occasional wealth and reach in science.

A broad literature exists that opines when modern science formed as a profession, what it looked like, and how it operated during its different phases. However, much of it has been written in the domain of sociology, not the history of science, and what does exist is sorely out of date: there have been almost no studies in the past quarter-century that explicitly study the issue of what constituted a professional or an amateur in any comprehensive way. Historian of biology Paul Farber has noted that although some very careful studies have been undertaken to distinguish just what constituted a profession among scientists in the nineteenth century, relatively little is understood today. "Although many historians, sociologists, and scientists agree that a salient feature of nineteenth-century science was its newly acquired professionalism," he notes, "the subject is one of which at present there is little understanding. There is no basic consensus concerning its causes, development, or even on the definitions of the basic terms with which to discuss it." Other historians have sounded an even stronger call to arms to better understand professionalization. Writing from a British perspective in 2001, Adrian Desmond wrote that "getting an embedded, localized definition of 'professional' in Victorian times is itself becoming *the* problematic in the history of biology."

I have resisted providing crisp definitions of "scientist," "professional," and "amateur" at the outset. In some ways, the entire book is a reaching toward descriptions rather than definitions because—as historian Paul Lucier has noted—pat characterizations of the terms are problematic. The actors themselves had contradictory notions of just what respectively constituted a scientist, a professional, and an amateur, which adds to the terminological difficulties. We thus have lacked a grasp of key aspects of these concepts, as understood in their era, and a gap in our historical understanding of the professionalization of science has remained.

In this book I address this deficit by using the historic study of birds to point to three new markers that distinguished professionals from amateurs. The characteristics I describe, and the subsequent creation of the modern zoologist, were set into motion in part by the arrival and subsequent wide acceptance of key aspects of Charles Darwin's theory of evolution. The publication of *On the Origin of Species* in 1859 and *The Descent of Man* in 1871 gave a pair of combination punches to the body of science that had for centuries been commingled with religious explanations and sentiment. Affinities among species had certainly been long recognized and described as a central part of classification efforts, but as British naturalist George James Allman noted in 1874, "It is only quite lately that a new significance has been given [to affinities between zoological objects] by the assumption that they may indicate something more than simple agreement with a common plan—that they may be derived by inheritance from a common ancestral form, and that they therefore afford evidence of a true blood relationship between the organisms presenting them."

The three markers that are threaded through this book were key influences, but often diffused. First, Darwinian evolutionary concepts finally gave scientists the ability to create accurate, nonspeculative classification schemas that brought the study of life into sharper focus. The relationships among groups of birds became a critical issue. Museums' long runs of series of the same birds became vital: for they allowed scientists to identify subspecies, catching birds in the act of evolving, as it were—making the transition from one form to another.

Second, the tightly related issues of accountability and authority came into play in the last quarter of the nineteenth century. If God and the Bible could not be used as guiding principles to order the natural world and scientific work, what could? The arrival of certain kinds of authority systems in the 1880s helped to provide a key definition of the professional. Not everyone accepted these authority systems, and the responses of individuals to them spoke volumes about how they identified themselves ideologically. Did scientists hew to a particular kind of authority—government commissions, checklists, organizing bodies, or classification systems imparted from above—or did they resist them and attempt to organize their attitudes and work in the natural sciences around different principles? Changing notions of accountability and responsibility toward a parent institution, the government, or simply the public correspondingly changed procedural aspects of dealing with collections of birds. Classification systems were in the air. Melvil Dewey's radical and proprietary new organizational schema for books, created in 1876 and pressed into increasingly wider service, attempted to organize all knowledge into ten top-level categories. Dewey quickly gained currency in libraries around the world, and another popular classification scheme for books, the Cutter Expansive Classification system, followed in 1880. Scientists working in institutions with their own libraries would have immediately noticed the changes taking place in the way printed works were being classified and placed in a hierarchy.

Third, one notable effect of Darwin's theory on biologists, and especially on ornithology, was a sea change in the use of language. The influence of Darwinian evolutionary ideas on language was much more diffused than the other two markers, because other elements also strongly influenced linguistic changes, such as the rise of the modern university and the creation of professional societies that wanted their own vocabularies. Language was a particularly visible element distinguishing those who identified themselves as professionals or as amateurs, and it would especially affect their publications. Professionals used increasingly technical terms and specialized vocabularies to discuss their work with birds.

On the other hand, amateurs fought to retain language that a much wider readership could grasp and appreciate. If they published works on the subject of their affection, their works were usually just that: filled with affection and sentimentality, often marked by lively readability, rather than the drier language of supposedly objective science. Amateur writers also spanned a very large language gamut. Their work ranged from technically accurate descriptions of birds at one end of the spectrum to literary efforts that anthropomorphized birds at the other.

These three elements all reinforced one another. Classification schemas required more detailed and technical language and a consistent hierarchy of information. Distinguishing fine differences among subspecies called for more precise terminology, and careful use of anatomical terms became essential. Classification schemas also implied authority: Who made the rules for them, and whose systems of classification or nomenclature would become the standard? When was deviation from these new norms acceptable, or was it? And individuals' accountability to other elements or organizations meant that discussion of birds was less and less about the rhapsody of nature and more about fitting into a shared set of technical criteria.

The relations among the actual players—the amateurs and professionals themselves—were complex. Frustration, division, and arrogance often flowed from the scientists toward the public. "We *are* everything and everybody in this country worth anything in ornith[ology]," crowed ornithologist and physician Elliott Coues to Joel Allen in 1883, referring to the newly formed American Ornithologists' Union, which had just twenty-one founding members. However, ornithologists working at full-time salaried positions in natural history museums or on behalf of professional organizations such as the AOU were also highly unsure at times about their own abilities—partly because their tasks were not all strictly ornithological, and they often veered into other areas in which they were less competent. The AOU's first treasurer, Clinton Hart Merriam, who would become the first chief of the US Department of Agriculture's Division of Economic Ornithology, had misgivings about his basic accounting skills, despite his official status overseeing the AOU's funds. "I am a wretched book-keeper and always get muddled up in simple accounts, and am afraid of doing some awkward blunder th[a]t will place me in a bad light before the Union—for I know that the average 'Yankee' is sharp on this sort of thing," Merriam complained to Allen in 1884. Professionals in the bird world thus had a relative position. At times, they were the alpha males in their field—they led the way in the areas I reinforce as key older markers of professionalism that we still recognize today: publications, full-time salaried positions, and a network of colleagues, among other things. But they were also made to feel like amateurs themselves at times, whether by physicists encroaching onto their territory or by famous people in other fields whose credentials as successful people often trumped their own.

I have used the terms "scientist" and "professional" somewhat interchangeably, because they were often used vaguely and contradictorily by the actors themselves. In this, I part ways with the supposition of the most recent scholarship on the professionalization of the sciences, which argues that "professionals were men of science who engaged in commercial relations with private enterprises and took fees for their services," whereas scientists "rejected such commercial work and feared the corrupting influences of cash and capitalism." The evidence I have found, at least from the last quarter of the nineteenth century, shows that the two terms were used by participants with nowhere near that kind of precision, and often in contradictory ways.

In laying out how the three elements—classification, language, and accountability—differentiated amateurs from professionals during the last quarter of the nineteenth century, and why it mattered, I have used Robert Ridgway, the Smithsonian's first curator of birds, as the focal point for this study. He was among the world's most famous scientists at the end of the nineteenth century but has slipped into relative obscurity, and his life has long deserved a fuller accounting. Ridgway certainly was not responsible for single-handedly leading the way toward the ornithology of the twentieth and twenty-first centuries. No one person was; the transition involved the labors of perhaps two dozen people—the "feathery tribe" of the title of this work. They worked sometimes in concert and sometimes in bitter and drawn-out conflict, driven not just by a desire to advance scientific knowledge but at least as much by ego, pride, honor, insecurity, religion, altruism, and a host of other often-clashing sensibilities. Some of these men were trained as doctors; many had no formal education in scientific endeavor; a number had clerical training; and others had nothing but enthusiasm and persistent drive. Still others were salaried bird workers in the United States and abroad. This latter group included Joel Asaph Allen at Harvard's Museum of Comparative Zoology and later at the American Museum of Natural History in New York; William Brewster of the Museum of Comparative Zoology; Charles B. Cory at the Field Museum in Chicago; Joseph Grinnell of the Museum of Vertebrate Zoology at the University of California, Berkeley; Henry W. Henshaw of the Bureau of American Ethnology and the US Biological Survey; Charles A. Keeler at the Academy of Sciences in San Francisco; Joseph Wade, publisher of the *Ornithologist and Oologist*; José Castulo Zeledón at the Museo Nacional de Costa Rica; Philip Lutley Sclater of the British Museum; and a number of others. But by virtue of Ridgway's position as the first curator of birds at the Smithsonian Institution, mandated by Congress as the National Museum, and by the imprimatur of his position and his connections with the most influential men of science and natural history of his day, he played the greatest role among these men in the formation of American ornithology and the professionalization of the field. Ridgway was a key transitional figure between the leisurely and aesthetic, sensory-driven pursuit of bird collecting and the later study of the living bird. His influence was enormous, and his story is overdue.

Ridgway also advanced the study of birds through more than five hundred articles totaling three thousand pages, as well as twenty-three books and other monographs totaling more than thirteen thousand pages. His magnum opus, *The Birds of North and Middle America*—eight volumes written by him and three by a successor working largely from his notes— stood as the definitive systematic work on birds from Canada through Central America for decades after his death in 1929. His ideas about subspecies and trinomials were a key part of the acceptance of evolution as a driving force in American zoology. In addition, he was a skilled illustrator whose representations of birds saw wide distribution in print.

Although American practice is the primary focus of this work, there is also an international component. Ornithological discourse never took place in a vacuum—60 percent of bird species are migratory and thus are often found intercontinentally—and their study and description has always relied on international efforts. British practices parted ways with American practices in the important realm of trinomials, with England a much more reluctant convert to trinomials than the United States. This has been a longstanding puzzle, because trinomials were a particularly potent marker of evolutionary change and divergence, and Great Britain had been the first country to be introduced to the topic via Darwin. My analysis resolves this apparent contradiction, and it relates largely to the differences in the size, terrain, and climate between British and American geography. Ridgway also had influential transactions with important ornithologists in other countries, such as his friend and collaborator José Zeledón of Costa Rica. Consideration of these factors provides a larger international context for the work being done by Ridgway and affords us a fuller picture of international influences on the evolution of American ornithological practice.

It is my hope that beyond an audience of scholars and scientists, this book will also be useful to modern birdwatchers and nature lovers who want a better understanding of the historical underpinnings of their passions. Modern biology is about the living bird, but the systematics work done by Ridgway and his ilk—the labors by which we understand bird ranges, subspecies, taxonomy, and other vital elements of birds—are the foundation on which modern ornithology has been constructed.

One of my personal predilections about scholarship is that historical arguments aren't worth a damn without the backing of primary sources as critical evidence. Throughout each chapter, I rely heavily on the actual words of the players themselves—extracts from correspondence, memos, marginalia, and other original writings of ornithologists and their circle of collectors, supporters, detractors, friends, and family—as a way of illustrating many points. Allowing the members of this tribe to speak in their own voices reveals a rich store of quotidian detail about their lives, consisting of their daily enthusiasms, worries, family and professional concerns, and hopes and dreams.

There is still a great deal more to be researched and written about every aspect covered in this book. I hope that beyond its contributions to the subjects at hand, it will motivate others to delve even more deeply into these subjects and to hold up different prisms to the people, places, and themes in order to reveal them even more fully and in greater dimension.

Acknowledgments

If gratefulness is a vessel, then mine is an outsized chalice filled to the brim. The extraordinary support and encouragement I've received from many people have touched me deeply and made this book a reality.

Working as a curator at the Huntington Library, one of the world's great research institutions, has given me access to rich source material of all kinds, along with the country's finest archival and curatorial colleagues, and I've mined these sources and contacts vigorously in my quest for the fullest possible version of the events that support my arguments about the nature of the study of birds in the last quarter of the nineteenth century. My greatest institutional debt by far is thus to the Huntington for numerous kinds of support at every level.

Among the Huntington staff who supported this book, my heartfelt thanks to two people in particular: my boss and mentor David Zeidberg, Director of the Library, and the inimitable Mary Robertson, my predecessor as Chief Curator of Manuscripts. David and Mary both enthusiastically supported three fellowships over the years, each of which involved time out of the office and away from my curatorial duties. I'm also grateful to Peter Blodgett, Curator of Western History; Alan Jutzi, Chief Curator of Rare Books, and his paging staff; the Reader Services staff (but especially Christopher Adde, who helped me pore over almost illegible microfilm copies of Illinois newspapers); and others in those departments. Melissa Lindberg compiled useful bibliographic details.

I have especially benefitted from conversations with readers at the Huntington, some of the country's finest scholars in a variety of fields, who have passed in and out of my office, the hallways, and the coffee cart and contributed to my understanding of the subject. Kathy Olesko of Georgetown University, visiting for a year as the Dibner Senior Fellow, gave me useful advice about the writing process. Bruce Moran of the University of Nevada at Reno, another Dibner Senior Fellow, read the entire manuscript and was especially enthusiastic about my book when I needed it most. He also encouraged me to think much more about what practice in science meant during the nineteenth century and why it was important. Diane Riska-Taylor, former professional ornithologist and now a historian of science, read the manuscript and helped make it better in a number of ways. Jane Smith of Northwestern University freely shared ideas about professionals and amateurs in the history of botany and even did some research in Illinois on Ridgway family members on my behalf after leaving the Huntington.

Several interns helped with research aspects ranging from the factoidally minute to the substantial. These included Melissa Ming-Hwei Lo, a doctoral student in Harvard's History of Science program who interned for me in the summer of 2009. She unearthed useful details in the secondary literature on my behalf and, in the course of discussions, led me to a better understanding of the role of professionalization in the sciences. Another Harvard history of science intern, Clare Moran, provided substantial research support during the same summer. A passel of my grad students also helped with research aspects over the years, most notably Kathryn Wolford, Roxi Opris, and Cassandra Ticer of Claremont Graduate University. Emma Wartzman of Oberlin College provided tremendous last-minute help with the details of Ridgway's bibliography that comprises the book's appendix.

The research for this book has taken me to Montreal, Cambridge (Massachusetts), Oxford, London, Berkeley, and Washington, DC. Along the way, many people gave me coffee and helped me. At the Smithsonian's Division of Birds, my thanks to Storrs Olson, who served as one of my fellowship sponsors and took a continued interest in the project over the years; James Dean, the Collection Manager; and Christina Gebhard, who provided me with data and wonderful images. Other Smithsonian staff helped, most notably Pamela Henson, the Director of the Institutional History Division, whose enthusiasm and deep expertise on numerous aspects of the book were material assets; Leslie Overstreet of the Cullman Library; Rick Stamm, the Keeper of the Castle (if that's not the world's coolest official job title, I don't know what is); and the archivists at the Smithsonian Institution Archives. The warm collegiality of the entire Archives staff sustained me through a very cold Washington winter and a brutally humid summer in two different years. My Huntington predecessor Ron Brashear and his wife, Madeline, also housed and fed me during both Smithsonian fellowships. At the Honnold Library of the Claremont Colleges, Carrie Marsh and Jennifer Bidwell of Special Collections provided nearly instantaneous service and support at every turn. At Utah State University in Orem, Utah, my thanks to Stephen Sturgeon, Manuscripts Curator, for assistance. At the Blacker-Wood Library at McGill University in Montreal, with its relatively little-known but jaw-dropping collection of original bird artwork and ornithologists' correspondence, Eleanor MacLean, the Biology Liaison Librarian, went to great lengths on my behalf. At the Harvard Museum of Comparative Zoology Archives, the Special Collections staff was extremely helpful. In Mount Carmel, Illinois, Ridgway's hometown, Claudia Dant of the Wabash County Museum sent me several useful items, as did Paul Wirth, the President of the Richland Heritage Museum. At the Libraries of the University of Southern California, archivist and manuscripts librarian Claude Zachary was a researcher's dream, immediately providing images, text, and other material.

The British archives I used during my research proved invaluable. While on a fellowship at Oxford University, it was a treat to spend a month at the Alexander Library, then located at the Edward Grey Institute of Field Ornithology. Not only is the Alexander Library—now part of the Bodleian Library system—the best ornithology library in Europe, but the staff is as outstanding as the books. My heartfelt thanks to Alexander Librarian Linda Birch, as well as Mike Wilson, who has cheerfully answered erratic queries from me for years. At the Natural History Museum in London, Assistant Archivist Polly Tucker helped me navigate the collections with British aplomb.

At the Los Angeles County Museum of Natural History, Collections Manager Kimball Garrett, who is also one of the country's finest field ornithologists, spent a number of days spread over six months patiently teaching me how to make study skins, as well as reading and commenting on the chapter on study collections. Lloyd Kiff of the Peregrine Fund took a great interest in Ridgway and provided much encouragement and a number of useful contacts. My steadfast friend Bill Wheaton, a scientist and geographer who has traveled the planet with me in search of birds for many years and on many other adventures besides, asked important questions and listened for years to my endless chatter about the subject.

Numerous people were incredibly generous with the results of their research on related topics. Kristin Johnson, at the time a National Science Foundation post-doc at Arizona State University, sent me her MA thesis on the history of British ornithology. Paul Farber of Oregon State University offered useful advice, as did Jared Diamond. Matthew Godfrey sent me his MA thesis on Ridgway. Mike Dooley, Smithsonian Fellow, offered up transcripts of Ridgway correspondence he'd found elsewhere in the course of his research, as did Betsy Mendelsohn of the University of Chicago. The late Marianne Ainley and her colleague Keir Sterling provided me with their unpublished manuscript of the history of the American Ornithologists' Union. Eric Ward, a doctoral student in history at the University of Missouri–Kansas City currently working toward a dissertation on Ridgway and the Smithsonian, provided enthusiasm and motivation.

The late Ernst Mayr, perhaps the twentieth century's most important thinker on evolution, also played a role. I had originally conceived of this work as a general popular biography of Ridgway. In 2003 I wrote to him asking for some suggestions about such a work. "I don't know what to tell you about Ridgway," Mayr's long handwritten response began, which was intriguing, because he had written the brief but authoritative account of Robert Ridgway's life for the *Dictionary of Scientific Biography* in 1975 and was well acquainted with Ridgway's work and his influence on the field. He noted that he had leavened his comments about Ridgway in his published account in order to be more "neutral" and intimated that the story of Ridgway's life and influence was considerably more complicated than he had let on in print. Thus began an intermittent correspondence between us that proved invaluable for Mayr's insights on a number of related points until his passing in 2005 and motivated me to expand the book beyond a general biography.

At Yale University Press, my editor, Jean Thomson Black, enthusiastically and skillfully untangled the often-torturous details involved in birthing the manuscript, was always a pleasure to work with, and has my endless gratitude. Michelle Komie, senior editor for art and architecture at the Press, also gave priceless advice. Laura Jones Dooley's thorough copyediting helped clarify and regularize countless elements in the manuscript. Two anonymous readers provided extensive critiques and suggestions, which have greatly improved the manuscript.

Throughout, my family has been a shining beacon of love and great good humor: my wife, Pam, flat-out the funniest person I've ever known, whose fierce intellect remains only slightly dimmed from birthing and raising our two children; and those children, Paxton Jay Lewis and Iris Palila Lewis, who have saved my life and continue to make it worth living every day. Having birdy middle names was somehow a necessity.

Transcription Notes

Because I quote so heavily from correspondence, a word about my transcriptions seems useful. Authors of handwritten letters often use the epistolary version of squawks, yelps, half-thoughts, and marks that are often untranscribable. For instance, the ways in which writers moved from sentence to sentence are myriad, ranging from double equal signs to slash marks to jottings of their own invention that have no typographical equivalent. In contested portions, I've simply made a sentence break where it seemed most logical. Writers also tried to emphasize things in different ways. To complicate matters for those attempting to channel them, many authors sometimes underlined the middle of a word rather than the whole word; unless it was clear that they meant to emphasize just part of the word, I've interpreted those partial marks as representing emphasis on the entire word and have treated them accordingly as italic text. Sometimes writers would underline a sentence for emphasis, only to reach the end of the line midsentence and continue on the next line with no underline. And sometimes writers would double- and triple-underline words for particular emphasis. These have all been rendered into less evocative but probably less distracting and more consistent italics. All obvious punctuation errors—for example, an opening parenthesis mark given without a corresponding closing parenthesis mark—have been silently corrected. I have converted most ampersands within quotations to "and" and made several other similar modest changes. Spelling is rigorously presented as written, and I've thus dispensed in almost every case with what would be a distracting flurry of "sic" notations for misspelled words. All items in italics are as written, with no emphasis added.

CHAPTER ONE

The Making of a Bird Man

I would not flatter you but I have always thought you had
the capacity of acquiring knowledge of the right kind, and
now I think the way is opened for you to make your mark
in *your* profession.

FANNIE GUNN to her sixteen-year-old nephew
Robert Ridgway, March 10, 1867

ROBERT RIDGWAY, who between 1875 and 1925 was one of the
best-known ornithologists in the world, is largely forgotten today.
Through a series of circumstances related to scientific authority,
the use of language, a bond with a senior scientist, access to a network of
other like-minded colleagues, and a few lucky breaks, Ridgway began his
career under the immensely influential imprimatur of the Smithsonian
Institution. Although he was at times painfully shy, the tremendously talented and driven Ridgway held a position that afforded him the credential
of full-time employment in a bird-related field. This entitled him to prestigious memberships, publication advantages, almost unlimited access to
bird study collections from around the world, and invitations on important field expeditions. These benefits, along with his native abilities, let
him wield an outsized influence. Ridgway became a giant in the world of
avian systematics, taxonomy, nomenclature, writing, and publishing for
nearly half a century. In the course of those efforts, he and his colleagues
both worked with and struggled against one another, as well as with and
against amateurs, in trying to create a number of standards for ornithological practice. In doing so, they helped to redefine just what constituted
a professional scientist.

To understand these dynamics, we need to look to events taking place
in the 1840s: most significantly, a dawning understanding about the mu-
tability of species. Brought into the influential climate of drawing rooms,
meeting halls, and newspapers through the publication in 1844 of Robert
Chambers's *Vestiges of Creation*, the topic of evolution (or "transmuta-

tion," as it was then known) began to enter the public discourse. Read by Abraham Lincoln, Queen Victoria, and tens of thousands of others, this clever book untethered the discussion of creation and change in the natural world from the oxcart of biblical scripture and hooked it up instead to the locomotive of the scientific underpinnings of the creation of species. The public's exposure to the subject grew rapidly, and the thinking of scientists almost everywhere was increasingly influenced by the publication in November 1859 of Charles Darwin's opus *On the Origin of Species,* which was promptly published in America in early 1860, passing through four American printings that year. Darwin's work was much more rigorous than Chambers's book, and in absorbing it, scientists began to take more seriously their suspicions and dawning realizations about geographical variations of species. No longer could they seriously consider birds as immutable, formed into their current precise shapes and colors by God, never to have a feather out of place, so to speak. Museums had long held many multiples in their collections, but their collecting efforts changed after Darwin as it became clearer that long series of the same birds—or similar birds—provided the opportunity to compare hundreds of examples of the same species in a way that would lead ornithologists to a variety of new understandings about birds that, in the words of one scholar, were "caught in the act of evolving."[1]

It was not just the existence of these long series, however, but their availability to researchers from other museums that gave these runs of the same birds their real power. Classification was not a new activity; it had been part of the triad of cataloging, description, and classification of zoological specimens for many decades, even centuries. But evolution began to provide the answer to the "why" of the natural world rather than the "what" of classification, and comparisons began to flourish because evolution provided a rationale for understanding differences and similarities. Bird specimens could be compared for geographic variability, so that, say, two hundred specimens of the same sparrow on the East Coast of the United States, when compared with another hundred specimens of that sparrow at another museum in the Midwest or West, would often readily reveal differences, leading to new subspecies, or even new species. Because Ridgway oversaw the country's largest series of birds, he was in an excellent position to provide these comparative studies, both nationally and internationally. He exchanged series of birds with countless colleagues around the world, and his publications brought discoveries gleaned from these series to a wider audience that was just beginning to absorb their implications.

Growing up in Illinois

Robert Ridgway was born on July 2, 1850, in Mount Carmel, Illinois, then a town of 935 souls, about 240 miles south of Chicago.[2] Some 159 families were counted in the 1850 census, and the town was moderately prosperous from its agricultural production. His parents, David and Henrietta Ridgway (née Reed), were Quakers and in their late twenties when they married in 1849. Henrietta Reed's maternal grandparents had emigrated from Gaithersburg, Maryland, to Mansfield, Ohio, where she and her eight siblings were born, and later from there to Wabash County, Illinois. David Ridgway, born in Harrisburg, Pennsylvania, in 1819, had come to the Mount Carmel region several years earlier seeking his fortune. Mount Carmel sat on the western side of the Wabash River, the boundary for much of the divide between Illinois and Indiana. The recent construction of the Wabash and Erie Canal—at nearly five hundred miles the longest canal ever built in North America—offered the promise of growth for the region's economy and offered Mount Carmel the potential for greater access to markets farther away. David Ridgway had high hopes for a successful business enterprise in Mount Carmel. He and his brother, William, quickly established a partnership in a drugstore business, and David and Henrietta began raising a large family. The town grew rapidly, and by 1860, the number of families resident in the town had grown by 70 percent.[3]

Robert was the eldest of ten children, six boys and four girls. Precocious in his careful use of language as well as in his artistic ability, young Robbie stood out immediately. The other member of the Ridgway clan to reach national prominence was his second-born brother, John Livzey Ridgway, an accomplished illustrator and artist best known for such popular bird artwork as the bird images on Singer Sewing Machine promotional cards distributed by the thousands in the early twentieth century. Another brother, Joseph, became a taxidermist and was employed by the Iowa State Museum and the Field Museum in Chicago.[4] Although not as talented an artist as his brother John, Robert Ridgway was nevertheless an accomplished illustrator. He followed squarely in the nineteenth-century tradition of the artist-naturalist, and many hundreds of his illustrations appear in his writings, often unattributed, as well as in a range of technical and popular works.

By the relatively scant accounts and correspondence available, Ridgway's childhood was reasonably happy, although his family was very poor throughout most of his early years. Various family members also suffered from health problems, and Robert often took to the outdoors—perhaps as an escape from family stresses, or perhaps simply because of the sheer magnetic pull in that direction for a boy being raised in a spectacular wilderness. His parents also encouraged his outdoor activities; armed with a gun, Robert was safe from most threats, and the gentle rolling terrain of rural southeastern Illinois posed few risks. The combination of a bucolic natural setting and two parents who stimulated an interest in natural history for their son proved to be a happy confluence of factors in Robert's development of an interest in birds. His uncle William was an additional influence, urging Robert to draw birds from life, and his aunt Fannie Gunn also provided crucial encouragement.

Initially, the bulk of Ridgway's interest in birds was stimulated by his time hunting. Even the poorest families owned at least one rifle, used for a variety of purposes ranging from obtaining food to keeping pest animals at bay to self-defense. Although he was a bookish young man from the start, the role of the rifle in ornithology—the study of birds being largely initiated by sighting carefully down the barrel of a gun—played an important if somewhat awkward role in Robert's youthful endeavors, as it did for virtually all naturalists of his era. His father's only recreation was hunting and walking in the woods (activities prohibited on Sundays, which were free of hunting or other chores due to the family's strict observance of the Sabbath). Father and son would spend a great deal of time together at both of these outdoor activities, sometimes going out at 3:00 a.m. to hunt wild turkeys. By several accounts young Robert was neither an expert shot nor initially comfortable around a rifle. He would often tamp down the charge into his father's gun with a metal rod, and the rod would jam inextricably into the barrel. The only solution would be to point the gun skyward, fire the gun to unclog it, and then track down the rod in order to try again. "Recollections of these difficulties bring to mind many incidents connected with my early shooting experience," he noted wryly near the end of his life.[5]

One afternoon, alone in the woods with his father's single-barreled shotgun, Ridgway heard a Barred Owl in the distance. A good mimic, Ridgway hooted back, and the owl shortly appeared in a nearby tree, landing amid a dozen Blue Jays. He aimed and fired, but to his dismay, the percussion cap went off without firing the shell, and the owl flew away, accompanied by most of the jays. He removed the gun from his shoulder, it suddenly fired belatedly of its own accord, and a single jay fell from the tree.

These sorts of vaudeville acts in the wilderness, if they were ever made known to his father, would probably have been disappointing, because his father seemed to have high hopes for a rigorous physical life for young Robert. When he finally got his own gun, it was a touching present from his father, shrewdly economical and imbued with a history of its own. In the summer of 1858 the steamboat *Kate Sarchet* sank in the Wabash River, and when salvagers rescued her cargo, it included a rifle. David Ridgway bought the badly rusted weapon for a fraction of a dollar, had the barrel shortened and the gun cleaned, derusted, rebored, outfitted with new hardware, and finished with a new wooden stock of wild cherry. Along with the gift of the gun came a powder horn, a leather shot belt, and other requisite accessories.

Although he would occasionally forget parts of the setup, such as the percussion caps, on his outings, young Ridgway was delighted with his new gun and eventually became more proficient. Hunting played an important role in his life as a budding ornithologist because at the time, the field guide as we know it today did not exist. The lack of portable and comprehensive printed guides was not the only restriction for naturalists interested in studying birds in Ridgway's youth and, indeed, for much of his career: the Galilean-style binoculars of the era were heavy and relatively low powered. The best-quality field glasses typically had a magnification of up to five power, half of that generally in use today, and because of their design, especially at higher magnifications, had a restricted field of view and let in relatively little light. Nor had the camera advanced suitably for photographing fast-moving objects in black and white, let alone color, out of doors. So the gun ruled supreme as the natural tool for studying birds in the nineteenth century and well into the twentieth century. It also allowed for much more leisurely hands-on examination of specimens, leaving future ornithologists with a three-dimensional record of the bird. Ridgway's visual skills were exceptional, and he quickly became adept at identifying birds on the wing. His keen eyesight would also later serve him well as a taxonomist. Successful taxonomic work requires excellent visual pattern recognition and analysis skills, especially in an era before one could easily photograph a specimen at one institution and compare that photo with another similar bird.

Robert mixed his own gunpowder in his father's store, and because he had already begun to draw and paint birds, he made his own paints as well. As a pharmacist's son, he was comfortable grinding his own pigments and mixing watercolors. Although frustrated with the quality of those paints and, to his embarrassment, sometimes forced to use toy paint, these pharmaceutical skills served him well in his later work on his books on color standards. He also collected and drew birds' nests, eggs, leaves, and anything else that seemed interesting from around the woods of his hometown. Mount Carmel in the early and mid-1860s was a far wilder place than it would be only twenty years later, before corn and other crops led to the destruction of thousands of acres of old-growth forest. The deep woods and larger expanses of land meant that Ridgway could see many species in different habitats, and this richness meant that he had a great deal to say and ask of Spencer Fullerton Baird of the Smithsonian Institution, with whom he would enter into written conversation for three years before the two met in person.

As the firstborn child, Robert bore the brunt of chores until his siblings grew old enough to help. Each Saturday he cut Sunday's firewood, a task he performed as quickly as possible to leave time for hunting in the afternoon. The household's Sabbath restrictions even prohibited whistling. This one stricture may have been enforced for more than the sanctity of not laboring on Sunday. Ridgway was apparently an annoyingly enthusiastic lifelong whistler, and colleague Elliott Coues would later reproach him for whistling when the two shared an office. "You're quite a whistler, aren't you?" Coues asked him one day, as Ridgway recounted later. "Feeling somewhat flattered, I replied, 'well, I can whistle sometimes,' but my vanity was crushed when he answered, 'how much will you take to stop!'" For ornithologists of the era, whistling must have gone hand in hand with their work, naturally enough, as they often had to rely on that tool to call birds within shooting or viewing range, and still do.[6]

Music was not a major theme in the Ridgway household generally; his uncle William disapproved of it, and his mother, while not minding "dancing to good music," noted while writing to her twelve-year-old son that "it is the associations that makes the harm." His interest in music came early, however, and at age nine or ten, Robert served as a drummer in a small three-person band formed with two cousins, playing patriotic songs while marching at recruitment events for the Northern cause at the dawn of the Civil War. This activity generated a well-known photograph of Ridgway, taken by his sister Fannie and widely published during and after his lifetime, in his uniform, "frown and all," as he described it in later life.[7]

His musical talents (or lack thereof) aside, Ridgway's relationship with his parents appears to have been good. He described his mother as "firm in her discipline but very affectionate." Judging from her regular letters to him while he was away on the government-sponsored Fortieth Parallel Survey expedition in 1867 and 1868, however, she was needy and insecure, and she often treated him like a peer, even at a young age. Perhaps this made him grow up more quickly, but it was difficult for him to have less time to be a son.

The family suffered a number of setbacks during Robert's teenage years. David Ridgway, a pharmacist, prospered well enough in the first few years of his business practice in Mount Carmel that he built the first modern business building in the town, a three-story brick-and-iron building that housed his drugstore. But the elder Ridgway, a soft touch, advanced credit and did not press the collection of accounts vigorously enough, and the store suffered heavy financial losses. To add to his woes, the uninsured building was destroyed by fire and upon being rebuilt (well after Robert had left town and was working at the Smithsonian) was partially demolished by a tornado. Fifteen people died in the storm, and more than a hundred businesses and residences were destroyed—a tremendous blow to the town. This tornado also destroyed the Ridgway family residence, along with most of the family's possessions. Seven members of the family were at home at the time. Miraculously, although a surviving photograph shows the house reduced to rubble and one sister suffered minor injuries, no one was seriously hurt.[8] Shaken by these reverses, the family moved to Wheatland, Indiana, about forty miles to the northeast, buying a farm from the sale of the remnants of the pharmacy business and the land on which their house had stood.

Despite the family's struggles, Robert continued his field studies and his interest in birds and natural history. At the precocious age of thirteen, at the urging of the mother of his good friend Lucien Turner, he wrote to US Commissioner of Patents David P. Holloway (having accidentally written to the previous commissioner, William D. Bishop, who served for less than a year between 1859 and 1860—Ridgway must have been operating on outdated information). In his letter, he asked for help identifying a bird he'd seen around Mount Carmel. Although the Patent Office had once owned an extensive natural history collection, the National Museum held most of those collections by the time of Ridgway's letter. Holloway didn't know "a hawk from a handsaw," so he sent the letter along to his colleague Spencer Fullerton Baird, then the assistant secretary of the Smithsonian Institution.[9]

Baird was one of the country's most eminent authorities on natural history, having donated to the museum his substantial collection of bird specimens (many of them from John James Audubon) and having focused most of his scientific life on birds. More than virtually anyone else in midcentury America, Baird promoted natural history as a science. He wrote some of the earliest monographs on the distribution of North American birds.[10] His mentorship and support of young scientists created a ripple effect, laden with the authority of America's national science museum. Baird assumed duties as assistant secretary at the Smithsonian as a young man of twenty-six in 1850, serving in the second-highest post for more than a quarter-century, to 1878, and then as secretary until his death in 1887. His boyhood desire to identify birds and his subsequent writing to Audubon for more information closely mirrored Ridgway's approach, which no doubt struck a chord with him.

Robert's letter to the commissioner of patents finally ended up in Baird's hands in the late spring of 1864. In it, he asked for help in identifying birds and included a full-sized color drawing of a pair of purple finches. Ridgway lacked access to any but the most general works on natural history, but the boy was careful in his written expression. Baird answered a few weeks later praising his artistic ability, identifying the bird, and asking him to continue to correspond with any questions. In an extremely telling comment, he advised the young man that he "must learn the scientific names of the birds, and thus be able to talk and write about them with persons not knowing the English names used in your part of the country."[11] This was advice Robert took to heart and quickly internalized; he began using scientific names as quickly as he could learn them from Baird and from a handful of published works.

In a subsequent letter, encouraged by Baird's kind answer, Robert excitedly tumbled forth a litany of questions: "I will also ask a few questions concerning the names of birds:—is the Imperial-Wood-pecker the large species 17 or 18 inches in length of black plumage dusky bill and scarlet crown and occipital-crest and also with white bands on head and neck."[12] Baird responded, "No. This is Hylotomus pileatus [the Pileated Woodpecker]." Ridgway, who had turned fourteen a month earlier and was working from a specimen in his fast-growing collection, used a clean, detailed descriptive language he must have mimicked from his tiny library. Asking about the Great Crested Flycatcher (now *Chondestes grammacus*), he described his unknown bird as "a large Sparrow with light draby-ash coloured plumage above beneath whitish with a dusky spot on the breast cheek and various marks on the head chestnut and with the tail (except two middle feathers broadly tipped with white; its song is soft and loud resembling in sound that of the indigo bird)." Baird scrawled "Chondestes grammaca" after the description.

In his missive, Baird managed simultaneously to convey helpfulness and busyness, answering Ridgway by crossing out his errors, writing corrections, and generally responding like the former college professor he was. One half expects to see a grade at the bottom of the letter. For his part, Robert made his case as a serious student of natural history, using careful language to describe the birds he'd seen, in a tone Baird would find familiar. He used technical terms in detached language, describing specific anatomic details of local birds with a brisk precision startling in a young teen: "A warbler, olive-green above cheeks yellow throat and anterior part of breast black with more or less of a grey patch belly white, wings black skirted with light blue," he wrote about one specimen. "Bill pea-green; iris yellow; toes yellowish gray; claws dark-horn; horns ¾ of an inch in length, reddish umber edged with white," he wrote about another.[13]

Ridgway's small circle of friends in Mount Carmel must have contributed to his inclinations toward nature. One of his closest childhood pals, Lucien McShann Turner, became one of the country's best-known naturalists. This was quite a coincidence, and one wonders what was in the water in the town, given its miniscule population of barely a thousand people during the Civil War, when Ridgway and Turner were gallivanting around the local countryside. Turner went on to work in Canada for the Smithsonian in the early 1880s and became a leading expert on the natural history of the Arctic. Ridgway and Turner would write one government publication together, *Contributions to the Natural History of Alaska* (1886). Ridgway did the artwork, Turner the text. It must have been a satisfying collaboration for old friends who saw each other rarely. Lucien's brother Granville Turner was Ridgway's other closest friend. He described Granville as having "under his seemingly rude, careless exterior . . . a warm and good heart, such as I know of, in none other—sympathetic, and he has often disclosed his secret troubles to me with moist eyes." Granville Turner doesn't appear to have particularly distinguished himself in later life, although in 1904, while living in Seattle, he patented an umbrella that could be rendered useless to anyone except the rightful owner. Perhaps brawls broke out regularly in the rainy Pacific Northwest at the start of the twentieth century owing to a shortage of bumbershoots. In any case, Ridgway's friends were loyal, and he was loyal to them.[14]

These home relationships helped to constitute Ridgway's personality to some degree, but it was to Baird that he was most firmly anchored by 1865, revealing more details of his life to him as the months passed. "I believe I have never told you of my circumstances and situation in life and think that it would be no harm to tell you," the young naturalist confessed. "I am a poor boy; the oldest of a family of seven children; my father is the junior partner in a drug-store, the slim proceeds of which are scarcely enough to supply us with the necessities of life." With the unknowing sweetness of a boy writing to a much older man about whom he knew little, Ridgway often addressed him as "Dear Friend." Baird, who wrote between three thousand and five thousand letters a year to correspondents around the globe, always found time to write to his young charge, although it was often some weeks between communiqués. Besides questions of identification, their letters often discussed bird illustrations. Ridgway detailed such difficulties as being able to work on his artwork only in the evenings on Saturdays, as well as a lack of proper paints for bird illustrations, complaining that he had to use toy paints rather than artists' paints. His early drawings were done both with these toy paints as well as with colors of his own manufacture which he ground, obtained from his father's drugstore. (Baird, always extraordinarily preoccupied, managed to find the time to procure some of his daughter's art-quality paints and send them along to Ridgway.) Baird also urged Ridgway to use the greatest possible precision in his own artwork—advice that would help him internalize the vital benefits of precision and accuracy in the study of birds. "The object should be in a drawing form to secure artistic elegance and at the same time a minute, almost microscopic, accuracy in matters of detail," he advised.[15]

Mount Carmel was far from Chicago, the nearest big city, and in his rural setting, Ridgway had only one precious book that was primarily about birds: a volume of Samuel G. Goodrich's popular book *Animal Kingdom*—unillustrated—which he would quote to Baird in his attempts to pin down identification of species he'd seen. Ornithologists in the nineteenth century generally did not consider illustrations of much use for identification purposes. In suggesting a book to help him identify the local birds around Mount Carmel, Baird wrote to Ridgway, "I am sorry you did not get the text of Birds N. Am. [*Birds of North America*, written by Baird, John Cassin, and George N. Lawrence in 1860]. Nothing will replace it as a means of identifying the species you are likely to meet with. The plates [which were bound in a separate atlas] are not at all necessary."[16]

The possibility of a different life was already tugging at Ridgway by the time he was fifteen. "My parents do not fully understand my views and motives and consequently I do not expect to receive that encouragement which I would wish to receive from them," Ridgway wrote Baird. "My mother is willing to do a good part for me, but my father says that my study and drawing is more expensive than beneficial and calls it mere child's play." He felt pulled toward Baird, and his gratefulness to the Smithsonian's assistant secretary was apparent. "Forgive me for not having before expressed my thanks to you for your kindness to me . . . remember that if my thanks have not been expressed they have a home in my heart, there being too full for expression," he wrote Baird in 1866.[17] The next year, Baird would secure Ridgway another kind of home: a place on an expedition to the western regions of the United States, removing him from Illinois and opening him to the possibilities of a wider world.

The Fortieth Parallel Survey with Clarence King, 1867–1869

The 1860s and 1870s were seminal decades in the growth of science and technology in the United States. These years included a complex interplay of social change, technological advance, and the influence of Darwin's writings on a wide sphere that included the natural as well as the social sciences. Education in science began to change and become more formalized. Technologies such as the railroad and other forms of industrialization had transformed the American economy as well as daily life, and as with countless other places, Ridgway's small hometown of Mount

Carmel was dramatically affected by the arrival of the railroad during and after the 1860s. The American editions of *Origin of Species* quickly entered into wider circulation than they had enjoyed in England, and Darwinian notions began to saturate more and more aspects of daily life.[18] America also remained relatively insular politically, with its extranational aggression in the form of the conquest of Hawaii, Cuba, and the Philippines some thirty years away. During Ridgway's youth the railroad was king, bursting with the promise of a transcontinental route to stitch the country together and to bring prosperity to the nation. The American West, filled with unexploited riches, beckoned, and a series of explorations— geological, zoological, and botanical—were legitimized by federally sponsored surveys.

The second half of the nineteenth century also brought what William Goetzmann has called "an intense preoccupation with the discovery of time." The trans-Mississippi West offered a clarion call to a group that historian Henry Adams described as "men with knives in their brains."[19] This preoccupation was driven at least in part by the railroad, with its elements of punctuality, schedules, and the annihilation of space and time. Notions of precision played directly into the professionalization of science, spilling over into identification and description tasks. As the world changed, increased accountability and precision became vital: without them, the new information about the world being amassed by explorers, engineers, and scientists would be a chaotic mass.

Into this milieu stepped a supremely confident young American geologist named Clarence King, pulling Ridgway and others along with him to survey the far western United States. A tireless advocate of western exploration, King was also an expert on and relentless advocate for geology and mining. He was no stranger to the western regions, having already served as a member of the California Geological Survey. The genesis for the Fortieth Parallel Survey came about under the administration of Abraham Lincoln, whose members had a great interest in a transcontinental railroad—not just as a route by which to quickly transport troops across long distances west of the Mississippi but also as a way to extend the nation's economic and political reach westward. This survey was designed to create a detailed map and scrutinize the resources along a hundred-mile-wide section of the proposed route of the transcontinental Union and Central Pacific Railroads. King asked Congress for a small appropriation to undertake a scientific survey of the Great Basin area in Nevada and Utah Territories. His efforts generated various allies, each with his own vested interest in the project, including Edwin Stanton, Lincoln's secretary of war and a former Californian; Brigadier General Andrew A. Humphreys, the army's chief engineer; and California senator John Conness. The senator attached a proviso for the survey to an appropriations bill already passing through the Senate. The federal statute that emerged authorized a geological and topographical exploration of the territory between the Rocky Mountains and the Sierra Nevada range, including the route of the Pacific railroad. Humphreys was King's boss for the undertaking, and he laid out the plan of action: the survey was to "examine and describe the geological structure, geographical conditions and natural resources of a belt of country extending from the 120th meridian eastward to the 105th meridian, along the fortieth parallel of latitude, with sufficient expansion north and south to include the line of the 'Central' and 'Union Pacific' railroads . . . [and] make collections in botany and zoology."[20]

The Fortieth Parallel Survey and others like it were designed to supplement and inform the railroad's progress, but in fact the railroad ended up serving as the handmaiden to exploration instead. The urgency for a rail route across the continent provided a rare opportunity to make extraordinary progress in both the practical and theoretical sciences. The collecting of physical specimens created a snapshot in time of the era: it would demonstrate what plants and animals existed in specific regions and under what conditions. This would prove invaluable in subsequent decades because they provided vital comparative data. The specimens from the trip, in fact, continue to provide priceless information for scientists today; nineteenth-century specimens have often provided information undreamed of by the original collectors.[21]

The body of people involved in the scientific discovery of western North America, working from the eighteenth century onward, was not a narrow group of elite workers. Rather, the task involved, as Goetzmann notes, "entire nations and whole societies, merchants, dilettantes, and even state legislatures [who] launched exploring expeditions to discover nature's wonders." In the King survey, all the participants were civilians rather than military men for the first time. The work of King, Ridgway, and their compatriots, along with the enormous body of material and information gathered by such earlier expeditions as the United States Exploring Expedition of 1838–1842, dramatically advanced scientific knowledge in the nineteenth century. Four great surveys of western North America—

the Hayden, King, Powell, and Wheeler Surveys—added to this store-house between the 1850s and the 1870s.[22]

The quantity of geological, botanical, and zoological specimens brought from remote regions and deposited in urban centers in the United States, along with the expeditions' cartographic, photographic, and textual records, meant that institutions had to find new ways to systematize, organize, and present these materials. Expeditions—usually on the move for months at a time—made it possible for specimens to be abundantly collected across a wide range of territories, and in biology, they filled in numerous gaps in the record. This huge influx of material was a kind of second information revolution, following on the heels of the invention of the printing press in the fifteenth century. The modern museum rose up and, with it, the ability to scrutinize physical materials in institutions vested in providing a systematic approach to information through objects. The Smithsonian, we will see, rose and prospered in significant ways because of these explorations.

At the end of March 1867, Baird wrote to Ridgway offering, contingent upon King's need for a zoologist, a position on the Fortieth Parallel Survey. "The Survey will last probably three years and be prosecuted with unexampled advantage for a thorough exploration of the country in every direction," he explained. "It offers more promise of novelty in Natural History than any other part of North America; while in ornithology it will afford a chance of completing the information respecting the habits of the western species." Baird also anticipated what he knew would be comments about the young man's suitability for the job: "You need not have any apprehension of your fitness to discharge the duties; you know more than enough to begin, and you will soon grow up to all the possible requirements." Some of his commentary was clearly meant for Ridgway's parents; he dismissed any dangers from Indians and—stretching the truth considerably—noted that the work would be "so near the railroad that all collections can be readily sent in."[23]

King had needed a naturalist before he knew of Ridgway. One leading candidate for the job was Ridgway's future confidant Joel Allen, then in his late twenties and serving as the curator of birds at Harvard's Museum of Comparative Zoology. Addison Emery Verrill, Yale University's first professor of zoology, wrote to Allen on March 21, 1867: "They are trying to find some young man to go as Zoölogist and Botanist and doubtless you can get the place if you desire . . . It seems to me that if you want to travel to the far west there could not be a better opportunity." But by March 27, the deal was sealed with Ridgway, and Allen, who had been interested, missed out. "We had a telegraphic despatch yesterday to the effect that a Zoölogist has just been engaged for . . . the Expedition. But I do not yet know who. So I suppose the matter is settled," wrote Verrill.[24] Baird's support meant that the job was Ridgway's for the taking.

Ridgway's parents had agreed to their son's place on the expedition after some prolonged hesitation. His father wrote to Baird, noting that the trip would be a great trial for one so young but that he thought it advantageous for his son. "I feel that it will be very hard for him to be set among strangers, and in better society than he has been accustomed to. But he can learn and will likely soon overcome his difidence." Ridgway's shyness would prove to die hard. "You will find Robert a very obedient boy, and he has no bad habits that I know of," he concluded.[25]

Baird offered Ridgway a choice: work for the Smithsonian in Washington or go on the King expedition, at a monthly salary of fifty dollars and all expenses paid. Ridgway unhesitatingly chose the expedition. His case was no doubt strengthened by his mother's sister Fannie Gunn, who wrote Robert:

> Should you go to Washington of course you will learn and see a great many things that would interest you but they are all the result of other mens ambitions, and not your own acquirements . . . I think the [expedition] would indeed be a great adventure for a boy of your age, and aside from the salary would be very advantageous to you both mentally and physically. I would not flatter you but I have always thought you had the capacity of acquiring knowledge of the right kind, and now I think the way is opened for you to make your mark in *your* profession. Now I ask why may not you be a discoverer of something great and wonderful as well as any other man, now that an opening is made for you. You have long wanted to find a way of following your favorite pursuit, and now that way is right before you . . . there would be danger and peril even hardship to encounter and endure, the most sanguine would not doubt, but not so many more so perilous as the discovery of our own continent.[26]

Leaving home was an exhilarating experience for Ridgway. His mother admonished him—futilely, as it turned out—to "have a good religious life as well as a usefull one . . . never forget to aknolage the power that . . . created you with such peculiar talent." His parents escorted him to neighboring Olney, he boarded the train, and arrived in Washington on April 18, 1867. The young country boy who had never been outside of Illinois was amazed by the nation's capitol. "Pa you could hardly believe that such an enormous city could possibly exist—indeed I never dreamed of so great and fine a city," he wrote to his father the evening he arrived. "I have not seen Mr. Baird yet but will go to the Smithsonian in the morning. Washington is a magnificent city and the cupola of the capital is larger than the Olney seminary, if you will believe me . . . Pa maybe you will not believe me, but the Ohio River at Cincinnati is about two thirds as wide as the Wabash at Mt Carmel . . . I am well contented and delighted with the trip."[27] Despite his initial awe, the teenage Ridgway hit the ground running hard. He spent his first three weeks at the Smithsonian, where Baird taught him how to prepare study skins, showed him the facilities and collections in great detail, and acted generally as a mentor and father figure. Baird, the statesman of national science, had treated other promising young naturalists with the same care, including Henry Henshaw and Elliott Coues, but none so young as Ridgway, who was still just sixteen.

Those first weeks proved instructive. A close eye to details of differences from bird to bird often bore fruit, because many species had never been compared across the country, and there was great knowledge to unearth. Ridgway actually made his first discovery of a new species not long after arriving at the Smithsonian—a bird he had collected in his hometown and had brought to Washington for identification by Baird. "Ma, tell Pa that I wish I had a hundred black-birds in my box instead of three," he wrote home. "Ours is a distinct species from the eastern one, and I have the honor of first pointing out the specific differences in the two."[28] At age sixteen, Ridgway had made his first scientific discovery. His find also underscored the key role that study collections with numerous examples of the same bird would play in the discovery of new species and subspecies in the coming decades.

At the same time he was quickly learning the ropes at the Smithsonian, Ridgway was also busy preparing for the trip, as was the rest of the team. The endeavor was ambitious. The survey members were tasked to study a thick strap of land stretching for about a hundred miles on both sides of the fortieth parallel. The region lay largely in the vast continental drainage called the Great Basin, located in actuality between 39 and 42 degrees north latitude. The survey would eventually encompass nearly a thousand miles, with the expedition beginning at Nevada's western border and crossing through Nevada, Utah, and Wyoming Territories to end just east of Cheyenne, Wyoming. Understanding the geology, and thus the mineral resources of the region, was a key component of the trip, and the reason for the presence of no fewer than five geologists, including King.

Clarence King was youthful and enthusiastic, described by the expedition's botanist as a man "of striking personality, wonderfully genial and sweet in manner." He wore a derby and lemon yellow gloves in the field, brought a black manservant along to the most remote parts of his expedition, and poured forth enthusiasm and drive. The federal commissioner of mining statistics, on dropping by King's camp one summer, was startled to find a "polished gentleman . . . in immaculate linen, silk stockings, low shoes, and clothing without a wrinkle." Ridgway wrote to his mother about King a week before he departed on the steamer: "He is a very kind, and agreeable young man and I can not help liking him." King was universally admired on the trip; he was described by one of his men as having a "musical, infectious laugh," and spent long hours holding forth on a variety of subjects to expedition members. He was also a quick wit, and occasionally hyperbolic in his descriptions of nature. As one admirer noted, "The trouble with King is that his description of the sunset spoils the original."[29]

But King was more than a likable dandy and a good raconteur. He was also a skilled organizer of large projects, which stood him in good stead in the late 1860s as he focused on making a detailed map and survey of the region's resources. Furthermore, he imposed a brand of discipline on his men that would serve them well as future scientists. An especially significant aspect of his personality and leadership intentions is revealed in a group of notes to survey geologist James Hague, written after the trip, in which he stated what he wanted to be remembered for as the group's leader. "If I succeeded in anything, it was in personally impressing the whole corps and making it uniformly harmonious and patient; and I think I did that as much as anything else by a sort of natural spirit of command and personal sympathy with all hands and conditions, from geologists to mules. ('Tis but a step from the sublime to the ridiculous.) Perhaps you know of my great struggle for discipline and the hand to hand capture of a murderer and deserter from the escort." A soldier on the trip had deserted, and in King's later telling of the story, he had chased the infidel for a hundred miles across the desert, grappled with him, arrested him, and lodged him in a Texas jail. King claimed that the event forever reduced the soldiers and the workingmen of the survey to obedience.

Fearlessness was another important trait of King's, which he encouraged in his men. "I care very little about my reputation as a geologist, but a good deal as being a fellow not easily scared," he wrote Hague. He concluded, "If there were any graceful and inoffensive way of doing it, I wish it could be intimated in my life and engraved on my tombstone that I am to the last fibre aristocratic in belief, that I think the only fine thing to do with the masses is to govern and educate them into some semblance of social superior." As a scientist leading his troops through the desert, he modeled superiority through mannered living amid hardship and served as the exemplar of the scientist as a member of a privileged class. (In a strange twist, King also lived out the last several years of his life with his common-law wife, persuading her he was an African-American Pullman porter named James Todd, despite his fair complexion and blue eyes. He pretended to be Todd while at home and continued to work as King, a white geologist, when in the field.)[30]

The Fortieth Parallel Survey team departed in two small groups from New York on May 1 and May 11, 1867 (King had fallen sick and had sent several members on the way early, recovering sufficiently to join the second group ten days later). The first group was an advance party consisting of Frederick A. Clark, a topographic assistant, and four geologists—Samuel Emmons, James Gardiner, and brothers Arnold and James Hague. The second group included Ridgway, along with William Bailey Jr., the expedition's slender and somewhat frail botanist, King, and photographer Timothy O'Sullivan, who had gained national fame from his widely published photographs of Civil War scenes. They were accompanied by Henry Custer, a topographic assistant to King. Ridgway came armed with a letter of introduction from Baird, noting his official capacity as the trip's naturalist and commending him to the attention of the friends of science generally.[31]

Ridgway's group traveled aboard the three-thousand-ton steamer *Henry Chauncey,* along with another 950 people, primarily soldiers going to garrisons at far western posts. Ridgway shared a room with O'Sullivan and Bailey. The boat passed Cape Hatteras, the Bahamas, and Cuba, arriving in Aspinwall (now Colón) on the Caribbean coast of Panama on May 19, where the party crossed the isthmus by rail and boarded the spacious side-wheeler steamer *Constitution.* The two parties would not unite until they met up in San Francisco some weeks later.

Ridgway was thrilled to be under way. Writing from the *Chauncey* to his father on May 17, he exultantly noted that he was "now on the bosom of the 'deep blue sea,' flowing sweetly along: the fragrant balmy air of the tropics dispels the intense heat which however on the sea is hardly felt." He described passing Cuba and Haiti, the ship's landing at Aspinwall, and the trip across the isthmus on the Panama Railroad:

> Day dawned, and unveiled to us . . . an earthly Paradise! Such as pen or words cannot describe . . . along the shore of the beautiful bay, not a spot of earth was to be seen—nothing but a dense, impenetrable growth of *the* most gorgeous, and rank vegetation on the earth . . . Now, while dreamily musing, I look back and find myself whirling along over the beautiful Isthmus . . . screaming parrots, and screeching monkeys scramble among the branches of the trees along the edges of the stagnant pools, at either side, are numerous regally-colored birds, the fiery scarlet ibis, and purple gallinules, snowy-white herons, and golden-plumaged stripes wading about in the water: curiously shaped and gorgeously painted birds, some of the large size lazily sit half asleep on the dead boughs while stretching out full length basking in the sun large iguanas and huge serpents are seen. Every now and then we pass the lovely country seat of some wealthy Spaniard, the yards and jardins profuse with the most beautiful flowers on earth, with beautiful green parrots living among the boughs. As I said before there are no words can possibly give even the faintest idea of the magnificence of tropical America.[32]

This brief exposure to the tropics made a deep impression on him. Ridgway—who had noted to Baird three years earlier that he didn't have much taste for "foreign species," preferring North American birds—was already undergoing a transformation. Decades later he would return to Central America for several extended stays. His collecting work in tropical America would also expand his vocabulary of classification, exposing him to species which were in the same families as US birds but far different in their appearance.[33] Ridgway studied birds every step along the way, observing several species of oceangoing birds following the ship between Panama and San Francisco, including the Black-Footed Albatross, the White-Headed Gull, and others. In his final published report for the federal government, he noted that "we reluctantly omit, as too far beyond the geographical province of our subject, some notes on the Isthmus of Panama," and then proceeded anyway to describe some of the species he saw along the Panama Railroad and the Bay of Panama.

From Panama the group traveled up to Acapulco, arriving in the early hours of the morning of May 25. They watched the sun rise the next morning, delighted to be in the tropics and immediately intrigued by the doings in Mexico. "A small fort commands the harbor and bore the *Mexican* flag. The Frenchies have skedaddled," noted Bailey, referring to the end of the three-year French occupation of Mexico.[34] Emperor Maximilian, in fact, had been arrested just nine days earlier while attempting to sneak out of the country, and was later executed. King's party stopped there for the night of May 25, leaving the next day for San Francisco, where they arrived on June 2 around 2:00 a.m. There the groups from the two ships were reunited and joined by their military escort.

The total size of the party, not counting the soldiers, numbered somewhere between fifteen and twenty men. Intermingled with more experienced scientists, were a number of young or inexperienced members who were somewhat surprised by the relative swiftness of their invitations to join the survey and subsequent departure. "Had I been visited upon suddenly by a delegation of politicians, telling me I had been nominated for the Presidency, I could not have been more appalled," recalled Bailey in later years. None other than the great Harvard botanist Asa Gray—another friend of Baird's—had written to Bailey, asking him of his interest in accompanying the expedition. It couldn't have hurt that Bailey's father was a professor at West Point. But again, somehow, youth trumped experience, although Gray didn't exactly rave about Bailey's qualifications in his letter. "Mr. King desires a young man who shall at the same time be an accomplished botanist," Gray wrote the twenty-four-year-old Bailey. "As the two things are incompatible, I think you will do as well as anybody."[35] Geologist Samuel Emmons, who was tight with Ridgway, was another young man, only twenty-six at the start of the trip.

No scientific expedition today would be likely to take relatively inexperienced youngsters along as members of the official scientific corps, especially on such a long trip. But in the nineteenth century it was much more possible. Perhaps enthusiastic, inexperienced young men were easier to locate than older trained ones, or perhaps the risks seemed more suited to the youthful and unwed. Or maybe the fact that young men had lived through the Civil War years meant that they were accorded more respect as part of a cohort that had lived, fought, and died together. These young naturalists were also probably less likely to censor their own newly forming scientific ideas because they had no fear of damaging their nonexistent reputations. These were amateurs doing some evolving of their own. Everyone starts somewhere, and these greenhorns of the 1860s would become the skilled, widely published professionals of the 1880s and

1890s, many working productively well into the twentieth century. But at the start of the Fortieth Parallel expedition, all Ridgway knew was that he loved studying birds, was thrilled to be on an expedition with important scientists, and wanted to do everything he could to stamp himself with the imprimatur of authority in order to gain acceptance.

The group was an army on the move. The party included an ambulance wagon, a portable darkroom for O'Sullivan's photographic work, and a supply caravan and mess wagon. King's surveying equipment included fragile and bulky gear: a six-inch theodolite, a number of compasses and chronometers of varying sizes, thermometers, and other glassware. The expedition even carried its own flag, sporting crossed hammers, which was mounted at camp outside King's tent. Ridgway's part of the load was substantial, for the supplies necessary to sustain a naturalist's long expedition to a remote region amounted to heavy quantities of equipment and supplies. In a letter to his father, Ridgway inventoried some of his six hundred pounds of gear:

> My outfit could not possibly be any better than it is. I will give you a few items of its character, to show you more fully how I am equipped:
> 2 Maynard guns (shot & rifle); these cost about $60 apiece and are the best guns in existence
> 25 lbs finest (no 12) dust shot
> 50 lbs no 10 shot
> 25 lbs no 7 shot
> 25 lbs no 3 shot
> Some larger sizes
> 20 lbs arsenic
> 30 lbs alum and saltpeter
> 50 gallons 98% alcohol
> 2 lbs Tartar emetic for medicating alcohol
> 1 lb Chrystallized Carbolic Acid, for strengthening alcohol
> 6 drachms Strychnine in clr bottles
> 1 gross homeopathic vials; corked;
> 1 gross larger vials (2 & 4 oz) corked
> 10 lbs raw cotton;
> 10 boxes for nests & eggs.
> Corked lined boxes for insects
> 10 lbs Plaster-paris,
> Dozen Mouse-traps
> Dozen Stub-traps for various animals
> With chests, Appurtenances, and everything my outfit will cost the Government about $500.00.[36]

Most of these supplies were for use in killing and preserving bird specimens. His fellow travelers would have taken special note of the contents of his outfit. The fifty gallons of alcohol were "doctored to lessen its attraction to nonscientific personnel of the expeditions," notes one biography of Baird, who oversaw almost all of the Smithsonian's own outings.[37] Ridgway also was responsible for collecting other animal specimens on the trip, although he collected a relatively small number of animals besides birds. He preserved some specimens in alcohol and prepared most of the dry bird specimens: skin, feathers, and most bones intact, the viscera removed, and the body stuffed with cotton batting. These study skins, as they are called, served as a kind of physical photograph because their preservation maintained the key morphological components of the living bird: its plumage, coloration, size, and a number of other details of interest to ornithologists and, later, to evolutionary biologists.

Ridgway and botanist Bailey both found a great deal to like about the undertaking. "We are living like belligerent roosters," Bailey wrote to his brother from the expedition's first camp on the outskirts of Sacramento on the eve of their departure into the wilderness in mid-June of 1867, as they waited for the snows in the Sierras to recede. The same week, Ridgway wrote home that he, too, had been "enjoying myself finely ever since I have been here." By the third day the party had reached the foothills of the Sierras, at about two thousand feet. Ridgway's first impressions included this description of the view: "The whole valley of the Sacramento, spread out like a map before us, fading away into hazy indistinctness until finally lost in the distance: the dim-blue outline of the far-distant Coast-range, terminating the view toward the west; the beautiful Sacramento river dividing the valley and flashing in the sun-light like a silver streak." Ridgway proclaimed himself very content and noted that his family would not recognize him at a fit and trim 140 pounds.[38]

Offsetting the difficulties of being away from home and family and in trying conditions were the aforementioned joys of spectacular vistas, along with the thrills of discoveries. The body of scientific work accomplished by the team was considerable. The survey, which made a total of forty camps along its thousand-mile route, collected thousands of specimens, including geologic samples, birds, plants, and sediment from bodies of water. Ridgway was helped considerably by his friend Hubbard G. Parker, an enthusiastic Nevada Indian agent who was an amateur taxidermist. Parker, who provided "frequent and gratuitous service" and is thanked at the out-

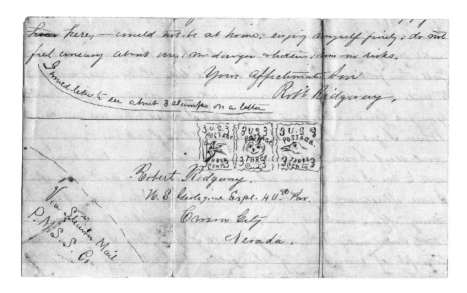

Robert Ridgway playfully illustrated three birds on stamps: a woodpecker, an owl, and a crow. Letter from Ridgway to his mother, Henrietta, June 24, 1867. Special Collections and Archives, Merrill-Cazier Library, Utah State University, USU Caine MSS8, box 1, fol. 4.

set of Ridgway's published report, often accompanied the young naturalist on long rides looking for birds as well as helping to prepare skins of many birds throughout the first summer. The agent also provided levity for the other survey members with his apparently exceptional sense of humor. "He is one of the most original and entertaining sticks I have ever met," Bailey wrote home. "He is an old Californian of vast adventure and a friend of dry humour which is inexhaustible."[39] Sometimes time crawled from the tedium of the desert, and at other times, with the help of people like Parker, it flew by.

After leaving Sacramento, the group worked its way eastward across the Sierras and into Nevada along the Truckee River for most of the summer of that year, including an expedition to Pyramid Lake, and then across the desert to Humboldt Lake, where Ridgway came down with a severe case of malaria. From there the team passed along the western slope of the Humboldt Mountains, through the Truckee Meadows in the early fall, and then overwintered in Carson City, Nevada, at the end of the year. The winters, particularly the one of 1867 and 1868, were sometimes rambunctious. King made regular trips to San Francisco, leaving most of the men behind, and drink and rowdiness were often the norm for many of the company. Bailey complained at the end of a long winter that "there is no head and no discipline—and I am not alone in my thinking." The lack of King's authority was keenly felt whenever he was away. Some of the members, such as Emmons and the Hagues, were regular church-goers. Others passed the time in more earthly pursuits, and the days were filled with activity. "The time flies like a Central Pacific locomotive on a down grade and the breaks off," observed Bailey.[40]

Ridgway worked throughout the winter. Two printed synopses of his findings later detailed the names, abundance, and habitat of seventy species he found around Carson City.[41] The team finally departed for Ruby Valley in late spring, and from there, worked their way north into southern Idaho, and then back down to Utah. By this time, most of 1868 had passed, and the group spent the fall of that year in northwestern Utah.

The new scientific information, specimens, drawings, and spectacular photographs accumulated on the trip were obtained at a high price. King was struck by lightning at least once, leaving him badly burned and briefly paralyzed; at least half of the party was often sick at the same time. Ridgway and others suffered regularly from fevers, delirium, and diarrhea, long days in the saddle over difficult terrain, swarms of mosquitoes in the summer and frigid weather in winter, and occasional desertions (four of the twenty-soldier party the first year) and theft by the very soldiers accompanying the group. As the first season in the wilderness drew to a close, Bailey noted, "considering the terrible drawbacks, the work has been done well and thoroughly—and no one will ever know what our maps and reports and notes have cost us." Another omnipresent stress was fear of attack by the Bannock Indian tribe. Bailey, who had espoused a number of ugly racist attitudes in his letters home during the Central American part of the trip, also revealed an extreme anti-Indian bias also present in most other members of the group: "The universal sentiment is for Indian extermination—and the troops don't kill them off fast enough."[42]

Ridgway had a considerably better attitude about the natives than did Bailey. During the trip he had close interactions with the Paiute Indians, whom he found friendly, resourceful, and helpful. He learned the names and pronunciation of at least eighty Paiute, Shoshone, and Washoe Indian bird names, including local synonyms for several, and described those names in detail both in his letters home and in his published report. He transcribed them, noting the emphases in pronunciation and elaborating how they were said. Woodhouse's Jay (a Nevadan subspecies of the Western Scrub-Jay) was "We'ahk"; Parkman's Wren (a subspecies of what is now the House Wren) was "Tuse-vot'-tse-pah"; and for Say's Flycatcher (now Say's Phoebe), "to-que'-ohk." These were likely phonetic descriptions of either vocalizations or nonvocal noises; Say's Phoebe, for example, makes a loud snapping sound by quickly closing its mandibles or, occasionally, when trying to capture prey. Despite his mother's fears about Indians, Ridgway reported that they were in their camp daily. The Paiute were very quiet and peaceable, and brought fish to trade for provisions.[43] The survey group had no distressing incidents with them, despite repeated warnings and concerns from various quarters.

One intense highlight for Ridgway was a trip with Parker and several other expedition members to Pyramid Lake in Nevada Territory in August 1867. Ascending out onto the lake and climbing up Pyramid Rock, a steep triangular shale formation, Ridgway found the view down "a frightful one" but kept his wits about him enough to spot Great Blue Herons and Peregrine Falcons breeding on the rock. He was stirred by the gigantic flocks of American White Pelicans, which roosted along the lake by the thousands, the noise of their wings so loud that the men could hardly hear each other speak. Always on task, he gathered more than a hundred of their eggs for shipment back to Washington.[44] Ridgway also reached the summit of Pyramid Rock. Although he became paralyzed with fright on the way down, he may have been the first white man to reach the top. He also suffered from a bad case of malaria following this trip to Pyramid Lake. He lost consciousness, had to be carried by litter back to camp, and caused great worry among the team due to the delay in returning. It took him a full week to recover, and soon after he became ill, most of the rest of the survey also contracted the disease.

Other members of the team noted the difficult nature of their outing; Bailey wrote his brother, "Mr. Parker and Ridgway returned some ten days since from their trip to Pyramid Lake, where they had a pretty hard time—but obtained many specimens of birds. They were much delayed in returning by the floods and the wretched state of the roads. They had to leave their team at the foot of the Geiger grade and walk into Virginia City." Again, the jovial Parker is mentioned as a foil to hardship: "Parker always enlivens us very much—and we regret that he will be away for a month in San Francisco. While he was here, I lived with him in his room—in jolly comfort—but have not returned to my own Greenland." The photographer O'Sullivan was a fine enough comrade to all but occasionally annoyed members with his constant recounting of his glory days from the Civil War. "His chief fault is his reminiscence of the Potomac Army," Bailey wrote to his brother. "One would think he had slept with [Generals] Grant and Meade and was the direct confidant of [Secretary of War Edwin] Stanton."[45]

Bailey marveled at the transformations wrought by the telegraph, wiring his uncle in Providence, Rhode Island, for money and receiving the funds three days later: "Isn't that annihilating space?" he asked rhetorically. But technology was of little consolation for most of the trip, for it could not alter the arid landscape. The accomplished landscape watercolorist and engraver John Henry Hill joined the survey in May of 1868. He left after just a season, and when asked what the scenery of Nevada was like, replied simply, "Hell." Interpersonal tensions also arose. Bailey described geologist James Gardiner as having treated him unkindly from the start. "Gardner the second in command hates me and would do anything he could to remove me," he wrote his uncle William in early 1868, as they overwintered in Carson City. "The mean slights, and total misunderstanding of my work and actions, have almost made me desperate of what I say or do, so that where I believe I would have achieved a good reputation for cheerfulness and kindness at first, I am eventually looked on now as peevish," he complained.[46]

Whatever their hardships and stresses, the men apparently ate well most of the time. A friend of King's later recalled, "One of the rules of the Fortieth Parallel Survey was that its members should be as civilized as practicable, especially at meals. The men believed in a good and varied diet, well cooked and served; and when the accounting officers of the War Department demurred at passing a bill for currant jelly, they were met with a threat to charge up at the rate of beef the venison furnished by members of the mess." Bailey wrote to his brother that "we have a jewel of a darkie cook, a Liverpool man. He can make the best dinner of few materials than any cook I ever knew." Ridgway, for his part, either eschewed the good food or wanted to convince his parents he was living a more rugged culinary life than he really was: "You would not know me [I] live on Plum-duff [a stiff flour pudding with raisins] and friolas [beans], canned fruits [and] condensed milk. The best tea and coffee to be obtained on the Pacific Coast. Have plenty to wear." We know something of Ridgway's physical makeup from his letters during and just after the trip. Never heavy, he weighed 134 pounds during his late teenage years— a slight build he maintained for the rest of his life. When he got sick he would lose weight precipitously, dropping as low as 119 pounds. He stood at just about five and a half feet tall.[47]

While Ridgway and his colleagues labored in Utah and elsewhere during his travels, his family back in Illinois struggled through the typical serious family illnesses of the time. Ridgway's mother was also constantly anxious about her son's safety—a natural response barely two years after the end of the Civil War, when tragic news about young family males had been a regular event. The Far West, too, would have seemed distant and threatening as a wilderness. "Pa is very much afraid you will be exposed to the Indians but I am more afraid of other things such as *tall trees* and precipices or wild beastes," she confessed.[48]

Ridgway proved to be a disciplined collector during the two-year trip. He reported gathering 1,522 bird-related specimens for the Smithsonian: 769 skins and 753 nests and eggs, along with a handful of other nonavian specimens. Ridgway observed a total of 262 species, about 90 percent of which he saw east of the western slope of the Sierra Nevada. These different birds accounted for approximately a third of all species found today in North America—a testament to the species diversity found in the Far West. Ridgway was also quick to notice key habitat details: the fact that some birds were found only in specific locations, and found near specific food sources with unvarying certainty. He took careful note of the types of regions where birds were found and organized his final report accordingly.

In the fall of 1868, as funding began to dry up, King narrowed the focus of the expedition to study only geology, dropping botany and zoology and putting Ridgway out of a job. He returned to the Smithsonian for seven months of work writing up his notes from the expedition and sorting through the specimens he had collected, then spent several weeks with his parents in Mount Carmel. In early May 1869, his friend Frederick A. Clark, one of the topographical assistants on the expedition, summoned him to Utah at King's request. Ridgway spent about six weeks near Salt Lake City, moving through forty-three camps and doing further collecting for the Smithsonian as the final part of his survey work, finishing up in late July. Even being close to home was not without its hardships.

One day the small four-oared boat he was in capsized on Salt Lake—the "Nettie," which he had also used on Pyramid Lake—and he narrowly escaped drowning. "Ridgway not appreciating a salt watery grave, got himself landed and has fallen in love with Terra," King wrote wryly to James Gardiner of the survey. As the summer wound down, Ridgway tired of being on the road and was eager to get back to Washington, where Baird had offered him a job as an artist.[49]

Because of his rapidly growing skills as a scientist and the seasoning provided by nearly two years in the wilderness on a major expedition, it is easy to forget that Ridgway was just nineteen in 1870. After a year at the Smithsonian, he spent the summer in Mount Carmel with his parents, turned twenty, and promptly got sick with what his mother described as dysentery and rheumatism. In September, his mother and sister jointly wrote a letter to Baird asking for his permission to let Ridgway remain at home for several more weeks to recover from his various ailments: "Our object is to ask you if it will be any inconvenience to you if Robert should remain a few weeks longer," the two women wrote. This epistolary equivalent of a note pinned to his shirt for the teacher, if he ever found out about it, would've embarrassed young Robert mightily, who tried hard to be older than he was when around Baird and other scientists. His mother's letters could be peevish and complaining at times. She would often begin by attempting to comfort him and quickly move to a scolding that had religious overtones: "I thought I would write again and perhaps might say something to cheer you up a little and drive away the 'blues,'" she wrote him in a typical letter. "I cannot see much cause for your being unhappy . . . Now my dear Robert you must not indulge in such feelings for it is not only foolish but it is *wicked* for one so fortunate and one who as so many interested friends as you have."[50]

Beyond their common interests in birds, another factor united Baird and Ridgway: neither appeared to believe in a supreme being. Baird and his wife stopped attending church a few years after moving to Washington, DC, and some of his students and colleagues anecdotally described him as either an agnostic or an atheist. It was possible to believe in Darwinian evolution and still believe in a Christian God, but in the nineteenth century, it was much less common, and downright impossible for biblical literalists. And Baird distinctly believed that all animals came from a common ancestor. "My object is to define the forms [that is, subspecies] that characterize certain regions, not only as a matter of ornithology, but of physical geography," he wrote to one of his field collectors in 1869. "If I can do this to my own satisfaction, it is nothing to me that others say *they* can't. I can with my series convince anyone, un-prejudiced, of the differences, and I don't care whether they constitute varieties, species, or genera. I believe all the small ones [he names a number of bird species] came from one ancestor in the first place, and that all thrushes had a common origin in the preceding period."[51]

This would have accorded with Ridgway's views, or perhaps influenced them, because Ridgway almost never mentioned a deity after his teenage years. As a result, religious dogma did not stand in the way of rapid acceptance of evolution for either Ridgway or Baird. William H. Dall, the dean of Alaskan explorers and later a biographer of Baird, noted somewhat more explicitly about Baird's views at his memorial service that "unlike some of his contemporaries twenty years ago, the views of Darwin excited in him no reaction of mind against the hypotheses then novel and revolutionary. His friendly reception of the new theories was . . . quiet and undisturbed." But although he took more than a passing interest in the controversy over evolution, Baird himself never overtly put forth his views on Darwinism, preferring to focus in Baconian fashion only on what could be scientifically demonstrated by comparison of specimens.[52]

Furthermore, although he hedged his bets on the subject more than Baird, Clarence King was also apparently an evolutionist. He gave an important address at the Sheffield Scientific School in June 1877, entitled "Catastrophism and the Evolution of Environment," as well as a paper on "The Age of the Earth," published in the January 1893 number of *Annals of Science*. The first of these papers was a protest against the extreme uniformitarianism of famous geologists like Charles Lyell, an intimate of Darwin's who nevertheless couldn't swallow evolutionary theory whole. In the second work, King estimated the earth's age at 24 million years—a far cry from its current estimate of 4.5 billion years, but nothing like the 6,000 years claimed by biblical literalists. To admit support for evolution was heretical, even into the twentieth century. As Darwin famously confided to a friend, to say that species were not immutable was "like confessing a murder."[53] Thus people like Baird and King, serving in highly visible positions, were required to maintain decorum by avoiding explicit statements about their views on the subject.

The final published report on the Fortieth Parallel Survey included Ridgway's section on the trip's birds. The six-volume set was published over an eight-year span between 1870 and 1878, and although the pain, difficulty, boredom, and travail of the work is not evident in the 3,843 pages of the report, which included hundreds of color and black-and-white plates of specimens collected, it does show what a monumental scientific undertaking the survey was. Prominent Yale geologist William H. Brewer, writing to King, called the multivolume publication "a work so generally and rightly recognized among scientific men as the greatest which this country has yet produced in this direction."[54]

King asked Ridgway to keep his report on the ornithology of the expedition as concise as possible, "within the compass of 200 pages," owing to the bulk of the multiauthor work. Ridgway's portion ended up being nearly twice that long, at 366 pages. He had almost all of the writing completed by 1872, three years after the end of his time in the field with the expedition. He had been very busy with his new Smithsonian duties, but as a single man with no family obligations, he had no particular difficulties writing and organizing his part. When Joel Allen reviewed the work in the April 1878 issue of the Nuttall Ornithological Club's *Bulletin,* he praised the work and noted that "the Report, as a whole, is quite free from strictly technical matter, and hence attractive to general readers and amateurs, as well as of great value to specialists." Other reviews of the Fortieth Parallel work appeared, and occasionally reviewers made blunders. For instance, a brief squib in *Forest and Stream* in the spring of 1878 about Ridgway's volume noted that the expedition had been done under the auspices of Ferdinand Hayden, rather than Clarence King. This was a big deal, as the Hayden Expeditions were highly publicized annual federal survey trips into the western territories, both before and during the Fortieth Parallel Survey. After the error was pointed out to him, Charles Hallock, the owner of the magazine, wrote Ridgway, cringing. "I am made by other's blunders to appear to be the quintessence of stupidity boiled down. Reynolds is the man who wrote the notice of your '40th Parallel'—a young college graduate whose carelessness ought to be recognized 'in orders.'"[55] A correction, longer than the original announcement, was promptly published in the magazine.

By the time his part of the report was published, Ridgway had been considering a number of issues on naturalists' minds, such as the ways in which birds distributed themselves across the country. He made observations about birds' abilities to cross the Sierra Nevada range, the extremes of elevation at which some species were found, and cross-continental mi-

gratory patterns, and teased out far more comprehensive range information for a number of species than had previously existed. His report detailed information on each species observed, noting the sex, number of
specimens caught, general physical characteristics, and, in many places,
notes on the local Indian names of the bird calls and songs.

If he had been fond of nature in his little corner of the world in Illinois, Ridgway's time on the King survey gave him an explosively expanded
view of a new region, new people, and, of course, new birds. It was also his
first chance to spend extended time with experienced scientists and men
who had traveled widely, and the trip introduced him to the world of careful scientific collecting and its constituent parts. It also provided him with
the accoutrements of professional scientists: their language, their work
habits, and their daily lives. Forced into intimate contact with others for
months at a time, Ridgway could not help but be shaped by his comrades.

Most young men like Ridgway also came from families who had been
touched in some way by the Civil War. Required to be men before they
had barely had a chance to be boys, these lads were often quickly turned
into young adults through lives marked by family deaths and economic
hardship. Even before they were forced into manhood, though, boys from
rural settings—the norm in mid-nineteenth-century America—had the
chance to know the natural world. Boyhood was often about content-
ment, at least in later recollections, while later life was about being a
worker for hire and about sacrifice and honor. "How keen the perceptions
of boyhood, and how great a boy's capacity for enjoyment! In those old
days I am sure that it would be impossible for a human being to be more
perfectly happy, than I was," Ridgway wrote to a friend many years later.[56]
This combination of youthful, spirited activities out in nature, combined
with desire for a scientific life and the shock of adulthood, was part of the
metallurgical process that forged boys into scientists after the Civil War.

Individual personalities and geography also shaped the contours of
scientific surveys. The aristocratic civility of King and his strong desires to
impress on his men a certain kind of decorous conduct were coupled with
the dawning age of working evolutionary practice. The rugged territory
of the Sierras, the Great Basin, and the arid regions the group traversed
set a physical stage for incipient professionalism because the expedition
was much more than a ramble in the countryside: it was a prepared, well-
armed, scientific society on the move. Ridgway in particular was helped

along by a sensitive ear for scientific language and a huge enthusiasm for his chosen field of bird study. The issue of the soon-to-be-completed transcontinental railroad, in the news and much discussed, also raised sensitivities about precision in daily life.

Ridgway's own personal history and abilities also had a good amount to do with his becoming that much-contested creature, the professional. First, he was required to grow up quickly. He was the oldest of what would become a family with ten children. Even as a young teen, his mother generally treated him as a peer, and not in a healthy way. She regularly asked him for advice on matters inappropriate for a boy of his age, sent him off to relatives to work for a year to help the family, and gave him free rein to spend time in the countryside by himself. Second, he had two important traits that were evident by the start of his teens: his budding skill as an artist, and what his mother called his "peculiar talent" with birds. The first required that he focus on the details of birds in order to draw well, even as a young boy, and the second put that focus into action as an ability to recognize what seemed to be slight differences among birds. Again, Ridgway's extreme visual acuity was a key asset. The latter quality was what would eventually help make him one of the world's best systematists and describers. Accompanying these native skills was his ability to mimic the language of other seasoned scientists, especially their careful, unemotional descriptive terminology.

Being a professional became more and more about being precise: with language, with descriptions, and with methods. As explorers, engineers, and scientists fanned out and either brought things back home or created a new world out of whole cloth via the railroad and other industrial innovations, accountability, precision, and careful description became vitally necessary, because without them, the new information being amassed about the world would be a chaotic and jumbled mass.

It is tempting to polarize Ridgway's experience as one of small-town upbringing and limited circumstances contrasted with exposure to other lands and travel across wide expanses involving experienced adult scientists as role models; home-grown enthusiasm versus formal national science; and even Baird's Gilded Age big-city life versus the "real world" Ridgway came from: rapidly vanishing natural America. However, there is a middle ground between these contrasts. Ridgway's abilities and circumstances intersected with his environment in a complex way that provided him with key elements of his future success: a natural training ground for his talents in rural Illinois involving careful collecting of birds; an uncommon ability with both illustrations and words that was not too far removed from that of much older professional scientists; and the availability of Spencer Fullerton Baird as a father figure who could lead him gradually toward a greater focusing of his interests. Baird's notions of precision in art and in his descriptions of birds also helped to shape Ridgway's sensibilities, letting him focus on the trees rather than the forest.

This father figure happened to not only be knowledgeable about birds—he was also the second in command at the country's national museum. And Baird's own lack of religious dogma, as well as his interest in classification and regional variation, would have been absorbed by Ridgway, who came to Washington already steeped in the requisite background to begin to understand evolution, regional variation, and the benefits of classifying and organizing birds. Indeed, Ridgway would note in the preface in the first volume of his magnum opus on the birds of North and Middle America that "it is difficult to understand how any one who has studied the subject [of evolution] seriously can by any possibility believe otherwise."[57]

Finally, the survey exposed the importance of collecting the same species of birds across a vast area. Biological surveys such as the King expedition provided a chance to discover birds with subspecific differences. As one historian has noted, very dense local collecting revealed a world not of discrete units but of fluctuating populations. Survey practice brought the gap between subspecies into view. Too few specimens, or too many, and the gaps were invisible to the observer's eye.[58] The broad geography of the West made it possible to search out similar birds in different regions, and this improved the chances of locating subspecies. Different populations of birds, which had become isolated from each other thousands of years apart, would have been subject to different evolutionary pressures—a wide mix of food niches, predators, and climates—and would thus slowly develop slightly different characteristics. For both the creation and the later identification of subspecies, geography was everything.

Many of the specimens from the western United States could thus be put to work to show some new evidence of evolution. These specimens came directly back to the Smithsonian. As the *Boston Daily Advertiser* noted, "All the surveys of our government have poured their accumulations into this storehouse, and the descriptive portion of our ornithology is now far more accurately portrayed than ever before." Following the Fortieth Parallel Survey, Ridgway's new home at the Smithsonian would continue to shape him, and he in turn would help to shape the emerging profession of American ornithology.[59]

CHAPTER TWO

The Smithsonian Years

While the ornithologist is engaged in working out his share
of the great problem of life and evolution, which of course,
entails much hard and dry study that has little popular
interest, he gathers around him much that is beautiful and
interesting from the popular standpoint. It is the fairy-land
of science.

VIRGINIA DISPATCH, on Ridgway's labors at the
Smithsonian, September 20, 1885

AFTER RIDGWAY finished his survey work in Utah in the late summer of 1869, he made a beeline for the Smithsonian and began paid
work as an illustrator. His parents were disappointed by his decision
to go to Washington without even a brief stop at home. Ridgway, however,
was delighted to get to the nation's capital. In landing a permanent job
at the Smithsonian, he owed everything to Baird. The assistant secretary
understood the value of an older expert's interest. He had been mentored
by none other than John James Audubon, who named a new bird after
Baird in 1843. A small, secretive grassland bird, Baird's Sparrow was discovered by Audubon on the plains of North Dakota, and he named it for
the promising ornithologist who was just out of his teenage years.

Ridgway was only one of many "Bairdians" mentored by the
ornithologist-turned-administrator. Others included Elliott Coues;
William Healey Dall, later a prominent naturalist; Henry W. Henshaw, a
scientist and collector who held several senior federal government posts;
William Henry Holmes, who would become a curator of anthropology
at the Field Museum, and then the head of the Smithsonian's Bureau of
American Ethnology; and Ridgway's later friend Leonhard Stejneger, a
herpetologist who would eventually work in the Smithsonian's Bird Division. Baird's influence on the practice of science after his arrival at the
Smithsonian in 1850 cannot be overstated. Even during his lifetime, his
approach was known as the "Bairdian School," marked by what Ridgway

described in Baird's eulogy as "a school strikingly characterized by peculiar exactness in dealing with facts, conciseness in expressing deductions, and careful analysis of the subject in its various bearings—methods so radically different from those of the older 'European School' that . . . conclusions or arguments can be traced back to their source and thus properly weighed, whereas the latter [European method] offers no basis for analysis."[1] These hallmarks of Baird's work would be adopted by Ridgway, who breathed in the atmosphere from his first day at the Smithsonian.

Washington was a lively place when Ridgway arrived. It was striving to become the intellectual as well as the political capital of the country, and city life during the last quarter of the nineteenth century involved a coherent group of scientists, educators, and literary figures, including Baird and Smithsonian colleagues such as ethnologist Matilda Coxe Stevenson, along with innovators like Edward Gallaudet, educator of the deaf; journalist and novelist Henry Adams; physicians Anita Newcomb McGee and Joseph M. Toner; Horatio King, the postmaster general; librarian John Shaw Billings; Clarence Dutton of the US Geological Survey; Clarence King; astronomers Simon Newcomb and Asaph Hall; and many others.[2]

Amid this mix, the National Museum was an important aspect of city life. Before construction of a dedicated building in 1881—then the National Museum building, and known today as the Arts and Industries Building)—the collections lived in the Lower Main Hall of the "Castle," as the magnificent brick building that long comprised the center of Smithsonian life is known. Done in a medieval revival style of architecture, the rich red brick building exuded associations with college campuses around the country. It was an immediate and distinctive landmark in the District of Columbia. Within the Castle, the Lower Main Hall was filled to overflowing with public exhibits almost from its opening in 1858, and a wide public came to see the institution's—and the world's—treasures.

Despite his enthusiasm for Washington life, Ridgway's Illinois roots still beckoned. His mother complained that he never wrote and that he seemed to have no interest in coming home to visit, but in the summer of 1870 he took a leave of several months, returning to the Smithsonian before year's end. Baird encouraged him to continue looking for species to fill in the Smithsonian's collection deficiencies even while he was at home in Illinois. Back at the Smithsonian, he stopped working as an illustrator sometime in mid-1871, and began working steadily and quietly among the Smithsonian's bird collections. He did so for two years, engaging in little correspondence home and making no return trips there until late August 1873, when he asked Baird for two months' leave to visit his family. We know that he made $840 per year during his first several years at the Smithsonian, based on Baird's covering his salary for one month during a visit home. In 1874, he was given a raise to $1,000 annually, with Baird describing him to Secretary Joseph Henry as "very important in the general museum work, in Cataloging and . . . the arrangement of skins and eggs of birds, and in making up sets for distribution."[3]

Ridgway's arrival at the Smithsonian also made for an important change in Baird's life. It freed the assistant secretary from managing the bird collections, which had greatly occupied him, to another great interest—the study of fishes.[4] In 1871, President Ulysses S. Grant appointed Baird as the commissioner of fish and fisheries (albeit without giving him a raise), which gave him much more time to work on fishes, as well as to recruit other fish-minded protégés such as George Brown Goode, who would oversee much of Ridgway's publications work later in the century.

By the time Ridgway began full-time work at the Smithsonian in 1869, the institution was twenty-three years old. It had been founded by James Smithson, a British scientist who had never been to America, and who had willed a large fortune to his nephew. However, Smithson's will stipulated that if his relative died without heirs, his estate would go to the US government for creating an "Establishment for the increase and diffusion of Knowledge among men." The nephew, Henry James Hungerford, was never in good health and died in his mid-twenties, a scant six years after Smithson himself. Smithson's mandate was encapsulated in a single sentence in his will.

The "increase and diffusion of knowledge" part was reasonably clear, if broad. But just what had he intended by an "establishment"? A laboratory? A library? A coffee shop? The amount of money, in today's dollars, was the British equivalent of $505,000, a tidy sum—depending on your calculations, worth as much as $156 million today. Waiting for the best time to convert Smithson's annuities to British gold sovereigns based on the market, the US government's agent responsible for getting the sum to America cashed in the annuities, picked them up in gold coins occupying eleven boxes and shipped them home in September 1838, a rough, monthlong sea voyage. The gold sovereigns were then converted to Ameri-

can currency. Science in America had been transformed forever by a foreigner with mysterious motivations who had no known connections to the
United States.[5] In 1846, nearly a decade later, Congress finally signed into
law the establishment of the Smithsonian Institution. The Smithsonian
did not begin receiving government funds until 1858, when it began operating as a hybrid public-private partnership, supported by private monies
(primarily Smithson's) as well as federal monies. Its functions included a
research-oriented museum, but it served primarily as the national library,
a role that ended in 1865 when the Smithsonian Castle suffered a major
fire, at the same time that the Library of Congress's role was dramatically
expanded. The Smithsonian then became a research-oriented museum,
open to the public and intended by Congress to be the nation's museum
of record.

Bird collections from a variety of government expeditions—most
famously, the Wilkes Expedition of 1838 to 1842, a naval exploration and
surveying expedition to the Pacific Ocean, and a number of others—were
housed at the Smithsonian almost immediately on its founding. A flurry
of midcentury expeditions quickly brought in more specimens. While it
had been preceded in prominence earlier in the century by the Academy
of Natural Sciences in Philadelphia, founded in 1812, the Smithsonian
quickly surpassed it in scope as these collections arrived after 1850. Baird
especially solicited specimens of fish and birds, along with cetaceans.

By the mid-1880s, Ridgway was engrossed in his institutional duties
and had built a substantial network of colleagues outside the Smithsonian. Using Bairdian precision, he was at work at nothing less than
untangling the relationships among birds. The dried, stuffed remains of
birds—study skins—were his thesaurus. He also made use of birds' eggs
in the collections, as well as skeletons and specimens preserved in alcohol, but for his particular purposes, his labors with study skins predomi-
nated. By 1882, the bird collections held some fifty thousand skins, and
these allowed Ridgway to see synonyms, antonyms, homonyms, and other
parts of speech among the morphological language of birds. Amid this intense activity, his actual relationship with Baird had trailed off. However,
their respective roles in each other's lives were clear when Baird died that
year. Ridgway was selected by the institution to write Baird's obituary and
give his memorial address. Baird's wife, Mary, wrote to Ridgway to thank
him for his comments, noting that "we know that you did appreciate and
love him."[6]

The Public

There was no one homogenous public that interacted with the Smithsonian's Bird Division, either in person or by mail. People had many different interests and goals. Some wanted nothing more than to thank
Ridgway for his influence on their lives. More demanding people desired
publications, identification help, information on starting a bird-related
organization, employment, and more. The ranks of visitors included the
curious, the idle, and the inspired. By the end of the century more than
two hundred thousand people visited the museum each year, countless
thousands passing through the public bird displays. Inevitably, a number of them worked their way to the Bird Division staff. Ridgway's availability ebbed and flowed depending on his schedule on any given day or
week, but more often than not he took the time to tour visitors around—
sometimes people who had come bearing a letter of introduction from
a friend or colleague in common but more often unexpected strangers.
The Smithsonian kept a sign at the base of the stairs in the Tower, warning uninvited visitors from allowing curiosity to draw them upstairs to
the private offices of Ridgway and other staffers. But in the days before
the telephone and instant communication from front desk to back office,
people still showed up unannounced. Ridgway's office was not easy to
reach, being on the fifth floor of the South Tower in the Castle. The room
was spacious but simple. It consisted of a desk with a series of cubbyholes
for filing correspondence and papers, a bookshelf atop the cubbyholes, a
worktable, high ceilings, big windows, and a large library-style stepladder
covered with books and publications.[7] The stairs to his office were apparently daunting—the fit and trim Ridgway recalled there being "about 87"
stone steps, and he enjoyed bounding up them several at a time.

Elevators were nonexistent in the building. One visitor, described as
a very large German man, came up to visit Ridgway one day. "Breathing
heavily, in fact audibly, he stopped, mopped his face, then placing his
right hand over his heart, as his chest heaved visibly, exclaimed, 'Shentle-
mens, my heart bleeds for you,'" Ridgway recalled.[8] Ridgway welcomed
this visitor and almost all others patiently and with his trademark cour-
tesy, giving tours when he had time. His attitude about visitors stood in
sharp contrast to the aging and more famous Baird, who remarked that
"I am obliged to resort to every possible dodge to avoid the crowd and fil-
ter it."[9]

Robert Ridgway in his Smithsonian office in the Castle, early 1880s.
Smithsonian Institution Archives, RU 95, box 19, fol. 25.

Some never forgot their encounters with Ridgway. "A few hours spent with you in the tower (Smithn. Inst.) in 1874 were of the most interest to me of any that I ever spent with any one man in my life," wrote Morris Gibbs in 1883. A physician who became one of Michigan's most prominent ornithologists, Gibbs noted that he had adopted Ridgway's recent schema of classification for his own work on Michigan birds. He corresponded regularly. "I bother you about twice a year, and have done so for about eleven years," he wrote Ridgway in 1888. "I date many plans from the time I visited you, and all have been more or less complete in their fulfillment. Every nook and corner of your room are plainly before me now as when I first left the city. I can remember your showing me the specimens of your collection, and can feel now the same sensations of unbounded gratification in the examination that possessed me at that moment. Neither have I lost one particle of gratitude for the kind attention shown me by you, and I think so often, how easy it is to influence a man's or boy's course and tastes by the treatment he receives from his superiors in his pursuits." Like most others, Gibbs wanted not praise in return but unvarnished opinions from an expert. "Do you read Forest & Stream? If so do you notice my article on Mich. Orni.? If so please criticise my efforts as severely as you can and let me know if you think it is worth while to continue them?"[10]

Most visitors came as tourists and never met Ridgway but walked through the public spaces as part of their visit to the museum's storehouses and treasures. The Smithsonian's bird exhibits were arranged with specific goals in mind. For instance, Ridgway wrote Baird in 1880 with suggestions about displaying duplicates from the research collections. Unique specimens would not be exhibited, he suggested. Mounted examples were most commonly shown, because they could be posed and given glass eyes and other accessories to make them resemble living birds. At the suggestion of newly appointed Smithsonian Secretary Samuel Pierpont Langley in 1889, the Bird Division also set out a very large glass case specifically for children that included birds most often described in popular natural histories. Ridgway made a point of including birds mentioned frequently by poets, so that the case would also have appeal to adults with no particular knowledge of birds. He included the Nightingale, the Skylark, the Condor, some eagles, a Bird of Paradise, a Flamingo, and a number of others. The children's exhibit case also included a few curiosities such as a wren's nest built inside a human skull. The Bird Division's exhibit areas could hold substantial numbers of birds, and in 1888, some 6,714 specimens were on public exhibit. Most of these specimens were mounted. Issues of cost, appeal, accuracy, and conservation of the mounted and taxidermied specimens in public spaces all arose with great frequency. Ridgway wanted these spaces to be inviting to the general public and recommended a number of popular texts on birds to be available to readers who could sit and leaf through these works. In the late 1890s, a number of these were suggested on the Smithsonian's behalf by Florence A. Merriam, a popular writer on birds, and Mrs. L. W. Maynard, another well-known author of bird books.[11] All of these efforts occupied a great deal of the division's time.

Bird identification for a wider public by mail was another common activity for Bird Division staff. Individuals as well as natural history institutions, writing from around the world, sent specimens for Ridgway to identify. Sometimes, as Ridgway himself had done with Baird, people simply described birds in a line or two and hoped for a response identifying them. Others sent specimens of birds and eggs in to provide more details in support of their query. Each of these required a response, and Ridgway answered almost all of them. Sometimes these requests strained the limits of credulity, but he remained polite with every correspondent. Writing to a New York correspondent, Ridgway thanked him for sending along the alleged egg of an African weaver-bird and coolly offered up some useful advice: that he "abandon any further effort to 'establish' your remarkable claim that these birds were really found breeding, in the wild state, by you in the Catskill Mountains." The egg was almost certainly that of a canary, Ridgway concluded.[12]

People often wrote to Ridgway asking if he could help them obtain a permit for killing birds in the District of Columbia. They sent specimens they hoped to barter with the Smithsonian in exchange for other birds. They also penned letters asking about breeding habits, song, nests, and many other field details of living birds. They requested advice about selling early editions of Audubon's books or other works. Many also wanted publications: the institution's official bird checklists were a common request. Some wanted to know what birds they'd find in a particular region they were planning on visiting—almost always for collecting purposes. Others felt bad for asking, but asked anyway. "Please do not consider me a beggar," one writer explained. "I only desire copies of works that I know

The Smithsonian Department of Birds on the southeast gallery of the
Lower Main Hall, National Museum of Natural History, ca. 1886.
Smithsonian Institution Archives, MNH-6063.

are given to so many of our Senators etc. who lay them aside and they are
never read. Is it not so?"[13]

Few topics were off-limits for those seeking information. A Baptist
minister wrote to Ridgway asking for his help with the habits of eagles in
illustrating a sermon, in which the eagle's nest equaled the church, the
eye of the eagle was equated with omniscience, the beak of the eagle sym-
bolized divine protection and defense, and the bird's wing was emblem-
atic of God's swiftness of flight. The institution's correspondents were also
not limited to a small class of well-educated professional males. As Daniel
Goldstein has noted in his seminal article about the Smithsonian's correspondents, they came from all walks of life. A woman artist in California, having read a book on color nomenclature written by Ridgway, asked
how to obtain specific colors of paint because she had nothing suitable
to paint flowers. "Their colors are so beautiful that it seems almost impossible to reproduce them in paint for the most beautiful colors that are

made seem dull beside the gorgeous colors of the wild flowers," she wrote. No doubt recalling Baird's kindness at sending him materials for his own illustration work when all he had was toy paint, Ridgway sent her the colors she needed.[14]

The Smithsonian not only sent publications and information (and art supplies) but also actual specimens, and not just to fellow museums. They aided schools—from kindergartens through colleges—by the donation of mounted specimens, information as to their names, and other useful facts for teachers imparting bird knowledge. Ridgway regularly offered to iden-tify birds for other institutions that called on his authority, such as the Portland Museum, the California Academy of Sciences, and many others. Even Philadelphia's august Academy of Natural Sciences did not satisfy its constituents, who were occasionally moved to call on the Smithsonian. "Our former strong corps of ornithologists at the Acad. of Natural Sci-ences, at Philadelphia, having passed away, we suffer great inconvenience in detirmining mooted questions now," a Pennsylvania collector wrote Ridgway, requesting aid.[15]

The general public and the scientific community were not his only correspondents. Ridgway occasionally heard from some of his friends from the Fortieth Parallel Survey expedition. Botanist William Bailey wrote him in 1873, apparently intrigued by a pair of articles Ridgway had recently written in the *American Naturalist* on the vegetation of the lower Wabash Valley, in Ridgway's native state of Illinois: "I received a docu-ment the other day in your familiar fist, and at once my mind went back to Carson City, Parker, the coyotes, the 'nice girl,' and the many other fascinations of that desert city . . . How long is it since you began to take Botany in yours? I did not know that you took kindly to the weeds."[16] He also counted other government workers among his correspondents, in-cluding members of the government's Division of Economic Ornithology and Mammalogy, which helped the government study and respond to wildlife and its impact on everything from agriculture to buildings to human populations.

The species-rich nature of the neotropics also meant that colleagues in Latin America played a role in developing Smithsonian collections, help which the institution reciprocated. International cooperation between the Costa Rican National Museum and the Smithsonian was especially frequent and quite involved. Both curator José Zeledón and his associate Anastacio Alfaro spent months at a time in Washington, working with

Smithsonian bird collections. This relationship provided a porous barrier for sharing techniques, supplies, and knowledge between the neotropics and the United States. Zeledón supplied specimens by the hundreds, and for his part, Ridgway supplied publications, guns, museum mounts for taxidermy uses, and more. Later, Zeledón also borrowed $4,500 from a Costa Rican lender to loan to Ridgway for the private publication of his 1912 book on colors. He had become a close friend, providing a refuge out of the country for Ridgway and his wife for some months shortly after the death of the Ridgways' only child, Audubon, in 1901. There is even a hint that Ridgway named a pet of his—perhaps a dog—after his friend. Writing to Ridgway in 1883, Zeledón closed, "I received the elegant fotographs of my namesake master Zeledón. He is a sweet and inteligent looking boy. Having my name could not be otherwise."[17]

The press occasionally visited Ridgway at work in the Smithsonian, and their reports bring some of his systematic work to life. "One who can get the privilege of a talk with Mr. Ridgway while he is deftly at work among the myriads of specimens from the feathery kingdom around him will discover that there are thousands of curious little facts about birds and their habits he never heard of before," a Virginia newspaperman wrote. "While the ornithologist is engaged in working out his share of the great problem of life and evolution, which of course, entails much hard and dry study that has little popular interest, he gathers around him much that is beautiful and interesting from the popular standpoint. It is the fairy-land of science."[18] It's easy to imagine the painfully shy Ridgway with his head down, being watched by the reporter. Ridgway, who—like his fellow curators—typically worked a six-day week, was known for his nose-to-the-grindstone attitude, even in a field with other prolific scientists. "I have just received and read my copy of your papers on the Crossbills," wrote his colleague Nathan Brown, the curator of ornithology at a museum in Portland, Maine. "What a worker you are! I hardly take note of one batch of discoveries, before you are out with another."[19]

Some of Ridgway's colleagues were neither quite amateur fish nor professional fowl but a bit of both, such as the members of the Ridgway Ornithological Club, founded in Chicago in 1883. Among other things, this club provided evidence of thirty-two-year-old Ridgway's status as an influential scientist in a different sphere beyond taxonomy, systematics, and collecting: he also provided institutional inspiration. After a flurry of informal meetings, the Ridgway Ornithological Club was incorporated on

June 6, 1884, and elected its officers. The club claimed nineteen members. The president was a twenty-six-year-old Illinois resident named Benjamin True Gault, an independently wealthy collector and writer. Named as vice-president and treasurer was George F. Morcom, an English expatriate who applied for work as a bookkeeper with a commission merchant for the firm of A. S. Maltman and Company in Chicago, despite never having ever entered a line in a ledger in his life. Morcom got the job and eventually became sole proprietor of the company. He also became president of the club in mid-1886.

The founder was Henry K. Coale, a young dealer in coal (what else?) and wood, who took the job as secretary and undertook most of the correspondence. Joseph Hancock, a physician and future author of a book called *Nature Sketches of Temperate America,* served as curator. The club had a large cabinet, six feet high by five feet wide, built for storage of up to two thousand study skins. Rounding out the group was F. L. Rice, the librarian, to help manage the little club's growing library. The group asked for — and received — an oil painting of Ridgway to grace their meeting room and looked to him for a variety of suggestions. The group's existence caught the eye of the British Ornithologists' Union, which reported on their doings in its journal *Ibis* in 1886, and the club lasted until at least 1890. When Coale's parents made plans to move to Washington, Coale sought to barter his personal collection of one thousand study skins for trial employment at the Smithsonian. Ridgway took the offer seriously enough to pass it along to Baird, suggesting that Coale be given temporary work prepping for a bird exhibit at the New Orleans Exposition, although nothing apparently came of it. Coale, who badly wanted a paid position in the bird line of employment, finally landed a job at the Field Museum in 1894, where he served as an assistant in the ornithology department.[20] Groups such as Coale's were semiprofessional in their makeup. While they were paid nothing, they collected birds purely for study purposes, were earnest in their pursuit of science despite a lack of formal training, and published articles in popular journals.

Bird conservation was an area of great interest to many that also cut across amateur and professional boundaries, and the Smithsonian supported conservation-related inquiries from both camps equally, even though they had somewhat different goals. Some correspondents voiced their concern about the killing of birds for plumage on women's hats and other decorative uses. They did so because they decried what they considered to be pure cruelty, decimating the birds they loved. Scientists and the collectors who worked for them, on the other hand, disliked the killing of birds for nonscientific purposes primarily because it gave a bad name to those who wanted to kill birds for scientific study. The Audubon Society was one organization that claimed both groups in its ranks, and Ridgway was a key contact point for its initial 1886 incarnation, the year before publication of its first short-lived *Audubon Magazine*. The secretary, Charles F. Amery, sent Ridgway fifty of the Audubon circulars and thanked him for being a valuable partner in their efforts. Ridgway took an active interest in the group and responded with a list of local names that the society might solicit for help. The region was a nexus for conservation: Amery remarked to Ridgway that "Washington is a large field for operations, and needs all the forces we can send to conquer it." Ridgway's sympathies for conservation work continued in future years, and in 1907 he was elected as honorary vice-president of the Washington Audubon Society. He also wrote the introduction to a popular conservation work by William Atherton DuPuy (1876–1941), his 1925 book *Our Bird Friends and Foes*, which remained in print until at least 1969.[21]

Conservation issues were often not straightforward at the intersection of amateur and scientific, because the term "scientific purposes" could be broadly interpreted to serve a number of competing interests. The officers of the Audubon Society, for instance, despite their distaste for the millinery trade, understood that there was a perfectly legitimate scientific purpose to collecting. Amery suggested a licensing system that would apply to everyone:

> There is not one case in a thousand in which the plea of "purely scientific purposes" has any meaning, but . . . it would perhaps be desirable to award licenses to all applicants, who have a certificate of fitness for the task of skin preserving, and who are prepared to give bonds that they will not abuse their permits by supplying milliners or exporting their skins etc. Permits for the collection of special specimens to complete a public museum, or the private collection of a distinguished naturalist should I think be granted without cavil, but applications for a general permit to shoot every thing for "purely scientific purposes" undefined should I think be subjected to the severest criticism.[22]

The very concept of just what constituted "science" was at issue in debates like these, which were typical. People used the term for different ends and understood it in different ways. To further entangle matters, a distinction between pure science and applied science began to emerge as actual scientific practices became more complex and more distinguishable from theoretical constructs. For some, the term "science" was simply anything that was not literature or art. For others, it had a more specific meaning. Definitions became important because many amateurs considered themselves to be scientists, while taxonomists, who also considered themselves scientists, would not likely attach that label to, say, a physician or an industrialist who collected birds.

The term's published definitions also changed as the decades passed. In 1819, a popular encyclopedia would characterize science this way: "A clear and certain knowledge of any thing, founded in self-evident principles, or demonstration. In this sense, *doubting* is opposed to science; and *opinion* is the middle between the two." This idea of doubt being antithetical to science would be stood on its head later in the century, when doubt would become a central element of science, and at least an implicit rationale for testing hypotheses with the scientific method. By 1867, a new definition of science began to emerge. "A science is a body of truths, the common principles of which are supposed to be known and separated," wrote the *Dictionary of Science, Literature and Art*, "so that the individual truths, even though some or all may be clear in themselves, have a guarantee that they could have been discovered and known, either with certainty or with such probability as the subject admits of, by other means than their own evidence." The element of discovery, or truth-seeking, was thus becoming a more central part of science. Science was also explicitly held up as the opposite of art. The aspect of actively seeking a truth became an increasingly unifying element in scientific thought after Darwin. "All mysteries yield to the solvent of investigation," noted the editors of the *American Naturalist* in 1882. By 1896, science would be described as "a rationally established system of facts and ideas which, over a given range of objects, confers certainty, assurance, probability, or even a doubt that knows why it doubts." Belief alone, the article continued, "founded on the authority of others, not regulated, and incapable of demonstration, or on the imagination or feelings to which a supernatural being is given . . . must be excluded from the domain of knowledge and of science in the broadest sense of the word." The supernatural was banished from scientific domain, doubt played a central role, and untested belief—no matter how widely held—was no longer sufficient to serve as science.[23]

Family Life

Young Robert Ridgway was handsome, slight of build, of erect bearing, with a shock of dark reddish hair and piercing blue eyes. John Muir, who accompanied him on a two-month expedition to Alaska in 1899, described him as having "wonderful bird eyes, all the birds of America in them." He sported a red moustache for virtually all of his adult life, and his brother John once described him, after his arrival in Washington DC, as a "dandy" who was "ever careful in his grooming." In 1871, those keen young eyes spotted a girl walking through the halls of the Smithsonian, and his confident demeanor must have caught hers. Her name was Julia Evelyn Perkins, and she was some seven years his junior—just fourteen upon their first meeting. Perkins was the daughter of one of the engravers who cut the woodblocks from Ridgway's original drawings for the printing for the first volume of a work called A History of North American Birds. They began a courtship, about which he wrote home freely despite her young age. His mother remarked, "Do you still visit the Miss Perkins and is she as facinating as ever I hope you will not get out of the notion of marying and conclude to be an old bachelder You will be much better off when you get a good wife." He waited until she was eighteen, and then asked for her hand. They married on October 12, 1875. The Ridgways' first house, which they rented, was a scant five hundred yards from the Smithsonian in what Robert Ridgway described as being by far the quietest section of the city.[24] In 1888 the couple bought a house in Brookland, a neighborhood in the northeastern part of the district, and settled in.

Back in Illinois, family members' skins were often thin when long intervals passed between letters, especially during Ridgway's first years at the Smithsonian, when his activities and busy life there would have made his time in Illinois seem far removed. "We begin to think it strange that we dont hear from you have had but one letter since Christmas and that was so short that it was not satisfactory believe you didnt even say whether you were well or not I hope you are not geting indeferant [indifferent] as to how we are geting along and have forgoten us but attrebut it to something else perhaps your letters have not reached us or you probabley are so busey you cannot find time to write," his mother wrote anxiously. As the months passed, the gap between city scientist and country family continued to grow. In the spring of 1871, Ridgway sent his mother

The Ridgways' residence at 1214 B St. SW, Washington, DC, ca. 1884.
Left to right: Robert Ridgway's wife, Evelyn Ridgway; Evelyn's youngest sister,
Belle Perkins; Robert and Evelyn Ridgway's son, Audubon Ridgway; Robert's
brother John Ridgway; and Hattie Perkins, Evelyn's older sister.
Smithsonian Institution Archives, RU 7006, box 194, fol. 52.

a copy of a recent paper of his, "A New Classification of the North American Falconidae," but noted wryly, "it's so dreadfully scientific that I'm sure it won't interest you." Ridgway visited his family that year but did not return home again for two years. He was given two months' vacation for the trip, having not used any time the previous year. His mother continued to urge her son in vain to take up the mantle of Christianity: "by the way when are you going to give your hart to the *Savior* Sooner or later we must do this in or[der] to secure our Salvation Dear Robie do not put of[f] this most important [of all] duties." She also urged him not to run with a bad crowd: "I hope and pray God every allmost every hour of my life that you may be able to keep out of the way of temptation and I would sugest that you *abandon intireley* the Society of intemperate persons Shun them as you would aney evil for the man that would tempt you to *drink* no mater what his profesion to friendship is your greatest foe and deserves your *everlasting indignation*," she wrote in 1874.[25] The urging for religion did no good. While his wife, like his parents, was a practicing Quaker, she made no particular religious or spiritual demands on him, and they apparently lived parallel lives, she attending Quaker meetings many Sundays, and him, never.

A number of family woes back in Mount Carmel brought him back home later in the 1870s. In June 1876, a tornado devastated the small town and destroyed both the family business and residence. Ridgway spent two weeks there helping his parents recover. On his other infrequent trips to Illinois that decade, he took his wife and young son, to the delight of family members. Despite his modest salary, Ridgway always tried to provide financial support for his immediate and extended family members when possible. He looked for jobs for relatives, sent money to his parents and siblings, and was generous, perhaps to a fault, with his and his wife's monies.

Not long after he began working for the Smithsonian, and no doubt thinking at times of the familial associations implied by his relationship with Baird, he began his own family. On May 15, 1877, the couple welcomed their first and only child. He was named Audubon Ridgway and was quickly dubbed "Audie." The event was a bright shining light for the Ridgways. Very little is known about Audie, and most of that from unpublished correspondence and other ephemeral source material. Like his father, Audie was interested in birds, accompanying Ridgway on at least one of his expeditions and starting his own collection of bird skins. With a wife, a child, an increasing roster of publications, and steady work, Ridgway was thriving.

The Bird Division

In addition to interactions with the public, a number of institutional elements shaped the reputation, attitudes, and work of the Bird Division's staff. Although he had been working there full-time from 1870 onward, Ridgway did not appear as an employee in the Smithsonian Annual Report until 1874, when he appeared on a list of assistants in various departments, all working under the auspices of Curator George Brown Goode.[26] Two years later, however, all the assistants in the different divisions, including ornithology, meteorology, mammalogy, and paleontology, were promoted to the rank of curator, and Goode was named assistant director of the National Museum. Ridgway was listed as "Professor" Ridgway, a title given in the annual reports to all the curators for several more years, despite their lack of teaching appointments anywhere. The curatorial staff now had official standing for the first time. This gave all divisions and their respective curators a new authority, and the more than ten thousand readers of the Smithsonian's annual reports now knew it.

Part of this refinement of organization was probably due to similar specializations taking place in science education across the country; by the 1870s, most students could take electives in science for the first time. The change in hierarchy was also caused in part by the rapid increase in specimens arriving from around the world, as collectors fanned out to collect and sell material to growing natural history institutions like the Smithsonian. Because Congress had stipulated that the Smithsonian, upon its founding, would be "the nation's museum," the institution's administrators also felt an obligation to be as comprehensive as possible. More material, and of more diverse kinds, meant that more formalized divisions of labor were needed. To give the lead workers in each division more professional status meant that they could also have people working under them, and a greater differentiation of staff was possible within each division in order to deal with the great influx of acquisitions.

As the years passed, Ridgway grew more comfortable in his own authority and ability. A number of his correspondents would address him as "Dr." or, more commonly, "Professor," the latter probably from his title in the Smithsonian's annual reports. But although he received an honorary master's degree from Indiana University in June 1884 for help he rendered in replacing bird specimens after a fire there, he had only a high-school education, never having attended college. Nevertheless, he was a very articulate correspondent, as well as an excellent speller and grammarian, and his writings convey the sense of a well-educated man.[27]

The Smithsonian's workers took pride in their role as stewards of many of the country's largest and most extensive series of birds. The collections grew quickly from the 1870s through the 1890s. The Smithsonian traded some of its duplicates for new birds and bought other collections. While acknowledging their shortcomings, staffers often positively crowed with delight when a rapid series of acquisitions built the collections in a particular area. After receiving several large foreign collections, Ridgway happily noted to Baird in 1884 that "with the exception of the Leyden Museum, we now have probably the best collection of Japanese birds extant, while two years ago we had no less than two dozen specimens!" One especially important acquisition of Asian birds, the Jouy Collection, was acquired in 1883 from Pierre Louis Jouy, who worked for Ridgway in the Bird Division and collected widely in Asia. Other colleagues regularly reinforced the Smithsonian's leading role. Bird curator Joel Allen, ensconced at the American Museum of Natural History in New York City, remarked to Ridgway in 1884 that "your Museum accessions for this month will far exceed ours, I fear, for the whole year! While it brings you much work, it also must give you great opportunities!"[28]

Despite these enthusiasms for their work and the renown of the collections, relatively few people at the Smithsonian made much money. Baird did his best to help with Ridgway's salary, agitating with Secretary Henry for a raise in 1874, describing Ridgway's duties in cataloging birds, arranging skins and eggs, creating sets for distribution to other institutions, and other work. Writing Ridgway several times in the following weeks about his possible raise, he told his young charge that "I do not see how I could get along without you" and to "keep up your heart about it. If you are in immediate want of money let me know, and I will send some."[29]

Bird workers at the Smithsonian, as well as at other museums, tried from time to time to sell their personal collections of bird skins to institutions for extra income. Ridgway had begun skinning and amassing a personal bird collection back in Mount Carmel as a teenager, and it numbered 1,040 by the time he was twenty years old. More than half of these (some 650) had been obtained from the Smithsonian itself over the years, apparently with Baird's blessings.[30] In 1875 he received Baird's permission to keep his collection at the Smithsonian for safekeeping, and in 1880, he offered to sell it to the institution, with the primary goal of buying a house in Washington. By then he owned about 2,400 specimens, consisting of nearly 850 species, of which 343 species and 503 specimens were neotropical—most of them probably obtained by trading with Zeledón. One selling point was that his collection included a number of birds not held by the national museum. Ridgway offered the collection to Baird for $2,000 or, for the neotropical species alone, $750, but Baird declined. He persisted, next trying to sell his collection to the Museum of Comparative Zoology at Harvard. Curator Joel Allen, still at Harvard, wrote Ridgway that its director, Louis Agassiz, would never compete against Baird for collections, whether or not Baird might want them.

Selling a personal collection to raise money was not uncommon; Allen had sold his own collection of 300 birds in 1861 to fund his education. Henry Henshaw would sell nearly 12,000 skins to the British Museum of Natural History in 1888, which made a dramatic dent in its desiderata of North American birds, as well as causing grumblings among his fellow Americans about a presumed lack of patriotism. Henshaw noted that although he had never planned on having his collection leave the United States, he had second thoughts about the relative scarcity of North American material in European institutions. He thus thought it "vastly more important" that his collection should go abroad than to remain in his own country. A number of his colleagues supported this idea after hearing his rationale and, while regretting the loss of the collection, agreed that it would be best served in Europe. Large transactions like this helped to boost the sheer numbers of specimens in the British Museum's bird collections past the Smithsonian's own holdings. Other prominent ornithologists who sold their personal collections included the legendary Philip Lutley Sclater of the British Museum, who sold his birds to the Natural History Museum in London in 1884 for £2,250, an enormous sum. Sclater's collection, however, was staggering. Among the 3,000 birds in his collection were 450 birds that had been new to science and had been described in print by Sclater himself.[31]

Ridgway finally donated his collection to the Smithsonian in the early 1880s, noting in an aside to prominent German ornithologist Count Hans Hermann Carl Ludwig von Berlepsch that he had made a voluntary and unconditional present of it to the National Museum.[32] Von Berlepsch had substantial inherited wealth, and it would have given Ridgway some distress to learn that on Berlepsch's death in 1915, his collection of 55,000 birds was sold to the Senckenberg Museum in Frankfurt, while the struggling Ridgway couldn't even peddle several thousand specimens.

Although he rarely complained about his income, Ridgway issued several mighty gripes about it at Baird over the years. In one blast from early 1881 he grumbled about how artists (and him in particular) were poorly compensated for their work and made other complaints that seemed generally aimed at getting things off his chest. Ridgway's first major work as an illustrator had been on the first three volumes of a five-volume work, *History of North American Birds*. Volumes 1 through 3, issued in 1874, covered land birds, and the last two volumes, detailing the continent's water birds, would later appear in 1884. The volumes on the land birds earned considerable fame for the authors, Baird and Thomas Brewer. Ridgway had been paid well for his artwork. But he had several bad experiences in getting paid for his artwork on other intervening projects, and when he was then asked in 1880 to hand color the plates for the projected last two volumes, he exploded. In particular, he railed bitterly about vague promises for possible payment for his artwork at some indefinite future time. "It seems to me hard to expect one who (excepting yourself) had by far the greatest labor to perform in the preparation of the work, and this in the midst of engrossing duties, to thrive upon the rather barren honor of authorship, and the still less nourishing food of possible future remuneration, the eager anticipation of which is apt to be the only reward." He was also simultaneously frustrated by his Smithsonian paycheck, and concluded his letter to Baird: "In short, I think that considering the vast amount of labor which I daily perform in my capacity of curator of a collection of 50,000 birds, and for which the salary is an altogether inadequate compensation, that it is hard to have to devote the little leisure time at my disposal to work which [will] not materially benefit me."[33] Although being paid a full-time wage was a key marker of the professional scientist, the amount of pay was not. "Poverty is about the only trouble I have," he noted to his close friend William Brewster in 1880.

These experiences left Ridgway somewhat cynical about naturalists' chances of financial success generally. "I would say that if one is without the means of living independently of the study of natural history (in any of its branches), he stands a very poor chance indeed of deriving the greatest benefit from such study," he wrote to one mother who asked for advice about her son's prospects for becoming a professional naturalist. "It is very necessary, in fact, to have some regular business, profession, or other occupation which shall supply the means of defraying one's expenses." The study of natural history was certainly a pleasurable undertaking, he agreed, but "as a *means of livelihood*, however, it must, in at least ninety-nine cases out of a hundred, prove a complete failure."[34]

Less ethical routes to boosting curatorial income were available, and unfortunately, even museum professionals were not immune from the promise of easy money, which could often be made by selling bird feathers. Zeledón remarked that along with specimens he collected and sold to institutions for scientific purposes, "a good deal of bright colored species can be sold to dealers for ornamental purposes. My mind is made up already to make this matter a business and give it a fair trial." This would have been horrifying to Ridgway to hear, especially with distaste running high in the United States for the millinery trade. However, the low salaries of curators, especially in Latin America, would have made such a proposition understandable.[35]

American salaries in the sciences generally were not very good, either in the rising universities or in other public or private institutions such as the Smithsonian. One writer described "the frugal table which the field of science has spread for ambitious young men who desire to live, or at least exist, on a purely scientific diet." But personal enthusiasm for the field, along with a sense of commitment to an institution's larger goals, made the issue of remuneration secondary, if frustrating or demoralizing at times. Although he was offered a position at least one other major museum, Ridgway threatened Baird that he was going leave only once, in the late summer of 1874, due to a difficult employee working under him — the aforementioned Pierre Jouy, who collected birds in Asia for the Smithsonian — as well as to his frustration over his salary. Money was often on his mind, and he agitated for raises for fellow workers in the Bird Division, from the taxidermist to his own assistant.[36]

The passing of years did not help much. The Panic of 1893, caused by railroad overbuilding and shaky railroad financing that led to a series of bank failures, meant that congressional allocations to the Smithsonian suffered, and staff had to tighten their belts even further than their usual lean operations dictated. In some cases, museum salaries were three and four times lower than those paid for similar positions at other institutions, and the effects lingered for years. Ridgway contemplated retiring from the Smithsonian as early as 1906. "The sooner I separate myself from the miserable existence of a 'Government Pauper' and become a free citizen . . . the better it will be for me in almost every way," he wrote bitterly to a friend. "It is a lamentable fact that *there is nothing to look forward to in the Government service.*" He remained in the Smithsonian's employ, but it was not until the last several years of his life that Ridgway's income became respectable by national standards. His salary and then his civil service pension during the last two years of his life went from $3,800 to $4,800 annually, and because these represented his many years of seniority, we can assume that this was the most he ever made per year. This was respectable pay in an era where Prohibition agents who engaged in often dangerous work for the federal government made a maximum of $2,800—but it had taken him a lifetime to get there.[37]

Despite the poor pay, various correspondents with Ridgway frequently sought work in the Bird Division. A typical letter along these lines was one from Christopher Greaves, a published author (via a little pamphlet titled *Half-Hours with the Birds,* 1900) who wrote Ridgway: "I also wish to ask you if there is any way of obtaining positions under the government in an ornithological line. I have long desired to take up work of this kind actively. I have long been an observer and writer on bird life since a boy, and have all my knowledge from practical experience[.] Are the[re] competitive examinations or in what way are the Govt. appointments made?" These frequent inquiries rarely resulted in jobs, but many felt compelled to ask. Family ties sometimes made getting work with the Smithsonian more likely. Ridgway's brother Charley found work as a telegraph operator there, and his brother John got a job with the Smithsonian as an illustrator between 1881 and 1882 after Ridgway recommended him.[38] John was also hired a decade later to do taxidermy work, at which he apparently already had some experience, in preparation for the 1893 Chicago World's Fair. He was employed for three months, at $1.35/day, for what would have been a measly $105 for a quarter-year's full-time labor.

Low salaries at least made it economical to engage more people in the work. By 1889, Ridgway felt confident enough of the Smithsonian's ability to support an enlarging staff for the Bird Division that he floated the idea of a full-time aide with Goode. He had in mind Charles W. Richmond of the Geological Survey. "The urgent need of an ornithological assistant is constantly felt, an undue proportion of my time being, under existing conditions, required for ordinary routine work and various details which, in justice to the department should not be done by one whose time is so valuable as that of a curator in the National Museum. By an *ornithological* assistant, I mean a person seriously interested in the study of ornithology, reasonably advanced in his knowledge of the subject, and sufficiently intelligent to render effective service."[39] Several capable men had worked in the Bird Division but kept moving on to other jobs. Goode approved Ridgway's request, and the twenty-one-year-old Richmond, had who started at the museum as a night watchman, proved an outstanding choice. He had first visited Ridgway as a young boy by making an unsolicited visit up the five flights of stairs to his office. He showed Ridgway his own collection of birds' eggs, watched him identify them with supernatural speed, and was hooked.

Perhaps overqualified by education, Richmond earned an MD degree in 1897 while working in the Bird Division in hopes of supporting his family by practicing medicine if necessary, but he never worked as a doctor. By the first decade of the twentieth century, Ridgway was heavily occupied with his work on *The Birds of North and Middle America* and relied on Richmond for many tasks. When he was ill, which was fairly frequently, Ridgway offered Richmond the chance to write reviews and other brief pieces for publication in his place. Steady, smart, and efficient, Richmond worked his way up through the division to become the director exactly thirty years later. Although he detested fieldwork, he thrived in the office setting, became an international authority on bird nomenclature, and kept an enormous, unique card file used by ornithologists around the world to sort out nomenclatural details.[40]

As a rule, both Ridgway and Baird were candid with colleagues and correspondents about what the museum lacked. They hoped that congressional funding would alleviate some of the financial stresses that were, in their opinion, becoming more and more obvious—the lack of a suitable working library, or convenient or comfortable laboratories and other work spaces. "May we hope that all these imperfections may soon be remedied, and that under the auspices and fostering care of our enlightened and generous government the United States National Museum may eventually become, in every department and in every feature, without a rival," Ridgway wrote in *Science* magazine.[41]

To complement their collectors in the field, and perhaps as a panacea for relatively modest salaries, scientists at most American natural history museums were allowed ample time in the field for collecting while on the job, typically working for four to six weeks at a time away. Often, these trips were as much of a vacation as work. Ridgway and William Brewster took a collecting trip of six weeks' duration to Illinois in 1878. The normally reserved Ridgway wrote to Brewster that on hearing the news that Brewster would accompany him back to Olney, "I believe I gave utterance to something very like an Indian war-whoop—much to the astonishment of the young man in the room above."[42] After a frenzy of planning, they met in Washington, took the train to Mount Carmel, and collected several hundred specimens. The two men reminisced about the trip for decades.

Ridgway's upbringing in a bucolic setting and his several years on the King Survey made certain aspects of city life in Washington difficult. "I get fearfully 'rusty' and discontented here, and probably will never get 'weaned' of my desire to live in the country, with *living* birds for my companions," he once wrote Brewster. Time out of the office naturally helped overstressed workers, and Ridgway was no exception. "In accordance with Article CLIV of Regulations governing the administration of the National Museum, I hereby respectfully apply for two months leave of absence, to date from April 15th or May 1st, as may prove most expedient," he wrote Baird in the spring of 1883. "My reasons for requesting *two* months leave are, first, that I have been unusually hard-worked for the past two months . . . so that I feel the urgent need of out-door recreation. In the second place, I am very desirous of making collections, *for the National Museum*, in a locality not previously worked, and which from inspection I am sure would prove by far more productive of desirable specimens than any place yet visited by me . . . upon my return I would, in consequence of renewed strength, be able to perform duties which, in my present jaded condition, I feel that I would be wholly unequal to."[43] His request was granted.

Earlier in his professional life, activities in the out of doors were usually happy events for Ridgway. Traveling to other locales on collecting trips was an important part of Smithsonian curatorial life. However, Ridgway balked more and more in later years at leaving home for the uncomfortable life required to get to remote areas for specimen collecting. This was not the gentler outdoors of his time in Olney, especially when he had to cook, camp, and sleep in the wilderness. Insects, poison ivy that rendered his hands unusable, undrinkable water, annoying companions, and other hazards were often present. Still, he understood the value of such expeditions for the Smithsonian, and during the 1880s and 1890s, he made collecting trips to Virginia, Florida, Maryland, Indiana, and Illinois, first at Baird's direction and then, after his mentor's death in 1887, on his own initiative.[44] He also visited Costa Rica several times, spending time with his friend José Zeledón of the Costa Rican National Museum.

A number of rare species now extinct or nearly extinct were the object of a number of Ridgway's collecting efforts. One especially memorable outing to Florida in February and March 1898 reveals much about the hardships, the efforts involved, and Ridgway's personality. He focused on locating specimens of the Carolina Parakeet and the Ivory-billed Wood-pecker, both already scarce and highly localized by the end of the nine-teenth century. Writing to Richmond from his camp at Lake Trafford, near a Seminole Indian agency in Lee County, Ridgway noted that three days of searching in the Big Cypress region twenty-five miles south of camp revealed not a single woodpecker but lots of turkeys. "We have prac-tically lived on turkey stew since coming to this place," he lamented. "It is very certain that the Ivory-billed Woodpecker does *not* occur in this por-tion of the State; neither does the Parroquet."[45]

This trip was particularly challenging. A severe drought had hit Florida in the fall of 1897, and in order to collect water, the team had to push aside dead fish floating in the shallow, muddy lakes. The water was virtually impossible to drink and had to be flavored with other elements. During this trip, however, the greatest hardship of all was his unnamed assistant. A more thoroughly disagreeable camp and traveling companion could scarcely be imagined, he wrote to Richmond. While the man had proven quite satisfactory as an assistant, Ridgway had not bargained for his profanity, unsolicited advice, and other unspecified difficult qualities. "All that I have gone through in the way of bad water, hard tramps, lack of success in finding what I was looking for, etc., have been mere trifles com-pared with the infliction just described," he lamented to Richmond. "We camped at Corbett's exactly one week during which time I went through more cussedness of all sorts than in any one entire year of my life."[46]

There were other opportunities to travel, and in 1899 he was part of an expedition of great importance to science. The outing was the Harri-man Alaska Expedition, funded by Edward Harriman, a railroad tycoon and one of the most powerful men in America. Heeding his doctor's ad-vice to rest, he planned to hunt bears in the far north. Never one to make small plans, however, Harriman decided to make the undertaking a major scientific one, and assembled a team. The size of the party put Clarence

King's earlier Fortieth Parallel Survey expedition to shame. It included twenty-three members in the scientific party alone, along with three artists, including the famed Louis Agassiz Fuertes; two physicians; a pair of taxidermists and specimen preparators; two photographers, including Edward S. Curtis, who would be influenced on the trip to begin a career as a portrayer of native American culture; naturalist John Muir; two stenographers; a chaplain; poet John Burroughs, who was tasked with writing a good part of the trip's narrative; eleven members of the Harriman family; and a complement of engineers and the ship's crew. It was a high-powered group, full of big thinkers. "Scientific explorers are not easily managed, and in large mixed lots are rather inflammable and explosive," warned Muir, the group's sage, "especially when compressed on a ship." However, Harriman's sense of luxury kept the stress to a minimum, perhaps due to the presence of cases of champagne and cigars, a piano, a five-hundred-volume library, eleven steers and other livestock, and a stateroom for each expedition member.[47]

Over its four-month tour, the expedition identified some six hundred species of plants and animals as being new to science (although many would later prove to have been discovered earlier), including thirty-eight new fossil species. The team also mapped the geographic distribution of many species—important information after Darwin, for the implications of evolution for biogeography were now clear. The group also discovered an unmapped fjord and named several glaciers. Although he had little to say about the trip in his correspondence and virtually nothing in print, Ridgway did note to Brewster that he was initially unhappy to leave his labors and his home behind. "It seems that at last I am to have an opportunity of seeing the North West Coast, having been favored by an invitation to join the Harriman expedition," he wrote his friend. "For many reasons I have hesitated about accepting the invitation, but have concluded that I ought to take advantage of the opportunity, which is one that may never occur again . . . Much as I desire to make the trip I shall very reluctantly leave my work, which is now progressing smoothly and expeditiously, and my place is growing more attractive every day." He described the birds and plants coming to life that spring around his Brookland home and remarked wistfully that he would leave them all with much regret.[48] He made no mention of missing his wife.

Although he had a relatively minor role in the outing, the trip gave the normally antisocial Ridgway the chance—or perhaps, the necessity, being shipbound with others for weeks—to interact with colleagues. He knew a number of the members of the scientific party well, either by reputation or as correspondents, including William Healey Dall, one of his fellow Bairdians; Daniel Elliot, renowned bird illustrator, curator, and American Ornithologists' Union founding member; and four old bird friends: Albert K. Fisher and Clinton Hart Merriam of the Biological Survey, George Bird Grinnell of *Forest and Stream,* and Charles Keeler of the California Academy of Sciences. It would have been an opportunity for Ridgway to exchange ideas, make scientific connections, renew friendships, and do important exploratory work.

These outings in the wilderness, all sanctioned by the Smithsonian and undertaken on Smithsonian time, served as a valuable counterbalance to the crowding staff often endured in the small workspaces in the Bird Division, where there was little privacy. Strangers and employees literally stood shoulder to shoulder at times. Ridgway's taxidermist, a Mr. Marshall, labored in conditions that were too small and *"and altogether too public,"* Ridgway noted emphatically to Baird. Ridgway described the public access to Marshall's space as being "one of the most serious disadvantages under which an artisan of his profession can labor. Many of the specimens which he has to mount try the patience and temper to the utmost, and to have strangers and others looking over ones shoulders at such times (and this occurs daily), is exasperating to an extreme degree. I venture to say that no artisan could do justice to his work under such circumstances."[49] Taxing conditions were certainly not limited to the Smithsonian. A first-person account in London's *Weekly Journal* in 1890, consisting of an interview with the British Museum's legendary ornithologist Richard Bowdler Sharpe, detailed the feeling of unhealthy bird spaces: "[I] quickly found myself in the 'Bird Room,' as it is called," the anonymous author wrote. "The atmosphere of this room seemed scarcely conducive to health, for the August sun beat fiercely on the skylights, and a strong smell of camphor and of spirits of wine pervaded the apartment into which I had been ushered."[50]

At the Smithsonian, the rapid pace of new birds coming in and the hectic nature of life in the Bird Division caused in part by these crowded conditions meant that Smithsonian security for collections could be lax. In the spring of 1884 a number of bird cases were left unlocked by an unknown staff person, putting the collections at risk of theft from visitors who strolled through the Bird Hall to view live songbirds on display and examine a number of mounted and prepared specimens. At other times, Smithsonian specimens wandered out of the building, either accidentally or willfully (although not under their own power). A Smithsonian employee wrote Goode in 1884 that he had "noticed yesterday in front of a shop on 12th St. between Penn. Ave. and E. St. a case of bird skins several of which bore regular Smithsonian labels. Thought I would call your attention to it." Upon investigation by Ridgway, these turned out to be cataloged duplicates either sold or given to collectors in exchange for other specimens.[51] Still, to have bird skins bearing Smithsonian labels out on the street for sale could not have been good for the division's reputation, and the occasional lack of communication among staff was troubling.

Even when they didn't leave the museum, specimens were mislaid regularly, often for years, due to the pace at which curators worked and the size of borrowed collections they reviewed. In 1900, Ridgway wrote to Goode's successor, Richard Rathbun, that the ornithological staff had to work so quickly and with so many different duties that things were getting lost. "At the present time, for example," he noted, "we have specimens belonging to the American Museum of Natural History, the Academy of Natural Sciences of Philadelphia, and to Mr. Joseph Grinnell, of Pasadena, California, which have not yet been found. There are not very many of these, but even one specimen thus overlooked may cause much trouble and inconvenience, not to say more serious consequences. In many cases, too, the labels of old specimens are so faded or grease-stained as to be almost illegible."[52]

Besides taxidermists standing toe-to-toe with the public and mislaid specimens, environmental controls in the era before air conditioning were, to put it mildly, erratic. Although the tall windows in Ridgway's office on the fifth floor of the South Tower commanded views of the Washington Monument to the west, the Potomac River to the south, and the Capitol to the east, the heat was excessive during the summer. Ridgway, writing to Joel Allen in June 1884 from his Tower office, remarked, "Hope you are standing the hot weather without injury to your health. The therm. at my desk registers 91° as I write this!" A curator would later crack that a scientist might be in danger of being roasted to death by the sun coming in through the tall windows.[53]

And with fluctuating temperatures, more than just the humans suffered; much could go wrong with basic materials. Eight of Ridgway's custom-made bird cases, kept in the basement, were apparently poorly manufactured out of unseasoned lumber, and they shrunk, cracked, and warped "to such to such an extent that not a single one of the sixteen double doors can be closed, nor scarcely a drawer put in or out without infinite trouble," Ridgway complained to his boss in 1884. Other aspects of the physical plant were less than ideal. Elliott Coues, used to spending weeks or months at a time out of doors and hardly one to complain about physical hardship, wrote to Chief Clerk William Jones Rhees to complain about the "very grimy walls" in his own office that were in need of cleaning, saying that "In its present state the room is scarcely habitable, and in a state far from being worthy of the Institution."[54]

Despite these challenging conditions, the Smithsonian worked hard to shine elsewhere, maintaining a presence at a variety of expositions and world's fairs around North America and the world, including Berlin and Paris. In the 1880s alone, it attended at least a dozen fairs, presenting exhibits and winning awards. These helped to make the Smithsonian a household name at home and abroad. The Bird Division had a presence at most of these venues. The popularization of scientific ornithology through these expositions bled a number of scientific issues into more public arenas and in the process raised the quality of popular discourse about science. One important aspect of these fairs was the presence of evolutionary thought as an undergirding principle in public science by the last decade in the century. The Chicago World's Fair, also known as the World's Columbian Exposition, held in Chicago in 1893 to celebrate the four hundredth anniversary of Christopher Columbus's arrival in the New World in 1492, was perhaps best known for its portrayal of architecture and urban planning. However, it also introduced millions of visitors to evolutionary ideas. Many of these ideas were couched in the fair's— and in the Smithsonian's—presentation of racial issues. They promoted American racial and ethnic hegemony and transmitted a new vision about American progress: the central role of evolution.[55] Many of the attitudes imparted by the Chicago World's Fair and others were Spencerian in nature—tuned to Herbert Spencer's idea of "survival of the fittest," a term Darwin had added with some reluctance to the fifth edition of the *Origin of Species* in 1869. Many implications of evolution were evident in specific Smithsonian displays. Attending a fair was a chance to learn about the distribution of birds, for instance. At the same time, the expositions helped the Smithsonian to coopt more popular and widespread public attitudes about science.

The Chicago World's Fair was a showcase for Ridgway. He was proud of his work, trying hard to make it interesting to both a scientific and popular audience, despite a shortage of institutional labor. Preparing specimens for the fair required creativity and attention to detail. Ridgway bought wild turkeys at market for one display. Trying to set up another exhibit involving a realistic setting for wild pigeons, he sent an assistant to Laurel, Maryland, to procure a branch of a water oak (*Quercus nigra*), which bore a particular acorn eaten by pigeons. Such tasks were highly labor-intensive, requiring a full day's work for an employee. But the rate of pay was so low that it did not seriously tax the institution's finances, and, in fact, was a form of efficiency allowed by the low salaries. Ridgway set out for Chicago in early April 1893 to supervise the work at the World's Fair. His wife and son visited the exhibit as well that summer, although Ridgway never saw it in finished form himself.[56]

At the Smithsonian, Ridgway was the undisputed king of birds—and for the American public, the Smithsonian was the undisputed national palace of science. The role of authority systems as a defining factor in the development of a professional scientific class was played out in complex ways at the Smithsonian. Along with scientists, the staff interacted with a large and broad public from all walks of life. The role and place of amateurs was thus influenced by their relationships with the Smithsonian, and amateurs' scientific work was often injected with a measure of scientific rigor through these interactions.

Other collective forms of American scientific authority certainly existed before and during the last quarter of the nineteenth century. The American Association for the Advancement of Science, founded in 1848, as well as the National Academy of Sciences, dating to 1863, had widespread effects of their own, through membership, meetings, and publications. But their impact was relatively limited compared to the Smithsonian. Created and supported by the federal government, the Smithsonian was partially funded by taxpayer dollars, had enormous collections, and was located within the crucible of American politics and history. Through broad interactions with the public, in person and by mail, via public exhibits in their buildings, and at fairs and exhibitions, the Smithsonian gave science a credibility and authority that was both institutional and national. Through these means, the Smithsonian's presence raised the nation's exposure to the new systematic science of the era. As

an authority on science, whether broadly or narrowly defined, the Smithsonian's Bird Division and its employees both assumed the mantle of authority and were given it by a broad general public. Ridgway's years with
the Smithsonian were in many ways the most populist aspect of his stint
as a scientist because he interacted with many thousands of people from
all walks of life, with only one commonality: birds.

Besides serving as a scientific touchstone for a lay public, Ridgway
and the Smithsonian facilitated much more technical work: that of bird
specialists, who contributed to the scientific reputation of the institution
and to their own profession. In 1885, Joel Allen, formerly at Harvard and
newly hired as curator of ornithology at New York's American Museum of
Natural History, described the Smithsonian as "the great ornithological
centre of America." Bird men from around the world used specimens at
the institution, received them by mail, and worked assiduously to sort out
issues of taxonomy and nomenclature, systematics, and bird biology. The
wide use of the Smithsonian's bird holdings was such that by 1932, Curator of Birds Herbert Friedmann was able to write to his counterpart Frank
Chapman at the American Museum of Natural History that more of the
basic systematic works on North American birds had probably been written on the basis of Smithsonian collections than any other single North
American institution. "Ridgway's great works, Coues, Henshaw, Nelson,
and others have all based their papers chiefly, if not wholly, on the collections in this Institution, and practically every worker on American birds
from the time of Audubon and Bonaparte up to the present is represented
by one or more types in our collection," Friedmann observed.[57] The role
of study collections, especially the Smithsonian's, is taken up in chapter 4, and was a key part of the fabric of professionalization because of the
collections' intersection with classification, language, and accountability.
The institution's publications, many stemming from this work, also radi-
ated out to a wider world and helped to extend its reach. Although these
writings were usually quite technical, they were printed in large numbers
and widely circulated.

Key individuals both inside and outside the Smithsonian helped to
give credibility to the institution's reputation and place. At the Smithsonian, Baird had vital influence on the growth and professionalization
of American science through his immense reach. He simultaneously directed the US Fish Commission, the United States National Museum at
the Smithsonian, and the entire Smithsonian during the last decade of

his life. Through these channels, he was able to exemplify the qualities of precision and his "peculiar exactness in dealing with facts" that Ridgway described in Baird's obituary. Ornithologist Leonard Stejneger also noted a specifically American attribute of the Bairdian school of ornithology, which Ridgway distinguished from the European approach: "The European school requires the investigator to accept an author's statements and conclusions on his personal responsibility alone, while the Bairdian furnishes him with tangible facts from which to take his deductions."[58]

As a scientist, Baird also focused keenly on systematic biology, and he transmitted many of these ideas to his staff. His key areas of interest became comprehensive for the institution as a whole, rather than simply representative. The value of long series of specimens and eggs, which allowed researchers to address Darwinian ideas of intergradations of species, became clearer with each passing year. These lengthy runs of birds "allowed for minute comparative studies, and yielded more exact biogeographic information, contributing to a more analytic approach to problems of speciation and evolution."[59]

Another element in building authority for American science was a group of scientists, some with Smithsonian ties, who dubbed themselves the Lazzaroni. Robert V. Bruce, in his 1987 Pulitzer Prize–winning book *The Launching of Modern American Science,* gives these men substantial credit for the rise of science after the middle of the nineteenth century. With a half-dozen scientists at its core, and others contributing from the periphery, the group "focused and projected the needs and aspirations of American scientists like a lens."[60] In some ways their institutional placement mirrored that of Smithsonian bird men like Baird and Ridgway and their Harvard counterparts William Brewster and Joel Allen. The four senior Lazzaroni were split evenly between Washington and Cambridge: Smithsonian Secretary Joseph Henry and geophysicist Alexander Dallas Bache and, at Harvard, naturalist Louis Agassiz and astronomer Benjamin Peirce. The group gathered together as scientific organizers and policymakers to influence scientific practice in America.

The Smithsonian's work in world fairs, especially at the 1893 Chicago World's Fair, also helped to embed scientific practice into American culture more broadly. The institution contributed "materially and conceptually to the organic whole of the exposition," as one historian has noted. Assistant Secretary Goode played a key role in shaping the exposition's message. He was an enthusiastic promoter of Darwin's ideas and would become an expert on Mendelian genetics and heredity. Years earlier, Baird had tapped him to oversee the Smithsonian's exhibit of animals at the Philadelphia Centennial Exhibition in 1876, and Goode had cut his teeth there. Goode thought of museums as systems of organization, and concepts of taxonomic order fit neatly into that idea. "The people's museum should be much more than a house full of specimens in glass cases," he wrote in 1896 in the Smithsonian's annual report. "It should be a house full of ideas, arranged with the strictest attention to system."[61] As the nineteenth century closed, no ideas of system in the natural world were more compelling to most scientists than Darwinian evolution, and if not all aspects of Darwin's work were accepted, many were already in the mainstream.

The Smithsonian became an authority in science in part because it tried to serve the widest possible audience, including both lay public and scientist. Goode noted, "The museums of the future in this democratic land should be adapted to the needs of the mechanic, the factory operator, the day laborer, the salesman, and the clerk, as much as to those of the professional man and the man of leisure." It was proper, he said, that there be laboratories and professional libraries for the development of the experts who worked organizing, arranging, and interpreting the contents of the museums, and museums that were not expert in their fields could not grow if they could not give proof of their claims to be centers of learning. On the other hand, he felt, "the public has a right to ask that much shall be done directly in their interest. They will gladly allow the museum officer to use part of his time in study and experiment," because scientific specimens were, he allowed, "the foundations of the intellectual superstructure which gives the institution its standing."[62] He supported science for all—a populist approach that left room for the casually interested, the untrained, and the dilettante, along with the systematist and nomenclator. This gave the Smithsonian a particular aesthetic mix that occupied a broad spectrum, as noted by the reporter who described the Smithsonian as "the fairy-land of science" that also supported "hard and dry study." It had a wide appeal in its displays in Washington, as well as in its presence in expositions and fairs across the country, and in its hands, the popular and the scientific could coexist happily.

For his part, the experience Ridgway gained at the Smithsonian helped him to serve as the institution's authoritative institutional voice about birds. A wider public recognized that he held special knowledge, asked for help, and received it. As a spokesman for science, Ridgway was ideal, because he was polite, knew people, and had substantial institutional resources at hand—if not monetarily, then informationally. He raised the quality of public understandings of the scientific study of birds and in the process challenged professional ornithologists to differentiate themselves from the larger public by reaching to greater heights. As we will see in the next chapter, his role in professional bird organizations allowed him to further this delineation, which would play out most dramatically on the pages of both popular and scientific journals about birds.

To Have or Have Not:
America's First Bird Organizations

The whole point of this particular kind of wise and
diplomatic humbuggery lies in its exclusiveness.

ELLIOTT COUES to Joel Allen, writing about the makeup
of the membership of the Nuttall Ornithological Club,
January 10, 1878

BEFORE 1873, America had no organizations or publications devoted exclusively to ornithology. That year, however, the
Cambridge-based Nuttall Ornithological Club (NOC) was born,
and eventually gave birth in 1883 to a new organization with greater aspirations and a national focus: the American Ornithologists' Union (AOU).
The founding of the AOU was a key moment in the professionalization of
science and receives the lion's share of scrutiny here. A handful of members in these organizations made fundamental decisions about who would
be allowed into the top tiers of membership and who would not. How
would the size and makeup of the founding members be established, and
who would do the deciding? The creation of a professional group played
havoc with some of the founders' insecurities about their status within
ornithology; at the same time, the founders' decisions created distinctions between amateurs and professionals that remain today. The professionalism they espoused was rough-edged and erratic at the outset,
often marked by differing ideas about what ornithologists' future goals as
a group should be. It was a period marked by bickering and often insular
thinking, as well as by grand plans.

As it loomed over the Smithsonian, the shadow of Darwin also stood
over professional natural science organizations after 1870. Because the
similarities and differences among and between species were no longer
something solely aesthetic or divinely ordained, their relationships became something worth studying in more detail, as well as more compara-

tively across institutions. But getting that information out of the study rooms of museums required publication in the pages of professional journals. With a broad scientific literature in hand that reported on discoveries and facts extracted from study collections, naturalists could do something new: they could think more and more synthetically about categories, and all that categories and subcategories meant to our greater understanding of the organization of the natural world. Darwinian evolution provided the "why" of biological change: plants and animals competed for food sources and reproductive success. From the 1870s onward, the close study of organisms began to provide the "how": the ways that their ranges, physical characteristics, and similarity to allied species interacted, thus providing a map to a greater understanding of evolutionary change.

The early 1870s were especially relevant to the formation of modern natural history organizations because of a key publication of Darwin's: the *Descent of Man, and Selection in Relation to Sex*, issued in 1871. The book spoke clearly for the first time about the mechanisms of sexual selection, and this was a topic of direct interest to the study of natural history. Darwinian thought continued to infiltrate society, but now it did so with greater urgency because it spoke to the human condition. However, the *Descent of Man* said volumes about speciation generally. In its review of the book, the *British Medical Journal* noted, "All recent research has an evolutional character. Everything tends that way in the present state of our knowledge." Closer to home, the *North American Review* stated that the *Descent of Man* opened the way for "a freer discussion of subordinate questions, less trammeled by the religious prejudices which have so long been serious obstacles to the progress of scientific researches."[1]

Scientific groups provided a way to speak the same language about science (or at least to argue in the same language) and through their publications could reach a wide audience. These groups began to form with greater frequency and specialization. Organizations like the Royal Society of London and the Linnaean Society, also London-based, had existed for a couple of centuries, but they were quite broad in their subject areas. More and more scientific groups began to form in America, with increasing specificity: the American Chemical Society (1876, now the world's largest scientific society), the American Microscopical Society (1878), the American Laryngological Association (1879), the American Association of Genito-Urinary Surgeons (1886), the Geological Society of America (1888), the American Association of American Entomologists (1889), the

Botanical Society of America (1893), the American Astronomical Society (1899), and many others. They served as a mechanism to protect against encroachment by sometimes-powerful individuals with different ideas than theirs; they provided for a sharing of public and private ideas about their fields through annual meetings and journal articles; and they allowed small clusters of scientists within those groups to generate a critical mass of opinion and credibility on a particular topic. In turn, these opinions and ideas—often backed by data but sometimes simply using the weight of an expert group's opinions—trickled into other spheres. They drifted slowly but steadily toward national centralization, which gave them broader authority than regional groups had previously been able to claim. Newspapers and popular journals reported on these groups and societies. People corresponded with their own extended networks of friends and family about them. And the scientists themselves produced numerous books about these ideas. Sometimes these volumes were produced by professional organizations and were intended for a narrow audience, but more often than not the expenses of publishing meant that they needed to reach as wide an audience as possible to succeed financially.[2]

Universities also began to serve more capably as training grounds for science after Darwin's works were in wide circulation. Elective courses—whereby students could select from a series of possible offerings rather than facing strict requirements to take a tightly defined roster of classes—finally came into widespread use by the early 1880s. This reduced the total number of students taking science courses—they could, after all, choose nonscience topics more readily than before. But it meant that those who were compelled to scientific education had more choices, and these students spurred their teachers to higher levels to match their interests. A tighter focus on scientific training also meant that more people identified themselves as scientists. In 1876, about three thousand Americans used some form of scientific training in their paid work.[3]

Other new elements helped the rise of organized science. The increasingly wide acceptance of Darwin's theories, the rise in US population after the Civil War and the need for cities to manage growth, and the impact of the Industrial Revolution spoke to the theoretical and practical benefits of a systematic approach to many daily things. The telegraph, which linked the country in 1861, along with the completion of the transcontinental railroad in 1869, helped to make America a truly national federation. After the Civil War steam-powered manufacturing surpassed water-powered manufacturing, freeing cities from the necessity of being near a waterway to make things. And as companies became truly national, stretching from San Francisco to New York, finances became correspondingly continental in scope. All of these things contributed to an environment that was ripe for the rise of national, discipline-specific organizations.

The Nuttall Ornithological Club

If it is possible to set a starting date—and even a time—for the actual beginning of the profession of ornithology in the United States, then the date of Monday, November 24, 1873, and the entirely reasonable hour of 7:30 p.m., at the residence of William Brewster in Cambridge, Massachusetts, serves as a leading candidate. On this evening the Nuttall Ornithological Club had its first official meeting. The gathering in Cambridge had its own antecedents. As was often the case for young naturalists in the later nineteenth century who would become leaders in the natural sciences, the founders' initial inspiration was naturalist and illustrator John James Audubon. Beginning in the autumn of 1871, Cambridge High School friends and constant companions William Brewster and Henry Henshaw—who would later become two of Ridgway's best friends and central members of the feathery tribe—began to get together once a week in the attic of the Brewster home at 145 Brattle Street in Cambridge to talk about birds, examine study skins, and pore over an octavo edition of Audubon's *Birds of America* owned by Brewster. (In apparent homage by a Cambridge city planner, a Brewster Street now crosses Brattle Street at one point, and a Henshaw Street can be found fewer than four miles away.) The two boys read from the Audubon volume aloud, comparing what they saw in the book with what they had found in their efforts to study birds in nearby Norton's Woods and elsewhere in the immediate Cambridge area. Soon three other friends joined the group: Ruthven Deane, Henry Purdie, and William Scott, later curator of ornithology at Princeton.

It is notable that the Metaphysical Club was just then being formed in the same small town. The Metaphysical Club's doings would have been no secret to the Nuttall Club's founders, some of whom were the same age and ran in the same circles. Cambridge residents Oliver Wendell Holmes, William James, Charles Sanders Peirce, and John Dewey developed a philosophy of pragmatism following the Civil War. As Lewis Menand notes, they were responsible for moving American thought into the modern world. Their club, formed in 1872, lasted barely a year. Although the group's accomplishments were somewhat ephemeral, Menand tries to quantify them: "They helped put an end to the idea that the universe is an idea, that beyond the mundane business of making our way as best we can in a world shot through with contingency, there exists some order, invisible to us, whose logic we transgress at our peril."[4] This was a Darwinian notion, and an important proto-professional one, because it spoke to emergent systems of organization in the natural world.

Some inspiration for the Nuttall Ornithological Club came directly from Ridgway, who was working at the Smithsonian after his time on the King Survey. He knew Brewster, then still a teenager, at least two years before the club's founding, and because he was already at the Smithsonian, answered Brewster's questions and received gifts of his duplicates. The other NOC founder, Henry Henshaw, began communicating regularly with Ridgway in October 1871, asking if the Smithsonian would be interested in trading bird skins from his collection. Young Henshaw—at age twenty-one, just four months older than Ridgway—had accumulated 140 species of birds, almost all of which he had more than two examples, and he was eager to trade. Ridgway and Henshaw thus began a long friendship lasting until Ridgway's death more than a half-century later. They exchanged bird skins, found a great deal to like about each other, and remained equals, particularly after the AOU's founding in 1883. Their correspondence shows a lively, feisty, and warm relationship. Even before the Nuttall Club was founded, Ridgway had in fact offered both Henshaw and Brewster jobs at the Smithsonian; both regretfully turned him down.[5]

Although the founders were all very different in personality, abilities, and ambitions, their intentions were unanimous: they wanted to start a society that studied birds, and although they began as dilettantes, they increasingly sought to legitimize the club as a scientific group. The first gathering saw Henshaw, Brewster, Deane, Purdie, and Scott joined by an additional four members, all by invitation: Ernest Ingersoll, later a distinguished naturalist and author; Francis Parkman Atkinson; Harry Balch Bailey; and Walter Woodman. Ingersoll proposed the name of the society as homage to naturalist Thomas Nuttall (1786–1859).

The name reflected the club's parochial origins and emphasis. For its nine young founders, Nuttall had been one of their own: he had lived in

their corner of Cambridge, taught as a lecturer at nearby Harvard, and
served as curator of the Cambridge Botanic Garden. Ironically, Nuttall
was more interested in botany than ornithology, although he made significant contributions to the study of birds, especially in his popular book
Manual of the Ornithology of the U.S. and Canada (1832, 1834). This work
mentioned localities in Cambridge, which appealed to members. A shy
and reticent lifelong bachelor, Nuttall was also a foreign import, having
emigrated from England as a young man. Nuttall had no particular fond-
ness for the Boston area. Claiming he was "vegetating at Harvard," he re-
signed his post there in 1834 and went west to Oregon on an expedition,
returning only briefly to Boston in 1836. Most of his remaining years in
America were spent in Philadelphia. So the name of the club paid honor
to a foreigner with no compelling ties to Boston nor a primary involvement or any lasting fame related to birds (although Nuttall's Woodpecker
bears his name). In any case, the name was rarely used in its fullest form
by the members, who called it "the Club" or "for a long period, the 'N.O.C.'
in days before overwork of the alphabet had wearied one's ears," as one
account of the group noted.[6]

By 1874, club members considered publishing a bulletin but decided
that the time was not right. Instead, they adopted the *American Sportsman* as a temporary medium of publication, generating eight articles in
1874 and 1875. *American Sportsman*, like many other young publications
of the era, struggled to locate its audience and stay in print, and in 1875
changed its name to *Rod and Gun*. Club members published another
seven articles in that successor magazine and several other publications.[7]
Eager for publication and for more control over their articles, members
agreed in 1875 to start their own organ, and the *Bulletin of the Nuttall
Ornithological Club* appeared in April 1876. The first issue consisted of
twenty-eight pages and a hand-colored frontispiece drawn by Ridgway.
The innocuous bird pictured, Brewster's Warbler, had been described just
two years earlier, and Ridgway's illustration was no doubt his choice, and
a tip of the hat to his friend. The accompanying article, in fact, rehashed
Brewster's description of the bird, written up in an earlier number of the
American Sportsman.

The bird itself serves as an exemplar of what systematics would be
like for many years to come. What in the heck *was* this little bird? Was
it a color morph of a similar species? Was it a well-known bird in immature plumage? Was it a hybrid? Not until 1911 was the bird pictured finally

determined, by careful use of Mendel's newly rediscovered and reformu-
lated laws of heredity, to be a hybrid of the Blue-Winged Warbler and the
Golden Warbler.[8] Indeed, Brewster, in his 1876 article, noted that "this
bird resembles most closely the Golden-Winged Warbler." This first bird
served as a kind of metaphorical frame for the longer discussions of sorting out taxonomic issues that would be a standard topic in pages of bird
journals from then forward.

Joel Allen was named executive editor, and Coues and Baird as assistant editors—Baird was apparently included strictly to press his reputation into service, as he appears not to have played any editorial role.
Although it was the first magazine in America dedicated solely to birds,
the NOC's *Bulletin* was not the only ornithological game in town. Other
magazines dedicated to natural history already existed in the country,
such as the *American Naturalist*. Conceived in 1866 and first published
in Salem, Massachusetts, in 1867 by four students of Louis Agassiz, the
Naturalist struggled for content for the first issue. Cofounder Alpheus
Packard, later a prominent entomologist, wrote Allen to ask for a piece.
The founders planned on a forty-eight-page magazine, and Allen agreed
to write a series of articles. Even at this early stage, the issue of technical versus popular presentation arose. "It is hard work writing for the
people," complained Packard to Allen. "I dislike it, and we cannot be too
careful to write in a lively style and use as few technicalities as possible.
Thus it takes three times as much room for a popular as a scientific article
on the same subject." The magazine laid out its intentions in the first
issue, describing it as a journal that would "popularize the best results
of scientific study, and thus serve as a medium between the teacher and
the student, or, more properly, between the older and younger student
of nature." This was one of the clearer statements of the day that explicitly noted a transition from an older generation of naturalists to a newer
one. The editors wanted to bridge what was clearly a growing generational
gap following the end of the Civil War. However, the *American Naturalist* also did something that both the NOC *Bulletin* and, later, the AOU's
Auk would consider inappropriate: they noted—as had countless writers
of earlier generations—that the value of the magazine would partially lie
in its "illustration of the wisdom and goodness of the Creator." This was
language that was never seen in the NOC *Bulletin* or in the *Auk*. For the
latter publications, mention of God was not scientific, and it was the rare

aspiring professional who would turn to a higher power in the course of
trying to gain scientific legitimacy.[9]

During 1876 the club added sixty-five members—hardly an avalanche
of interest, but enough to keep the publication going. The membership
size for the eight years of the club's existence was roughly on a par with
other newly formed scientific societies; the American Chemical Society's
membership, for instance, fluctuated between 192 members during its
founding year of 1876 to a high of 243 members in 1881. Although the *Bulletin* was regional in its subscription list, some members of the club had
high hopes for a more national organization. Generalist publications like
the *Naturalist*—which had relocated from Cambridge to Philadelphia by
1879—had large national readerships but were broad in their subjects,
while the *Bulletin* editors only had birds on the brain. The first number
of the twenty-four-page *Bulletin* was in the mail by mid-February of 1876,
with several hundred copies going out—some to subscribers but most as
complimentary copies to potentially interested parties. Numerous writers
sent in articles, and by 1877, the *Bulletin* had lots of material in hand. "It
is a great misfortune that the Bulletin cannot be at once permanently enlarged," Allen wrote Ridgway. "More excellent matter comes in to us than
we can possibly use, with our present limited space. If we had the means
we could easily find matter for an annual volume nearly or quite as large
as the '*Ibis*'!" The proceeds from subscriptions were not sufficient to allow
for expansion. Still, by then the club was debt free, and the future was "as
bright as we could expect," noted Allen. "Our first aim has been to show
that it is possible to publish in this country an ornithological magazine
worthy of support, trusting the rest to the friends of ornithology."[10]

The goals and personalities of the members of the Nuttall Club played
key roles in dictating the course of ornithology in the United States, because the overarching aims of the club, as well as the ways in which the
group was and wasn't exclusionary, set the tone and a precedent for the
future formation of the American Ornithologists' Union. The differing
views of the NOC members about the role of an ornithological society
were a preview of similar differences of opinion that the AOU would
later struggle with. Ironically, and in the opposite vein of the subsequent
AOU, the NOC began with the stated goal of *excluding* professionals. Despite having provided artwork, Ridgway otherwise ignored the first issue,
contributing nothing until a plea arrived from Allen asking him to write

something for the second issue, which he did. The second issue brought big changes. "It was decided to make a new departure in regard to the character of the Bulletin," he wrote Ridgway. "We wish to make it a general medium of publication for all the ornithologists of this country, and to increase its standing." He noted that the publication wanted to include "leading ornithological writers" in its pages. Allen echoed this private sentiment publicly in the second issue of the *Bulletin*, discussing the nascent group's start and remarking that "heretofore [until publication of the second issue] the Club had pursued the policy of excluding professional ornithologists, rather, however, from a feeling of modesty rather than from any motive of exclusiveness."[11] This would change dramatically from the second issue on.

Joel Allen was a key player both in the *Bulletin*'s relatively short life and subsequently in the AOU. He finished his schooling late, at age twenty-four, and went on to serve as a student of Louis Agassiz at Harvard from 1862 to 1865. He leavened his museum experience with fieldwork in Brazil, Florida, and the American West. Back at Harvard, he worked at the Museum of Comparative Zoology as an assistant in charge of the bird and mammal collections until 1885, when he took a curatorial position at the American Museum of Natural History in New York. Allen was an ornithologist and mammalogist with broad interests, many of them concerning biogeography: the study of the distribution of species and the explanations for related geographic patterns. A good deal of his work throughout his career, however, was unattributed, appearing in journals and magazines in the form of brief notes, reviews, biographical sketches, and other content.[12] Allen's role with the *Bulletin* was critical to its successes. He was a superb ornithologist, and usually a diplomat. He often served as an intermediary between fractious members, soothing over arguments, gently boosting a point here or there, and striving most of the time not to take sides and to remain a neutral editor, both at the helm of the *Bulletin* and later as the editor-in-chief of the *Auk*.

Though he was named the first editor-in-chief of the *Bulletin* in the inaugural issue even before he knew of its existence, Allen did not become a subscribing member of the club until just after the first issue was out. And beyond contributing his illustration of Brewster's Warbler, Ridgway was also slow off the mark, publishing nothing in the first issue. It was probably his friend Allen's active taking of the editorial reins that caused him to begin contributing to the publication. "Short articles from you,

Joel Allen, ca. 1876. Archives of the Museum of Comparative Zoology,
Ernst Mayr Library, Harvard University.

either of a technical or popular character, we shall always be grateful for,"
Allen wrote to Ridgway in 1876, shortly after the publication of the second
issue. The periodical's appeal to the amateur was also a calculated one as
well as the "feeling of modesty" expressed in print. "We are just now catering somewhat to the popular tastes, hoping thereby to increase the more
rapidly our subscription list, on which the life of the magazine at present
greatly depends," Allen noted to Ridgway in the same letter. Indeed, the
subscribers already included amateurs of all stripes, including at least one
daughter of John James Audubon on the subscription rolls in 1880.[13]

   Coeditor Coues, who certainly considered himself an insider despite
his permanent lack of an institutional position as an ornithologist, tried
to convince Allen to reorganize the club in 1878. He suggested trying to
control membership by category while still allowing for paid subscribers.
"The whole point of this particular kind of wise and diplomatic humbuggery lies in its exclusiveness," he snorted to Allen in 1878. Coues suggested
active, corresponding, honorary, and foreign members. The active category, he proposed, would include residents of Boston and vicinity. The
next category would be corresponding membership, which Coues de-

scribed as "any body in U.S. in good standing, whether obscure or not so he be a birdman—no *compliment,* no fees, no obligations. Any d.f. [damned fool] can 'correspond' you know." Honorary membership, the third category, would be more than just honorific; in fact, it would be the most prestigious. "Any *U.S.* ornithologist of *distinction,*" suggested Coues to Allen, "drawing the line pretty closely to make it a real honorable recognition of the comparatively few 'leaders'—men who have obviously set their stamp on the science in writings." The fact of publication was the key thing that made a bird man a real success, implied the well-published Coues, beyond any other facts related to employment, ability, knowledge, or social class. His final suggestion was for a foreign membership category—"limited to other than U.S. ornithologists, of cosmopolitan fame . . . men who have ear-marked ornithology at large."[14]

Coues's suggestions were only loosely taken up for the Nuttall Club. Resident and Corresponding members made up the bulk of the small group, accompanied by a category for honorary members, numbering just six: Coues, Allen, Baird, Ridgway, artist Daniel Giraud Elliot, and George N. Lawrence, the latter a venerable author on bird subjects. There was also a designation of honorary foreign member, a category that encompassed about a dozen prominent foreign ornithologists. Coues's thinking, however, says a lot about what he felt was a proper hierarchy within the group and what categories of membership he considered a "compliment."

By 1878 the *Bulletin* was holding steady, although its attempts to meet the needs of the smaller scientific community while risking alienation from the larger popular-science community meant that it had a hard time growing. "I think the number [the forthcoming issue of the summer of 1878] will be an excellent one—rather more popular in character than *subscription list!*" wrote Allen to Ridgway. That year the organization boasted ninety-eight subscribers.[15] Many of these were members of the old school of "cultivators" as well as the younger scientists working as "practitioners" and "researchers." These useful terms were coined by historian Nathan Reingold in 1969 as a means of differentiating among those contributing at different levels to the advancement of science. Cultivators were those who helped to develop a science primarily as attendees of meetings, discussers of issues, and expressers of opinions. They spanned a wide range of socioeconomic backgrounds, and it is difficult to generalize about them. However, they typically did not publish and were generally from an earlier school of ornithology that appreciated the beauty of birds, often reached conclusions about birds as evidence of a higher religious power, and emphasized the aesthetic over the systematic or taxonomic as their primary motivator. Practitioners and researchers, however, were generally workers with full-time paid positions in their field, demonstrated a high level of scientific knowledge and competence, and had a desire to integrate the field. Researchers in particular were, by Reingold's definition, the elite investigators into topics, producing a disproportionate share of both qualitative and quantitative research. However, as he notes, elites such as Coues were largely self-anointed, and accounts of their power and influence usually stemmed from themselves.

The existence of a new ornithology journal that had pretensions of professionalism brought the issues of what was scientific and what was popular into wider view, because now both camps had a potential forum by which to be heard. One mark of an amateur was someone who simply wanted to present a testimonial of the "This is a great bird" variety. The tensions between the two camps arose very early in the *Bulletin*'s existence. Allen wrote to Ridgway in 1877: "I am somewhat between 'two fires.' There is a pretty strong 'home' [that is, Cambridge] influence exerted to make the Bull. '*popular*,' with the hope of thereby increasing our subscription list and saving the thing from financial failure. My own preference would be to make it strictly scientific, and to allow the widest range in respect to subjects.—that is, not to limit its scope in any way beyond, of course, declining worthless articles. As it is there is a pretty strong feeling in favor of limiting it to N. Am. Orn. [North American ornithology] and to avoid as much as possible strictly testimonial papers."[16]

Allen would remain between these "two fires" for some time. It was the schism between the "lovers" who were typical of the club's earliest members and the scientists who had specific technical aims that marked the Nuttall Club's desire to move from amateur to professional and its eventual transformation into the AOU. But the ambiguities of personal position could make it difficult to tease apart what constituted "amateur" or "professional." Many of the amateurs were active and well known in other professional fields such as the law, the clergy, or medicine. But more substantially, it often seemed that simply being on the losing end of an argument with someone holding a prominent position in ornithology or being ridiculed in print or privately to others meant being not worthy simply by virtue of having a different opinion. More than just the self-anointing of experts that Reingold describes was at work. Authority in the field, if not credibility, came from a complex mix of institutional placement, gender, publication, personality, and self-promotion.

The financial aspects of the club's journal were important, because beyond any philosophical issues of audience, the very life of the *Bulletin* might hinge on whether it could meet its production costs. The editors fretted constantly in their correspondence about their financial status, the number of members, the cost of printing and mailing, the print quality of the publication, and other aspects of keeping a publication viable. Still, the *Bulletin* was succeeding, if not spectacularly. At the start of 1878 it doubled in size to forty-eight pages and continued to add subscribers. A new notice also began appearing inside the front cover. The publication still described its aims of serving its several audiences simultaneously, but it now prioritized them as being for the scientist first and the amateur second: "This Journal forms at present the only serial publication in America devoted to general ornithology. While it is intended to serve primarily as a medium of communication between working ornithologists, it also contains matter of sufficiently popular character to interest all who take an interest in the subject of which it treats."[17]

Just two bird publications aimed primarily at scientists existed outside the United States when the *Bulletin* began. The first was the German *Journal für Ornithologie,* which began in 1853. Both the NOC *Bulletin* and the *Auk* would be regularly compared by their founders and subscribers to the other precursor—the venerable *Ibis,* established in 1858 by the British Ornithologists' Union. The *Ibis* would be a constant touchstone for nascent American bird publication efforts. In 1878, Allen wrote to Ridgway, "The Bulletin seems to thrive, and I sincerely hope its life may be a long one, and that eventually it will be entitled to recognition as the American '*Ibis.*'"[18]

With Coues, Allen, and others at the editorial helm of the publication after 1878, the *Bulletin* took on a number of issues much on the public's mind, and potentially of interest to an audience much wider than scientists. However, they had difficulty proceeding with any cohesion because internal squabbling by prominent members caused the NOC great consternation. The most notable example of these conflicts revolved around the "Sparrow wars" which raged from the early 1870s for much of the following decade.

The House Sparrow (*Passer domesticus*) was first imported from Britain into the country in the early 1850s, providing a form of comfort for immigrants and others who missed the birds they had seen and heard regularly in Europe, as well as ostensibly helping to control insects. The sparrows spread rapidly, damaging agriculture, displacing other species, and becoming highly visible in the eastern United States. Most American "scientific ornithologists," as journals of the time called bird professionals, were against their presence, although a vocal and often older group of ornithologists, notably Thomas Brewer, supported them for sentimental reasons as well as defending them as destroyers of pest bugs. The conflict pitted the nostalgic views of an older school against scientific ornithologists who detested the bird.

The Boston press covered the flap for several years. Although many Nuttall Club members were said to dislike the sparrows, most were careful to avoid taking sides in print, privately calling it a "taboo subject" because of its potential to antagonize the public. However, they almost all wrote in private against the sparrows. William Brewster called the diminutive bird a "wretched exotic." Ridgway noted privately that "the presence of the European House Sparrow in this country is, in my opinion (and also in the judgement of others) a National misfortune which is likely to assume calamitous proportions. All sentiment in favor of the bird is, I am quite sure, misplaced." He also called it "an evil which is probably now too late to remedy." The introduction of the bird into the United States was at first appealing, he wrote Allen, but it soon crowded out other species. He also saw no evidence of it as a control on insect populations and described the bird as "notoriously deficient" in "beauty of plumage or gift of song." Aesthetics died hard, and scientists of the era, while clearly working in a scientific vein, could not help referencing the aesthetics of birds—an attitude that they simultaneously disdained in the amateur populations because it was unscientific.[19]

The issue of sentiment was one that especially annoyed some ornithologists. Writing a piece on the House Sparrow in the *American Naturalist* in 1878, Coues wrote that "it is very regrettable that the 'sparrow question,' which has already become such a matter of national moment, should have degenerated into such a miserable personal controversy between the sentimentalists who misrepresent the facts and the ornithologists who understand them, that a prudent person, whatever his views, might refrain from having anything to do with it." He described, with a misogynist twist, the birds themselves as "wretched interlopers which we have so thoughtlessly introduced, and played with, and cuddled, like a parcel of hysterical, slate-pencil-eating school-girls."[20]

Thomas Brewer was another prominent ornithologist and a physician like Coues. But the similarities ended there. Brewster was old enough to have been a personal disciple of Audubon's. He defended the sparrow: it was not only a useful hedge against injurious insects, but it was also "frolicsome and entertaining" and "a great source of amusement to the children."[21] Perhaps Brewer felt obliged to defend the entire taxon: John Cassin of the Academy of Natural Sciences in Philadelphia had named Brewer's Sparrow in his friend's honor in 1856. Over the next five years Coues and Brewer fought bitterly over the issue in print and in private. "What an inexcusable petulant one-sided dear old fou-fou he is, to be sure!" Coues wrote to Allen about Brewer in 1878. Always the manipulator, he noted, "Brewer and I met very cordially, but next day in my office, he could not keep his temper entirely to my aggressive imperturbability." He continued:

> I think in the end the "veteran objector" will be left pretty much alone. I *am* right, and I expect to see you and others of renown and calibre backing me fully. I am very well satisfied with the state of the case; it is a very pretty quarrel as it stands. I want to impress upon you the duty of the Club to take up the matter formally, discuss it, and announce results of their deliberations. They owe it to *themselves*, if not to the subject. They are taken and held to be unpublished *professional* ornithologists—and people would say what's the use of ornithologists if they can't settle something that is right under their noses, and of great economic and practical importance as well as scientific interest. I wonder the public has not called on the Club to rise and explain. Some might think after all, that there was fire to Tho. B.'s smoke—or that the Club were afraid to tackle the redoubtable objector. You owe it to yourselves to show clean skirts in the matter, and you must *rise* to explain! Tell Boston the truth, boldly and candidly, and d—— the odds.[22]

Again, the tensions between amateur and professional reared their head. But this quarrel was an internal one between self-described scientists of different generations. Brewer's attitude was indicative of the fuzzy transition to greater professionalization of the field before and into the 1870s. Of an earlier generation, he was considered an authority in the field who also wrote popular accounts of birds. He was perhaps best known for having written, with Baird (who cowrote the text) and Ridgway (who did the illustrations), the physically and technically monumental 1874 three-volume work, nearly three thousand pages long—*A History of North American Birds*. This was the first attempt since Audubon to present an authoritative account of the continent's avifauna, and it was famous. But things were changing. While Brewer's understanding of the avifauna of North America was excellent and exhaustive, his attitudes about beauty and aesthetics as a part of his work were coming to seem antiquated to a new generation of ornithologists who found the sentimentality of no use to their science.

Despite his difficult personality, Coues was a force to be reckoned with, and even senior men in the field, Brewer notwithstanding, were loath to antagonize him. He published definitive, groundbreaking checklists and articles, led the way with trinomials and other systematic notions, and had many friends as well as enemies. As William Brewster noted to Allen over a typical editorial dust-up involving Coues and another member's article in the *Bulletin*, "The only thing that really weighs against this course is the danger of antagonizing Coues." Although they could override him, and did so frequently, they usually did so with considerable caution. But they had things to say about him in private. Ruthven Deane, writing to Allen, noted that Kansas naturalist Nathaniel Goss was "down on Coues for his evident self importance and remarked that when Coues took snuff he expected every naturalist to sneeze."[23]

Coues's ambitions and personality are most important here for what they led to: the formation of a national professional organization of ornithologists. He was probably the most vocal and enthusiastic supporter of such a group. He had no salaried position in the field and no formal education in zoology. But his ambitions were greater than the geographically localized Nuttall Ornithological Club could provide, and he saw the club as a useful stepping stone to a larger and more influential organization almost as soon as it was founded. Agitating against the sparrows could be an explicit move to gain scientific credibility. He made this clear as he continued his letter about sparrows to Allen:

> Such shrewd action [taking a firm stand against the House Sparrow] would at once *put* the club anywhere it wants to stand. The world is busy, and you have only to *assert* yourselves to be what you please, in order to be really taken for such. You can be an "American Ornithologists Union," just as easy as you can be a set of young gentlemen who need to smoke and powwow. You have an "organ" (and I think *now* as I never did before that the *Bull*. is established, and a straight road before you!).[24]

This letter—which contains the first known usage of the term "American Ornithologists Union," nearly six years before its inception—shows Coues in full form. He was very persuasive, and to his credit, Coues sometimes helped his colleagues feel they'd thought of an idea first, even though he may have been the initial spark for an idea.

Issues of gender also intersected with the tensions between amateur and professional. The overwhelming role of men in the NOC Club and the AOU—not a single woman is listed among either group's founding membership—was informed by popular attitudes about women's relationships to the natural world and their differing sensibilities about the sanctity of bird life. Women, with their culturally supported lack of desire to kill animals, were not even allowed a toehold in the new profession. One typical piece in an 1886 issue of *Science* magazine was entitled "An Appeal to the Women of the Country in Behalf of the Birds." Women, so the article noted, were responsible for the craze for feathered hats that led to the destruction of countless birds for their plumage. The issue was couched in almost biblical terms: "They have therefore sinned, for the most part, unwittingly, and are thus not seriously chargeable with blame," the author remarked. Nowhere in the piece, however, was there any mention of the culpability of the men who killed, transported, dressed, sold, or advertised the dead birds used for scientific or collecting purposes. However, the article noted, "ignorance can no longer be urged in palliation of a barbarous fashion." The article attempted to enlist the sympathies of women in reducing the demand for feathers. "Let our women say the word, and hundreds of thousands of bird-lives every year will be preserved."[25] You've gotten the bird world into this mess, the article seemed to be saying, and you should now take responsibility for it. Killing birds by the hundreds of thousands for science was acceptable to scientists; killing for women's fashion was most certainly not.

Where was Ridgway during these years? Although he wrote articles for the NOC *Bulletin,* he was busy with other matters. In fact, in early 1877 he confessed to Brewster that he hadn't even subscribed to the *Bulletin* yet.[26] The young ornithologist was everything that his alter ego Coues was not: shy, modest, embarrassed by ostentation, and often socially stiff and awkward. But Ridgway was not interested in machinations of the kind that drove Coues, and he was very busy with birds. His duties at the Smithsonian were intense and engrossing. Working there, with access to a rapidly growing body of specimens to sort through, he wrote a fair

amount about his discoveries. In the NOC's eight years before the founding of the AOU, Ridgway penned forty-eight brief pieces for the *Bulletin*.
These totaled 148 pages but were only a small percentage of his publishing
efforts. He also wrote another seventy-five articles for other publications
during the *Bulletin*'s existence. These totaled nearly a thousand pages,
appearing in *American Naturalist, Forest and Stream, Field and Forest,
Harper's*, the *Ibis, Ornithologist and Oologist*, and the Smithsonian's own
Bulletin, along with the US Geological Survey's government publication,
as well as the *Journal für Ornithologie*. This did not include his labors in
illustrating other bird publications, which he did on the side.

Unlike other key members, Ridgway was not from New England, nor
was he an officer of the club despite his friendships with its founders.
But Ridgway still served as an authority that the club members looked
up to, largely because of his Smithsonian status. They sometimes called
on Ridgway to help them resolve arguments, sending him specimens and
asking him to pronounce upon them, noting that he would serve as the
final word on a bird's contested identity. The beak, if not the buck, stopped
with him. His professional duties intersected with the NOC in other ways
as well. Allen sent him a plea on behalf of club members in 1881: "Several
members of the N.O.C. have asked how it will be possible for them to get
a copy of your new catalogue. They find it useless to write to the Dept. of
the Interior for scientific reports of any kind, a direct application almost
always failing to secure them. I think you have no idea how difficult it is
for an *outsider* to obtain such things. If you could send 10 or 12 copies of
the 'Nomenclature' to the Club for distribution among the members here
you would confer a very great favor upon quite a number who wish a copy
and know of no way of getting it."[27]

The eight years of the *Bulletin*'s existence provided authors with a
chance to air issues that had arisen in recent years and were gaining traction after Darwin's initial publication of the *Origin of Species*. One issue
much on Ridgway's mind, as well as that of other scientists, was the issue
of trinomials, or the splitting of Linnaean two-part Latinate names into
three names: genus, species, and subspecies. This was a vital development
for science generally. It spoke volumes about the recognition of evidence
that birds were evolving, and trinomials were an explicit acknowledgment
that intermediate forms of species existed. Issues such as trinomials, geographic range calculations of species, taxonomy, and other increasingly
technical matters meant that the character of the *Bulletin* continued to

change from its original amateur emphasis as the decade progressed. By 1880, editor Allen could note with pride about the forthcoming spring issue of the NOC *Bulletin* that "it will be the most technical and *scientific* no. we have yet issued."[28] The stage had been set for a ready transition to a larger, more ambitious national bird organization.

Formation of the American Ornithologists' Union

The American Ornithologists' Union—from the start a considerably more influential body that would eventually gain thousands of members—was founded in 1883, with the explicit goal of sorting out nomenclatural and taxonomic issues. More than any other single person, Coues deserves credit for starting the AOU, agitating for it intermittently between 1878 and early 1883, pushing his own ideas hard and claiming credit enthusiastically, all in his usual mildly paranoid manner.

Although the *Bulletin* had been somewhat too marked with a New England flavor to constitute a truly national group, there was no doubt about its attitude toward science—an attitude that it would pass to its successor, the *Auk*. Joel Allen, writing to Montague Chamberlain on the eve of the founding of the AOU, observed that there were plenty of complaints about there being too much science in the *Bulletin*. "As to the feeling of unfriendliness towards the Bull. for being 'scientific' and towards 'scientists in general as such' it seems to me this arises from a spirit to which we cannot cater," he noted, determined to continue the approach the *Bulletin* had set early on in catering to scientists first and amateurs second. "We publish primarily . . . in the subject of science, and secondly to interest and enlighten those who are not scientific, but not for those who do not care to learn and who are so ignorant of and indifferent to the matters treated as to harbor prejudice against science and its promoters." Although he thought the idea of an amateur department for the new AOU was smart, he also felt that no good would come from catering to a more common denominator. "I should say we had far better not lower our standards as to meet the wishes of so narrow-minded and intolerant a class," he concluded.[29]

Coues peppered his main coconspirator Allen with advice about how the transition should be made from the Nuttall Ornithological Club to the AOU, just how it should occur, who should be invited, and every other conceivable detail related to its new life on the national and international stage. "If you find it convenient to suicide [end the *Bulletin*] by July 1st, better do it, and say it has been done," he wrote Allen in early June 1883. Working behind the scenes, Coues attempted to move levers in order to get things done as he wanted, and Allen generally instituted Coues's suggestions—although often checking with both Ridgway and William Brewster. But the events of 1883 generally went as Coues laid them out to Allen:

> I have great faith in publicity, as helping things along, and know the value of horn-blowing. You could readily write a couple of pages, stating the facts, and explaining the nature and purposes of the new departure, while the call itself is kept perfectly private, and in print nothing need be said as to *who* has been invited or is to be invited. Probably the list I gave you could be raised to 50 desirable or at least unobjectionable names, and need not be larger. Out of fifty, we might count on 20 or 25 of the most earnest ones to be present—and that would be ample basis for *founding* the A.O.U.—large enough to be strong, small enough to be manageable. See Brewster, and let me have the documents back with all your emendations and criticisms, or prepare better ones as a formality. Then we should be able to touch off the machine and see how it fizzes![30]

Ridgway, who—based on his prominence in the field—should have signed the privately mailed call, was prevented by doing so through Coues's desires to keep the list to just three people: himself, Allen, and Brewster. Coues always held Brewster in high regard, although Brewster privately detested Coues. Brewster wrote to reassure Ridgway and to encourage him to attend the inaugural meeting: "We cannot do without you. There is a feeling, I find, that because you did not sign the call you are opposed to the whole scheme. This is unfortunate but if you failed to be at the Convention the suspicion would be confirmed. Now the call was simply an invitation to *meet* and *organize,* and it was signed by those who devised the scheme. It carries no authority and we all go into the Convention on equal terms and are equally founders of the future A.O.U."[31]

Although he had strong ideas about the role of amateurs, Ridgway preferred to leave the organizational scheming to Coues and focused on the technical tasks at hand. "I received yesterday the circular inviting me to attend 'a convention of American ornithologists' to be held in New York City on the 26th of September, prox. I need not inform you that I rejoice at the prospect of gaining more unanimity in the matter of nomenclature—one of the stated objects of the convention," noted Ridgway in the summer of 1883. He suggested that the Nomenclatural Committee that would be formed at the inaugural meeting should gather several series of study skins to begin working through taxonomic issues. Writing to Allen, Coues sniffed at the idea. "Bobby here is apparently all agog. Wants you to bring the M.C.Z. [Museum of Comparative Zoology] series and W.B. [William Brewster] to bring *his* series of all 'disputed forms' here, and have a go over. That is well—but all in good time—there's lots of time and lots to do, once we get the baby born, baptized and dressed."[32]

As planning continued, the issue of the amateurs versus the professionals promptly reared its head. Ridgway opined: "The great difficulty which I see in the way of having the question of nomenclature discussed by 'the convention as a whole,' is that this would involve the value of the views of a considerable number of amateurs, some of them with the crudest possible information upon the subject, as opposed to the intelligent discussion of the subject by a few well informed specialists." Allen responded that this was unlikely to be a problem, as he thought that since the inaugural meeting was by invitation only, with invites sent to "a limited and select number," that perhaps twelve to fifteen people would find it convenient to be present. He also thought that a spirit of compromise would generally prevail. "I feel that we shall all be willing to waive our preference on many, and all major points in the interest of harmony and unanimity," Allen told Ridgway, and thought that the outcome of their deliberations would go a long ways toward giving stability to ornithological nomenclature for a considerable period.[33]

Ridgway's personality contributed to his desire to focus more on systematics and less on the orchestration of a new group's goals, which would have involved interactions with a number of people he did not know well. Although he was proud of his achievements, Ridgway's modesty was informed by an almost pathological desire to avoid the limelight. He detested public speaking and normally could not be dragged to a podium to speak at a gathering of any kind. This nearly crippling shyness almost caused him not to appear at the inaugural meeting, although he fought that impulse long enough to attend. Writing to Brewster two weeks before the gathering, he admitted that "the only considerations which would lead me to prefer not being present and taking an active part in the convention are such as arise from my unfortunate diffidence, which I find it impossible to overcome. I shall, however, sacrifice personal consideration and perform what I consider to be my duty by being present and taking as active a part as may be required in that convention."[34]

That didn't mean, however, that Ridgway wanted any part of decision-making about what the group's membership would look like. That task was taken up by Coues, Allen, and Brewster. They began to try to determine who would be included, at what level of membership, and why. Coues opined to Allen that "there are several black sheep we wot [know] of, which few are the real occasion of making the call private after all. I think the three signers [of the call—Coues, Allen, and Brewster] would spot every one of them alike, and for the rest, supposing the signers agree here and now, to exclude any body that *any two of them* may wish?" However, as he quickly pointed out, "at the same time, we must have some definite theory of who are eligible and who not, to avoid any imputation of personalities. *'An author of recognized works of good repute'* might be about the theoretical gauge." In all of their eyes, publication was a key marker of standing in ornithology. The "good repute" aspect could be much debated, but the sheer fact of getting something into print meant a great deal to the fledgling group. But how to deal with the less significant members? Make them corresponding members—a category typically used for those without the credentials for being founding members. "For all the little fishes that fringe about the affair, we can provide easily enough with a fairly elastic list of 'Corresponding Members' to be elected by the founders and subsequently," Coues suggested.[35]

Despite his claims to want to set issues of personality aside, Coues's objections seem to have been primarily personal rather than professional. One possible invitee he suggested rejecting was Charles Bendire, an army captain who had a notable career as an ornithologist. "If there be any one I should want to strike out, it would be Capt. C.B.," he wrote to Allen. "He is . . . likely to be a chronic objector to things at large and general principles." But despite Coues's objections, Bendire made the cut, was invited, and attended the first meeting. An experienced field collector and, after 1884, honorary curator of the egg collections at the Smithsonian, Bendire wrote a popular book on the life histories of birds, made new discoveries about the migration habits of various, birds and discovered several new species, including Bendire's Thrasher (*Toxostoma bendirei*).

The actual venue, selected just a few weeks beforehand, was the lecture hall at the American Museum of Natural History. The group could not afford to hire a hall and had to find one that would allow them to meet without charge. The inaugural meeting, held over three days, September 26–28, 1883, was a complete success from the perspective of the organizers and attendees. Twenty-one men attended. A number of prominent naturalists issued expectant wishes. The bird illustrator and author Daniel Giraud Elliot, one of the founders of the American Museum of Natural History and famous for his large color-plate bird books, remarked that he could see no reason why a society couldn't be formed "to rival if not excel the BOU [British Ornithologists' Union] of eminent standing." Joel Allen, though absent due to medical or mental woes, was elected president of the new group. On the first day, Brewster wired Allen a shorter-than-haiku telegram, using just seven words to convey his busy joy: "Everything lovely. As lively as can be."[36]

The meeting went well into the evening of September 26 and reconvened the next day. The second day of the three-day meeting brought election results: Allen as president, Coues and Ridgway as vice-presidents, and Clinton Hart Merriam as secretary. The group's first council consisted of Spencer Fullerton Baird, George N. Lawrence, Henry Henshaw, William Brewster, and the one foreign officer, Montague Chamberlain from Canada. Most of these men would eventually serve as president of the AOU at some point in their careers (including Ridgway, who served in 1900 despite his shyness). Brewster, dashing off another quick update to Allen, wrote, "General feeling one of great and entire satisfaction. The thing is bound to succeed if no hitch comes to-morrow." He also favorably vetted the two lesser-known officers: "C. Hart [Clinton Hart Merriam] a brick and our right hand man, wise, prudent, quick to see and improve an opportunity—in short *two times* the man I had thought him. Chamberlain also a trump and as true as steel."[37]

After the officers were elected and a constitution adopted, the founding members held an election for active, foreign, corresponding, and associate members. Gone was the Nuttall Club's resident category. The associate member category was a new one, serving as a nod to amateurs they wanted to include but not at the highest level. (Coues once referred to the associate—the only category not limited to a set number—as a grade of membership requiring only being "out of jail and $3.00" [the cost of membership].)[38] Allen and Baird were both added as founding members, even though they were not present. Twenty-four active members were added to the original twenty-three, for a total of forty-seven inaugural active members. The constitution dictated that new active members—limited to fifty people total, and all US residents—would need to be confirmed by three-fourths of the other active members. Only active members could vote and be eligible for office.

In suggesting a list of founding members, Coues had been distinctly worried about British opinions, given their weight on the international stage. "I am impressed, like you, with the desirability of keeping the original members or 'founders' as few as practicable," he had written Allen, "but we *must* be Catholic, geographically speaking at least, or the howl of 'clique' will go abroad." Nearly half of those nominated to foreign membership were thus British, a fact that *Ibis* founder Alfred Newton praised: "The A.O.U. is entitled to and will receive the best wishes of all on this side of the water, you and your coadjutors must already feel sure—indeed, if there could be any doubt about it, that would be repelled by the *extraordinary honor* you have paid the British Naturalists in electing 10 of them out of your 21 original Foreign members! I need not say that I find with peculiar pleasure my own name in the honored list."[39] As with active members, foreign members—limited by the constitution to twenty-five—got to vote for their own kin, requiring a three-quarters majority to add new members. Corresponding members, one tier down from foreign members, were limited to one hundred, and added with the requisite three-quarters vote from other corresponding members.

Coues, writing to Allen the next week, noted that "I can hardly realize as yet how great and good a thing we did, in the premises, and in N.Y. it is more than a success—it is an exploit—and I am proud of my share in the general result." Ridgway chimed in as well, noting that "the result of the convention cannot fail to be productive of vast good to the ornithologists of this country, who certainly have reason to congratulate themselves on the harmony and good feeling which prevailed."[40] A major feat had been accomplished for American ornithology.

Despite their enthusiasm, however, the founders were not experienced in the ways of establishing a professional group. This was partly because there were relatively few such scientific organizations, but it was also because of the organizational inexperience of many members. Attendee Daniel Elliot wrote to Allen shortly after the meeting, and his comments show some of the ways in which the highly status-conscious AOU leadership skewed the typical organization of a scientific group by an officer-heavy roster. Elliot observed, "I cannot see any necessity for the existence of these officials [dual vice-presidents]. I cannot recall to mind any purely Scientific Society anywhere which has even one. The President is sufficient for all business purposes and in his absence it is customary for one of the Council or the most eminent member present to preside at the meetings of the Society."[41] But the issue of "the most eminent member" was one of the most contested notions of the group, and the very reason for having both Coues and Ridgway on equal footing as vice-presidents. Working as they were to pacify and please amateurs as well as serve what they perceived as the scientific community, they also feuded among themselves about who in their midst should have higher rank than other professionals.

Some of their selections, however, were quite strategically sound. Montague Chamberlain's selection as a member of the council was a useful attempt to close the divide between amateurs and professionals, or to at least draw in a wider amateur element. The founders of the AOU decided at their initial meeting that there should be someone on the editorial staff of the *Auk* to cater to the amateur element—"one who should be one of them, thus having their confidence, and in close correspondence with them . . . and draw them [amateurs] into the support of our publication." A self-identified amateur, Chamberlain was somewhat concerned at having been selected. "Considering that I am an amateur and just the amateur that I am known to be, it occurs to me that my appointment would be offensive to many of the members," he worried to Allen. "Tho' I am an amateur I am a member of the A.O.U., and I apprehend that it is possible the interests of the two may sometimes clash." Still, he promised to work diligently for the best interests of the union and did so for many years until his death in 1924.[42]

Like most people venturing into new public territory, the feathery tribe expressed confidence and bravado, but these were leavened by more private worries and insecurities. Indeed, the general tenor of the group's core members in the early 1880s is that of a bunch of anxious, unhealthy men, nervously parsing extremely fine distinctions of purpose, approach, and meaning in the AOU's founding and early months, and very often sick with various physical and mental ailments. Each key person had uncertainties as to what he understood was a big and important undertaking. Secretary and Treasurer Merriam, a rosy-cheeked young man of twenty-nine in 1883, and trained as a physician, complained that "the part of my duty as Sec. and Treas. that I most dread is the *keeping of the accounts*. I know nothing about bookkeeping and cannot afford the time to learn." Ridgway was also concerned about his perceived shortfalls—in his case, his perennial shyness. After the meeting, he wrote to Allen to say that he was "in all respects unfitted for taking an active part in matters of this kind, and sincerely trust it may never be necessary for me to exercise the functions of the office other than in a latent manner."[43] Part of the reason for these insecurities may have been the lack of an existing roadmap and a lack of institutional authority to look to. Then, too, they arrived at the AOU's founding with the imperfections all humans carry, hidden to a greater or lesser degree, but moved to confess shortcomings they would otherwise have kept to themselves. They worked for different organizations, came from varied walks of life, and were in fact the crucible from which their own authority would be founded.

Another notable element emerges about these men, especially during the AOU founding period between 1882 and 1884, and it relates to their health. Joel Allen complained to Ridgway, "I am still a sufferer from nervous prostration and to such an extent that I am incapacitated for any consecutive work"—something which would become a regular complaint of his. Allen had missed the inaugural meeting because of his health. Their narrative, in fact, is often about collapse. Brewster bemoaned his "excessive nervousness" and in August 1883 noted, "I am all broken up— literally good for nothing. In fact not strong enough to pack my trunk and start in search of health." Foreign members of the AOU also complained of their mental and physical health: Costa Rican naturalist José Zeledón wrote Ridgway that "my health is poor, very poor suffering constantly with a most tantalizing nervous irritability, poor digestion as well as mental prostration." Several years later he would also gripe about his debilitating "brain weakness." Ridgway also had the same kinds of problems; he nearly missed the inaugural meeting in New York for health reasons as well as shyness, noting in late August that because he had been breaking down physically over the past few months, it was "rather doubtful" he would make it to New York.

Not even the extraordinarily energetic Coues was immune. About to leave for England, he wrote to Allen in the late spring of 1884 that "I do not think I have any actual disease, but am just on the verge, apparently of a serious organic difficulty. My mind, too, is lethargic, and even writ-

Ridgway at his desk, ca. 1900. Note his arsenic-blackened middle and
index fingers. Smithsonian Institution Archives, RU 7006, box 194, fol. 51.

ing a few letters is a task." Even when feeling better, Coues remembered
feeling very bad: "I feel as if I had long lived half asphyxiated in some dark
cavern whence I had emerged to the light of day."[44]

One wonders how these men actually functioned, given their regular
complaints about diminished capacities. Mental health may have played
a larger role than any physical infirmities in these cases, because their

overarching self-descriptions at times are of men generally laid low and unable to work—classic symptoms of clinical depression. Although this depression may have been psychological in its origins, this seems unlikely given the very different life circumstances of the men. Rather, it seems more likely to have been a direct effect of the physical environment. Almost all of these naturalists' complaints were made during the late spring or summer. The timing points to a strong possible cause: arsenic poisoning. It was during the warm months that collection and creation of fresh bird skins took place, and the substance of choice for preparation of study skins was arsenic. It prevented rotting of the skins and helped to resist insect infestation. The neurological effects of arsenic are now known to include severe depression and cognitive impairment.[45] It could well have been a key factor in the mental and physical condition of the AOU's founders and their subsequent behavior. Because they were often rendered unavailable for meetings, travel, or consultation due to their state of health, mental or physical, it is more than a passing curiosity. These difficulties reduced their schedules and their effectiveness at times to where they could do no work at all.

Apart from these mental or physical distresses, but related in terms of effectiveness, perceived or real, was the issue of capability and self-confidence. Ridgway's shyness was well known among his colleagues, and a distinct source of anxiety. Merriam was especially uncertain of his own abilities. Besides lacking confidence in his financial abilities as treasurer, he also worried about his facility with language: "In my remarks upon *genera* in general (before the committee) I fear I did not make myself understood, and ever since have had an idea that in some way I made an ass of myself. The truth is that I have no command of language in speaking, and often either say what I don't mean, or wind up in an inextricable snarl so that neither I nor my audience know what I have said. This is a failing which I feel most keenly, but hope that time will remove."[46] Such were the burdens of leadership.

Amateurs and the AOU

The officers' insecurities about their new duties were exacerbated by challenges from outside elements. Despite their stated objective after the inaugural meeting of having an amateur department, the issue of excluding amateurs continued to arise, and a fairly broad public was not shy in complaining about this perceived slight. Merriam wrote to Allen after the first meeting that "certain parties in Boston have addressed personal letters to a large number of collectors and others, and have put to them questions like these: 'Did you attend the *Coues* convention? How absurd to attempt to force a *personal* nomenclature down the throats of Am. ornithologists' etc. 'The Nuttall Club died under false colors' — 'The organization was a put up job founded on fraud and *cannot live*' etc." But although some members of the public felt the new group was exclusionary, AOU officials felt that perhaps it was not exclusionary enough. Some wanted it both ways. Merriam spoke for most of the founding members when he wrote to Allen in mid-October that "the existence of an amateur dept. will certainly cheapen the publication in the eyes of foreigners while on the other hand how are we, without it, to retain (or acquire) the patronage and support of the Amat. element? This matter, it seems to me, is the most serious of all the questions pertaining to the welfare of [the] Union."[47]

The issue of how much actual content to aim at an amateur element was a key point of tension for the first years of the publication. During the last year of the *Bulletin*, Allen had made what he called "a faint and futile effort" to establish departments in the journal that would appeal to amateurs. Upon the founding of the AOU, he wrote to Ridgway to state his intentions to "lighten up" the section called "General Notes" in the new periodical, although without actually using the term "amateur" in any of the sections. "It may be safe to lighten up the department of 'General Notes' somewhat by admitting a small percentage of less formal and rather lighter matter," he admitted to Ridgway, "and to further 'popularize' the journal by establishing a department of 'Correspondence' or 'Letters to the Editors,' somewhat after the style of the Ibis, or perhaps of *Science, Nature*, etc."[48] These larger journals had realized that a broad audience helped with their most immediate need: a healthy circulation.

Some of the more popular bird publications aimed at the amateur element reached a large national audience. Frank Webster, the new owner of the *Ornithologist and Oologist*, distributed nearly forty thousand copies of his publication between 1883 and 1885. The magazine, which succeeded the earlier journal the *Oologist*, lived from 1880 to 1893, and was an important element in public perceptions of the amateur versus the professional in the last quarter of the nineteenth century. Webster had purchased the magazine from Joseph M. Wade, who was for a time an archenemy of Coues. Wade was a very important figure in American ornithology and natural history. He received formal training in natural history at Harvard's Museum of Comparative Zoology for several months as a young man, purchased a controlling interest in the *Ornithologist and Oologist* and undertook its transformation to a larger publication, and eventually turned his love of natural history into a profitable business as the country's largest specimen supplier during the last two decades of the nineteenth century.[49] If the AOU was the relatively dry, scientific side of birds in the public's eye, Wade's establishment was the stuff of fabulous fantasies for thousands of Americans, who had a fascination with their own private collections, as well as those using birds and their feathers for clothing, room decorations, and other aesthetic uses. Wade no doubt felt that the existence of the *Ornithologist and Oologist* would help his own business by reaching a larger audience regularly through print. But his primary goal was to promote birds to as wide an audience as possible: "My Idea has always been to popularize ornithology and avoid the Dry Sciences as much as possible," he wrote to former Nuttall Club member Ernest Ingersoll. "It can be made palatable to the Million[s] — at least I am not yet satisfied otherwise."[50]

In the fall of 1883, Wade penned a caustic editorial attack on the AOU that never saw the light of day. Because he had friends among the officers of the newly formed group, he wrote to them in private to complain and to note his forthcoming blast. Several people tried to dissuade him from the editorial. John Hall Sage, a banker who would serve as the AOU's secretary for nearly three decades, asked him to desist from publishing it. Clinton Hart Merriam also saw the article and worked to dissuade Wade. "Am having an awful time with Wade trying to get him not to print an article he has in type against the A.O.U.," he wrote to Allen, and echoed that attempt in a letter to Ridgway the next day. He also noted to Allen that "you are doubtless aware of the opposition . . . as well as of Wade's intention to continue the O. & O. another year in order to *hurt* the A.O.U. as much as possible." He also suggested a covert solution to the problem: "It has struck me that it might be best for *Chamberlain* to buy out the O. & O. (you know Wade offered it to him) as a private or individual enterprize, while in reality it would be under the control of the Council of the Union."[51]

The AOU members who knew of Wade's antagonisms realized that they needed to be careful with him; condemnation in print would probably backfire, given Wade's far superior number of readers across the

JUNE, 1886.

VOL. XI.　　No. 6.

Ornithologist

AND

Oologist

Established 1875.

Published by
FRANK B. WEBSTER,
409 Washington St., Boston, Mass.

Entered at Boston Post Office as Second-class Matter.

Cover of the *Ornithologist and Oologist*, the popular ornithology journal created for wide circulation. Huntington Library, San Marino, CA.

country. Coues, using the analogy of subordinating a horse to its rider, thought that the best course would be to ignore and thus isolate Wade: "If we agree to let him severely alone, it will be further necessary to let every body know the fact. I suppose that is the only course. If we undertake to cinch him, he would only buck worse, and we dont want anything of that sort going on." Popular science was popular for a reason. Scientific ornithology's technical language, focus on nomenclatural reform, and taxonomic goals put its adherents in a minority. However, Wade apparently had no personal animosity toward Allen or the other founders, and wrote to Allen on the founding of the AOU to state his case: "A word about the 'Union.' In behalf of our beautifull science and the boys and ladies. I hope you will remember that it is always desirable to simplyfy as much as the science will admit, and remember we are entering a Practical age where plainness of definition in all things is so very desirable. The easier understood the more students there will be—but you are so thoroughly imbued with science that I do not suppose that you can realize why we are not all Allens in that respect but we are not, and cannot be . . . I run [the magazine] for the best of the sciences—and if I make any mistakes it is from ignorance and not because I mean to."[52]

Wade had an excellent point—that by introducing the widest possible generation to ornithology, the science would be best served. The "Practical age" he mentioned was that of a country increasingly focused on business and technology and all that was implied by those activities: less time for theorizing and more time for doing, as he saw it. He was a pragmatist to the bone. And as it turned out, most of the officers wanted to coopt Wade into the group. He had been nominated to be a founding member of the AOU, on a four-to-one vote (Allen, Brewster, Ridgway, and Merriam voting for him, Coues against), although he did not attend the first meeting. Coues wrote to Allen about Wade after the inaugural gathering: "Do you really mean to say that that vulgar crank is going to *attack* the A.O.U. in his contemptible little sheet? Is he a fool? Has he declined his election? I hope he is ass enough to accept it, and *then* abuse the Union! It would be just like him! He may do so, as the parting *whiff* from the moribund sheet, and a sweet smell too. Didn't I foresee a scent in that quarter? I am sorry now that I did not oppose his election more strongly."[53]

For his part, Wade was smart enough to understand the value of a good publication, regardless of its origins. He praised Coues's *Checklist of North American Birds* extravagantly in the fall of 1883, noting that it was "the best thing since Audubon," and Coues promptly used it in the list of endorsements of his book, despite his private disdain of Wade. Not long after, Wade sold the publication to a kindred spirit, Frank Webster. Merriam found it a useful tool for promoting the AOU's own work on bird migration, and even though by January 1884 it was owned by Webster, Wade had produced that month's content the previous fall. "Have you seen the Jan. O. & O. published by Webster?" Merriam wrote to Allen. "In it is a most valuable article on bird migration by W.W. Cooke who incidentally mentions that *the Union* will go on with the work . . . In response to this inconspicuous notice in this insignificant journal I have already recd. two dozen letters asking for circulars and volunteering assistance."[54]

At the start of 1884, with Wade out of the publishing business and no longer a threat to the AOU, Coues changed his tune. His ego ever responsive to admiration, he responded to Wade's attempts at détente and told Allen that he had "had Wade in training for a month. He now writes nicely, and sends me an original drawing of [legendary American ornithologist Alexander] Wilson's for a X-Mas present! We had better take him into the true fold. Suppose you drop him a pleasant line?" But despite patching things up with Wade, Coues remained perturbed at the *Ornithologist and Oologist,* which after all was a competing publication with views that were still often at odds with those of the AOU. In the very first issue of the *Auk* he had an unpleasant dustup with Clinton Hart Merriam's uncle Augustus Chapman Merriam over Latin terminology for birds. Writing to Allen in early February 1884, he complained about the magazine's taking Merriam's side: "Doubtless you have seen the O&O for February and have noticed the tactics of the trashy little sheet. I suppose we shall have to take their small sauce for awhile. What do you think had best be done. I am never in the humor to enjoy small jibes and dirty flings, even for the most contemptible source. My irritation is of course at the service of the Auk, at any time and in any way that you think anything from my pen will do good. My present notion is to be contemptuously silent for awhile, then if the thing goes on, to stamp on it with full force." In the January issue of the *Auk,* Allen had penned a reasonable notice about the *Ornithologist and Oologist,* remarking that the January 1884 issue contained "the usual number of good articles and notes" and that there was doubtless room for a distinctly amateur journal. "We cordially wish it success," Allen wrote. This prompted a note to Allen from Coues: "Didn't you treat the O. & O. rather too well?" he wrote Allen in mid-January. "All things considered, a sneer or an eloquent silence might have been more apt."[55] Coues may have disliked being picked on, but Allen understood the benefit of supporting a large popular publication, or at least not antagonizing it.

In at least one arena, the AOU neatly sidestepped these squabbles by directly recruiting amateurs and putting them to work for scientific ends. To understand more about Darwinian speciation, ornithologists found that issues of geographic distribution were increasingly important to grasp and quantify. And to understand distribution, the new organization's officers reasoned, migration needed to be determined with far greater precision. They created a migration committee, headed by Merriam, almost immediately after the AOU's founding. This group was responsible for tracking the patterns of migration of North American birds, and Merriam deftly recruited a large body of amateurs to make and record observations.

Merriam had high hopes for the undertaking that were specific to capturing the amateur element. He told Allen that "I am led to venture the prophecy that the Committee on Migration is going to do more toward building up the *Union* and making it what it ought to be amongst the amateurs than any other one thing that has been, will be, or could be done." Merriam assigned experienced field ornithologists to oversee the efforts in specific states, and the effort may have been the first systematic state-by-state survey of individual species of any kind, and almost certainly the most detailed zoological migratory survey in the country. Founding member and conservationist Albert Fisher got Florida, Alabama, Georgia, the Carolinas, the Virginias, Tennessee, Kentucky, Maryland, Delaware, and New Jersey. Allen was assigned all of New England. Merriam himself took New York, Pennsylvania, Ohio, Indiana, Michigan, Wisconsin, and Minnesota. By late March he had added Alabama, Idaho, and Montana.

Merriam also wrote to the US Lighthouse Board to see if lighthouse keepers along the country's coasts could be pressed into the service of the AOU to do migration work. "Hope to receive substantial aid from them, and still control results *in full*," he noted to Allen. He promptly received the OK from not only the Lighthouse Board but the Canadian marine and fisheries minister as well. Baird offered to have the Smithsonian distribute the forms, but Merriam politely declined and enthusiastically began the work by himself. It quickly went from a trickle to a torrent. By early February he was mailing out between 80 and 125 migration circulars a day. By late March, he received a dozen or more letters on some days from lighthouse keepers alone. In early spring 1884 he wrote to Allen that the job was too much for one person. "For sometime it has been growing clearer every day that it will be an utter impossibility for me to do anything with the migration returns without a corps of assistants."[56] Amateurs, who had contributed by providing data, were thus also pressed into administrative service—all efforts that served as a form of citizen science.

The *Auk*

Although Ridgway's writings, and the role of publications generally in the transformation of ornithology, are treated in much greater detail in chapter 6, the formation of the AOU's quarterly journal, the *Auk*, needs to be treated here, because its formation and development was so strongly tied to the AOU's founding. The journal was critical to the group's future successes because it was to be, as Allen noted, the organization's principal work, and the element on which its reputation would chiefly rest.[57]

Allen served as the first editor of the *Auk*, with Coues, Ridgway, Brewster, and Montague Chamberlain as the four associate editors. Selection of the associate editors of the *Auk* had fallen to Allen, who told Ridgway that the task was "a matter of great delicacy and not easy to make." The original configuration of editors consisted of Allen and Coues as the two coeditors, following the NOC *Bulletin's* approach, but Allen was persuaded to add Ridgway, in part because Ridgway was perceived as more evenhanded. Chamberlain was added because of his alliance and self-identification with amateurs. Coues was not happy with the choice of Ridgway; the jealousy that he found difficult to shake made him respond negatively to any attempt to put Ridgway in a position of increased prominence within the organization. "Has R.R. any ideas whatever on editing, or conducting a periodical?" Coues wrote to Allen in October 1883, ignoring that Ridgway had written and revised four books and 190 articles of his own for publication. "I suppose not: if it is pure policy to include him, that could only be justified by expecting to get something out of him. But he is mortgaged and preempted, for practically all his matter, for the publications of S.F.B. [Spencer Fullerton Baird] since he scarcely touches any but S.I. [Smithsonian Institution] material."[58]

In fact, Ridgway touched a great deal of material from other institutions, and only one of his books and twenty-five of his articles to that date were published by the Smithsonian. Other leading members wanted Ridgway as an editor as much as Coues did not. His name recognition with a wider public, as well as positive perceptions of the Smithsonian, became a factor. Merriam wrote Allen to complain about the initial exclusion of Ridgway from the editorial staff. "In regard to the Editorial Staff I *decidedly object* to E.C. and W.B. *without* R.R., *because* of its effect on the Amateur element, and in fact, upon almost everyone outside of the Council. This is a critical period and R.R. must *appear* on a par with E.C. in the eyes of the public. Policy requires this now, I believe."[59]

Although the typographic elements were a carryover from the Nuttall Club's *Bulletin*, the group wasted no time in formalizing a new cover design, masthead, and name for the new publication in order to cast a new identity. Merriam sketched up a letterhead design a week after the September meeting and had it lithographed. Coues set up publication arrangements with the Boston publishers Estes and Lauriat, with whom he already had ties, as they were in the process of publishing a new edition of his *Checklist of North American Birds.*

The end of the Nuttall Club's *Bulletin* and the changeover to the AOU's *Auk* has often been described as a transition, with the *Auk* called a new series of the *Bulletin*. However, at least one of the club's early members denounced this idea vehemently. "It may never be known from whose misguided imagination first came the statement that the American Ornithologists' Union was in some fashion the offspring of our Club," wrote Charles Batchelder many years later, in 1937. "It is time that the myth should be exploded. The fact is that the Club had nothing whatever to do with the inception of the plan of a national organization or with the actual birth of the Union. Not only was it not consulted by the men who originated the [AOU] scheme, but the members as a whole were entirely unaware of it." Indeed, only about 60 percent of the club's membership was asked to join—and then only by entering what Batchelder called the "outer doors" of the organization as corresponding members. Coues worried that club members would try to start a competing bulletin, telling Allen that although the possibility was low, AOU members needed to remain vigilant. "If such a thing does start up," he cautioned, "we can see to it that it flops down promptly."[60] However, the AOU maintained the old series numbering from the *Bulletin*, noting it as such and even continuing the original *Bulletin's* sequential volume numbers on the front cover until the January 1937 issue. The words "Continuation of the Bulletin of

the Nuttall Ornithological Club" also graced the front of the cover, beginning with the first issue and finally ending in 1950. The club lives on today
in Cambridge as a separate entity and has published a number of memoirs
and other monographs.

The difficulty of coming up with a suitable name for the AOU's official new publication generated extraordinary correspondence between
the meeting in September and the publication of the first issue in January. Some officers hoped to avoid a single bird name, because they did
not want to sound like they were mimicking the British Ornithologists'
Union's esteemed *Ibis,* founded in 1859, the same year that Darwin's *Origin of Species* first appeared in England. Others found the allusion to
the BOU's name useful. But as was usually the case with AOU firsts,
Elliott Coues was apparently the one who came up with the name that
was used and survives to the present day. Writing to Allen in October
1883, he asked,

> Do you really like the name I proposed? The *Auk* is probably the best
> *bird-name,* if we must have one. It is brief. Even abrupt, raucous, dis-
> tinctive,—many things in its favor. But can you think of no better? There
> was a German "Ornis" once—not bad. "Auk" gives a chance for some
> wretched witticisms—as "awkward" &c. But I like it,—"brevity is the soul
> of wit," and in citation there is nothing like a short name of a periodical.
> I have had it in mind for many years . . . There is a Latin motto, half a
> play on words, which I always associated with the name. "Mortua alca
> inpennis, in pennis alca rediviva"—"The extinct great auk still lives in
> the writings of The Auk." Several here [at the Smithsonian, where Coues
> kept an office but had no formal position] seem to fancy it. It copies the
> Ibis motto-ing, too.[61]

This new name was at first blush received warmly and generally accepted by Allen and several others. Other prominent members, however,
soon objected. Two weeks later, Clinton Hart Merriam wrote to Allen to
complain. "Did it ever strike you that 'the Auk' (or any other bird) would
look like a servile imitation of our British friends?" he asked. Ridgway
was another objector, preferring either "Bulletin of the AOU" or "American Ornithologist" as a title for the new bulletin, as did Merriam. Canadian member Montague Chamberlain wrote to both Allen and Ridgway
to complain about it, calling the name "in very bad taste."[62]

Getting wind of these criticisms, Coues scolded Allen. "[The name]
I had supposed settled, and accordingly had so spoken of it in many let-

ters . . . What are the objections, and who objects? . . . it would seem to be for *you* to settle it. I thought you liked it very much, and had decided upon it . . . Don't you allow yourself to be talked over too easily, perhaps?" A week or so later, armed with more information about the source of the complaints, Coues wrote Allen again to make fun of both names proposed by Ridgway: "If the name 'auk' is to follow the bird to extinction, I should favor some other 'short sharp and decisive' bird name. With due respect to my esteemed friend 'Bull. A.O.U.' would sound like an octogenarian, as name, when the subtitle is to be "organ of the A.O.U.["]! 'American Ornithologist' is not bad: it might be improved by making it the 'American Ornithologist, Oölogist, and Taxidermist, a Quarterly Journal of Amateur Bird-Science, permitted to be run by a Society styling itself the A.O.U. in the interest of the Soreheads who were not invited to the First Congress.' !!!!! That would be neat, not to say gaudy!!!!"[63]

Allen, who as editor-in-chief of the new publication was the one receiving most of these opinions, was, like Ridgway, no fan of Coues's proposed name, despite his initial acceptance of it. He wrote to Ridgway in mid-November 1883: "I am greatly pleased to find you can so heartily approve my suggestions respecting the character of the new magazine. The name 'American Ornithologist' I find meets with general favor, and 'Auk' with very little. 'Bulletin A.O.U.' stands second in favor, and would be really *un*satisfactory to very few, which would not be the case with 'Auk.'" Despite Coues's protestations to the contrary, Allen had taken the opinions of many members into account and had his own objections. The history and character of the bird was not auspicious, he felt, besides its having been chosen as the trademark of a guano company that used both the name and figure of the auk on a label for its packages of bird poop. Allen thought these objections might have been somewhat trivial and unworthy of serious consideration, but "there remains still the fact that its adoption meets with opposition and would be unpopular," he observed to Ridgway.[64]

But sometime in December, the pendulum swung back to the original name. The venerable Baird was one of the few who liked Coues's proposed name, despite its relative lack of popularity, and he was probably the final arbiter in the founders' minds. He wrote Allen to say that "the name which you propose is a very quaint one, and I should be very well satisfied to see it adopted. It constitutes a good antithesis to the London Ibis." Allen was relieved to have the matter resolved and out of his hands.

"We are finally indebted to S.F.B. for the name!" he wrote to Ridgway.[65] It did have benefits. The associations people made with London's *Ibis* were probably more positive than negative, given the latter's established nature and good reputation. One significant precedent had already been set many years earlier by the founding of the American Association for the Advancement of Science in 1848; the AAAS had mimicked the British Association for the Advancement of Science by choosing a nearly identical name.

The *Ibis*, in noting the new journal's existence, was lukewarm about the name. "'The *auk*-ward organ, etc.'—I see our 'Ibis' friends take this view of the matter!" wrote Allen to Ridgway after the first issue was reviewed in England. Coues, with his usual wit, reported that "I hear the Auk called 'Ibis occidentalis' in England." What the *Ibis*'s review of the publication actually said was, "It is not for us to criticize the wisdom of this title, although, in reply to the Shakesperian question 'What's in a name?' we might say that there is a good deal in its appropriateness, or the reverse. But the reasons for its adoption are given; and we cordially desire for the new periodical a better fate than that which has befallen the most distinguished member of the family Alcidae."[66]

Once the journal's name was set, the membership took it in stride. William Brewster wrote to Ridgway to say that he was less than satisfied with the journal title but that it was only a name, not the journal itself. Chamberlain wrote Allen in a similar vein: "Of course the name of 'Auk' made me draw up my face a bit, but as the majority have decided so must it be, and the name may prove less objectionable to our friends than we anticipated." The editors of the first issue of the *Auk* anticipated complaints about the name by taking a defensive crouch, proclaiming that "the outcry from all quarters excepting headquarters of American ornithological science against the name of our new journal satisfies us that the best possible name is The Auk. Were the name of this journal one which anyone could have proposed and everyone liked, it could not have been an 'inspiration.'"[67]

The article then listed a long series of puns on the name, relating to "awkward," "New Yauk," and the "Auckland Islands," and noted that "the Auk could dive deeper and come up drier than any other bird, as Baird says." Then, with a whiff of desperation disguised as mild humor, it noted that because the title was a single syllable and only three letters long, bibliographers would appreciate its brevity. It also addressed the obvious positive comparisons with the *Ibis:* "We should like to 'ape' or otherwise imitate 'The Ibis' in sundry particulars."[68]

Recipients of the first issue couldn't resist writing in with their puns. Wrote Samuel Wells Willard, a young expert on bird migration and distribution: "The name is surely quite a novel one, and like the Englishman I was rather at a loss to know, whether it was the Hawk that ate fishes, or the 'awk that ate 'em." Walter E. Bryant, curator of birds at the California Academy of Sciences, wrote to Ridgway after the first issue with a crack echoing poet Robert Burns's dying words: "Don't let the Auk(-ward) squad fire over my grave."[69]

At Allen's request, Ridgway created two versions of artwork of the Great Auk for the cover of the first issue. As the press deadline grew closer, however, the editors expressed unhappiness with them. Ridgway deemed the first version "an entire failure" and suggested a professional artist in Cambridge or Boston to make a new drawing. Coues, writing to Allen, hissed, "*That* picture was certainly the sickest and most forlorn!" Because time was tight, they turned to an existing image that they could reduce. Allen suggested an image he had just seen, in a recent work by Field Museum curator and AOU founding member Charles B. Cory, called *Beautiful and Curious Birds of the World,* self-published in seven parts between 1880 and 1883. "It has a bit of ice-berg scenery in the back-ground on the right, and a group of Auks in the distance, on a hummock, on the left," he noted to Ridgway. "The figure of the bird is excellent, and the whole picture would make, so far as the two subjects can be someplace, a very fitting counterpart to the 'Ibis' vignette." The Library of Congress provided the copy, and the image that graced the cover of every *Auk* until 1913 was born. Six hundred copies were printed, on "medium-cheap quality white paper that will not tear easily," per Merriam's request, for subscribers, newsstands, and giveaways, with the remainder kept on hand for any unanticipated demand. This was three times the circulation the NOC's *Bulletin* had enjoyed and thus a significant increase in the dispersal of bird information over its predecessor.[70]

By mid-December, Allen had enough material in hand for two issues, the typesetting began, and the first issue was in the mail by early January, at a cost of three dollars for an annual subscription. However, the newborn journal had difficulties getting copies to subscribers. Nor was everyone pleased. Ridgway wrote to Allen in late January after receiving his copy to note that he was "rather disappointed" in the magazine's ap-

Cover of the first issue of the *Auk,* which graced the front of the magazine until 1913. University of Southern California, on behalf of USC Libraries.

pearance, "and feel more than ever dissatisfied with the name adopted. The plate and wood-cuts accompanying Mr. Cory's article also go far towards producing an unfavorable impression of this first number. However, I trust future numbers may look better." Ridgway was particularly incensed that spring when, after several months and several inquiries, his copy of the second issue had still not arrived. "You may perhaps imagine the state of supreme disgust that I am in at not receiving my copy of the April Auk," he wrote to Allen. "I have written you twice about it, and once to [publishers] Estes & Lauriat, and unless I hear from some one soon about the matter shall come to the conclusion that the publication is like its namesake—defunct." The fact that all the key players in the founding of the AOU had day jobs did not help in the smoothness of the operation. "This A.O.U. business *takes too much time.* Have done no writing of my own worth mentioning in over a month," Secretary Merriam groused to Allen in the fall of 1883.[71]

Bird professionals certainly existed before the last quarter of the nineteenth century. In the period marked by the *Bulletin* and the *Auk*'s publication, however, the scientist was becoming a new kind of professional. As Nathan Reingold has noted, "Defining professionalization is a thankless task. Many writers start out by warning that fashioning a definition is very difficult because there exists no ideal profession; one would be better advised to talk about the position of any particular profession on a kind of spectrum leading to an ideal."[72] The Nuttall Club, despite its strong regional origins, brought its participants under one national roof for the first time in the country's history. Its *Bulletin* extended the group's reach to a wider audience. Both the NOC and the AOU created categories that excluded some and included others and, in so doing, attempted to identify professionals or amateurs. Coues had described the most desirable members in 1878 as "Leaders—men who have obviously set their stamp on the science in writings." This was clearly open to interpretation; what constituted "science" differed from group to group, and amateurs certainly co-opted that term, as evidenced by Joseph Wade's description of "our beautiful science." The quantity of publications about birds wasn't a useful marker of scientific rigor or authority, as many literary bird lovers such as John Burroughs had written many widely published works on birds. Coues tried again in 1883 by describing desirable members as being authors of "recognized works of good repute." Again, these terms were open to wide interpretation. Who would decide the reputation of an author's work? In reality, those weren't the metrics the founders used at all. Rather, scientific ornithologists were defined as those who were unsentimental (as in their unanimous dislike of the House Sparrow), technical in their language (using Latinate names or anatomically-precise terms not in wide public use), had published in a narrow sphere of technical literature, and were respected by others who knew them, or knew someone who knew them. These were the qualities that helped people join the upper tiers of membership in bird organizations. Ultimately, membership in all categories for both organizations was decided by a small group of founders.

Beyond the founders' beliefs and arguments about what classes of people were most worthy of admission to membership, one group was made nearly invisible by broad assumptions: women, who faced an almost total lack of acceptance as worthies. Issues of gender were also a very important aspect of attitudes towards sentimentality. As Margaret Rossiter has noted, the stereotype of "science" stood at the opposite end of the spectrum of "womanly" activities. The former was tough, rigorous, impersonal, masculine, and competitive. Female activities and interests were seen rhetorically as the opposite: delicate, yielding, emotional, and nurturing. Any women interested in science were thus viewed as "unnatural" because they went against this grain. Women who liked birds for sentimental reasons were, by these descriptors, utterly removed from science. Their roles in some branches of science advanced considerably between 1880 and 1910, although ornithology appears to have lagged behind other professions in employing women. Astronomy was perhaps the most successful profession in recruiting and using women for scientific work. Numerous women were employed at the Harvard College Observatory in the 1890s, working at classifying stellar spectra. Prominent female astronomers including Henrietta Leavitt, Williamina Fleming, and Annie Jump Cannon did seminal work at Harvard during these years. On the West Coast, the Mount Wilson Observatory used women "calculators" who performed similar tasks designed to quantify the new informational bounty gleaned in the night sky in the first quarter of the twentieth century.[73]

Identifying just who amateurs were, what they were like, and what their views were about birds is an important task, although it is made considerably more difficult by the fact that they were a much more diffused group than those working for institutions or publishing widely. Their views in popular publications—particularly their letters to the editor— are probably the best window into their perceptions of themselves and of the scientists. Ridgway himself shows up in the *Ornithologist and Oologist* as a contributor just four times—in every case a correction he had written to the editor about some error he had noticed in the journal.

It is important to understand that many self-described amateurs still considered themselves—if not actually identifying themselves as scientists—to be interested in the study of ornithology as a science. "We do not mean youth who think that to possess a drawer or two of eggs in a fancy case constitutes a *naturalist,* or who think that to have a room ornamented with bright plumage renders them *graduated ornithologists,* nor do we mean others engaged in collecting *only* for the purpose of having a collection 'as large as any in the state,' wrote an *Ornithologist and Oologist* reader to the editor. "We refer to those who have either a natural or an educated taste in the study of ornithology *as a science* and who desire to assist in its development, and not those engaged in it as a pastime or play."[74] There was, then, often a spectrum rather than a sharp delineation among just what large groups of people considered to be scientific.

However, professionals considered language and tone to be a critical hallmark of the modern study of birds. While the *Ornithologist and Oologist* would run a multipage testosterone-saturated article called "Gastro-oological," about the virtues of actually eating the insides of eggs collected ("Barbarous, you say. Well, try a little savagery yourself"), the *Bulletin* and the *Auk* would have considered such an article wildly off-topic.[75] The presence or absence of sentiment was even more important, and it was almost entirely lacking in both the *Bulletin* and the *Auk*.

Another hallmark of this new professionalism was the distinct lack of sentiment in both journals. The *American Naturalist*'s stated goal of illustrating the wisdom and goodness of the Creator gave way to a lack of religiosity. Birds, with their often melodious songs and their many beautiful plumages, were far easier to wax rhapsodic about than, say, the families of barnacles that Darwin had studied. So it was especially unpopular to push back against deeply established sentiments about birds in order to talk about them as technical things that were worthy of comparative study and analysis. But push the NOC and the AOU did, and by the mid-1880s, the AOU was off and running, gaining subscribers at a steady rate, and never looking back. Its four-tiered membership structure was set and would remain in place to the present day with only minor tweaks. Although research agendas in ornithology changed dramatically in subsequent decades—most important, changing from the study of systematics to the study of the living bird—the group remained the hierarchical professional entity it had become during its evolution in its Nuttall Club days and through the establishment of the AOU.

CHAPTER FOUR

Bird Study Collections

I do not care for naming species, but I do greatly
care for having them in the Museum!

RICHARD BOWDLER SHARPE of the British Museum
to collector Charles Walter de Vis, April 3, 1897

I F COMPREHENSION of the scientific value of study collections of birds was akin to knowing a language, these understandings
would be broken up roughly like this: groupings of birds before Audubon's death, or perhaps just up to Darwin's writing of *On the Origin of
Species*—until the late 1850s—were like snatches of conversation in a broken patois, half-audible down a long hallway. Collections existed, and
sophisticated classification and morphological work had been undertaken
with them for decades, but the tales they told about evolution were more
or less incomprehensible, and furthermore, the debate about just what
constituted a species was essentially nonexistent in the face of broad acceptance of the immutability of species. After Audubon's time, the mut-
terings became firmer, more audible voices that spoke more and more
clearly because they began to resemble a vocabulary. Collectors ranged
more widely, more species were known, and birds were beginning to show
up in collections with other similar examples and in greater quantities.
The voices were increasingly crisp and sophisticated, with many accents.
To untrained ears, they might have sounded forced, at times unintelli-
gible. Not until the 1870s, however, did these collections begin to make
sense of the real workings of the natural world. By 1880, Ridgway—trying
to sell his personal collection to the Museum of Comparative Zoology
at Harvard—could note to Joel Allen, its curator of birds, that "there is
not a single duplicate in the entire lot, but the series showing individual
variations are very complete—in fact, the collection may be styled 'Darwinian.'"[1] In this third era, study collections, in tight conjunction with

the names assigned to birds, finally came to resemble a thesaurus. In the last quarter of the century they provided a sophisticated, highly detailed, and contextual body of material for use in biology, and the creation and use of these collections is an important part of the professionalization of ornithology because of their ties to issues of classification, language, and accountability.

New Species

As collectors fanned out around the world and museums grew in size and scope, the number of birds thought to be new to science rose quickly. Their discovery was sometimes so steady that ornithologists became blasé about new finds. Writing to William Brewster in 1886, Ridgway asked his friend to describe a newly discovered subspecies of grouse, noting that he was "much too busy to bother with the job." Indeed, this was an era in which Ridgway raised few eyebrows with his 1894 article titled "Descriptions of Twenty-Two New Species of Birds from the Galapagos Islands."[2]

Just how common were discoveries of new species in the late nineteenth century? Apparently common enough that in one case, Richard Bowdler Sharpe of the British Museum turned away an entire shipment that included a bird entirely unknown to science because of the payment desired by the collector who offered it. "Dear Mr. Haddock," his missive began. "I have sent the box back. Your letter has fairly taken my breath away. Seventy-five pounds! If the skins were in the finest state of preservation they could not be worth £10, and as it is they would be very dear at £5 . . . I am very sorry not to have had the Little Honey guide which I should have described as new and named after you, but of course I cannot give more than things are worth."[3]

Although collectors were active around the world, they did not restrict themselves to remote forests from distant lands. New species could pop up very close to urban areas, somehow having avoided detection for many decades. In mid-June 1881, Eugene Bicknell, an experienced amateur ornithologist, shot a bird in the Catskills not far from urban New York, at an elevation of about four thousand feet, near the summit of Slide Mountain, after hearing its unfamiliar song. Confused as to its identity, Bicknell sent it along to Ridgway for identification. Bicknell had decided, very cautiously, that it was a Gray-Cheeked Thrush (then known as *Hylocichla aliciae*, now *Catharus minimus*), although far south of its normal range. He hoped to publish a small blurb about the bird in a regional journal, as birders often did on finding a species outside its established locale. He asked Ridgway to "kindly give me your opinion of it, that my authority for its identity can be unquestioned." He noted that he had seen the same bird about a year earlier but had not had any success in catching one. In late September he caught a second specimen and send that along to Ridgway as well. Bicknell's collection of bird skins included nineteen specimens of *H. aliciae* with which to compare his new find, and he carefully noted the differences to Ridgway. Between Bicknell's suspicions about the bird and his own ability to compare them to the Smithsonian's examples of Gray-Cheeked Thrushes, Ridgway quickly pinpointed the thrush as new to science. This was a remarkable discovery so close to such a big city and in such well-traversed territory. Ridgway promptly named the bird Bicknell's Thrush, in honor of its discoverer.[4]

The bird quickly became well known, because its appearance so close to an urban center was widely commented on in print. Poet John Burroughs waxed lyrical in talking about the bird. "The song is in a minor key, finer, more attenuated, and more under the breath than that of any other thrush," he wrote. It seemed as if the bird was "blowing in a delicate, slender, golden tube, so fine and yet flute-like and resonant the song appeared. At times it was like a musical whisper of great sweetness and power." Ridgway was both more resolutely technical in his description and, importantly for science and speciation, much more comparative. What was most significant about the bird for his purposes was that it was different from other similar birds. "Similar to *Hylocichla alicieae* Baird, but much smaller and (usually) with the bill more slender," his priority-claiming description of the new bird began. "The general aspect of the upper parts approaches more closely that of *H. ustulata*, but the shade is much darker and less fulvous, which, as in typical *aliciae*, there is no trace of a lighter orbital ring." The bird's beak, which in Burroughs's telling was most significant for the song it poured forth, received attention from Ridgway for its peculiar shape, "being much depressed basally, with the middle portion of the culmen somewhat concave." The respective descriptions of the literary naturalist and the technical ornithologist could not have been more different.[5]

Despite his confidence that he had found a new species, Bicknell himself remained unfailingly modest. Such was the province of the amateur in the face of authority, particularly in the domain of collecting. Often, even after making a spectacular discovery, amateurs practically groveled.

On hearing that Ridgway planned to name the bird after him—fairly standard practice for new discoveries, but by no means universal—Bicknell wrote him to say that he felt very grateful and that his only objection could be that "I feel myself scarcely intitled to the honor." It never went to his head. Writing to Joel Allen a couple of years later about publication of an article, Bicknell noted, "I should be glad to have your advice on this matter, if it would be convenient for you to give it. If ever I tend to overstep the limit of treatment, which you may regard suitable for subjects treated in *The Auk,* I would feel it a favor if you would interfere, and give me the benefit of your opinion."[6] Many supposedly new birds from the nineteenth century turned out to be subspecies. Bicknell's Thrush, however, was an entirely new species—though the diagnosis was not proved with certainty until 1995.

For zoologists, reaching agreement on the definition of "species" and the criteria by which a species can be distinguished has been a long and freighted conversation that continues today. There is more than one species concept. The first version dates back to Aristotle and was known as the typological species concept: birds were grouped by physical traits, with, say, big beaks or certain colors of plumage or similar types of feet. The biological species concept arose in the mid-nineteenth century and is still in wide currency. It applies most effectively to vertebrate zoology, although other disciplines have tried to find an equivalent theory. It says that a species consists of individuals that can *usually* breed among themselves (one imagines the minority of these birds with headaches or suffering from overwork) and produce fertile offspring. Other less restrictive definitions exist today, including the phylogenetic species concept, which defines a species as a group based on a shared and unique evolutionary history: interbreeding is less relevant than the bird's place in an ancestral chart. For instance, although the Pacific Slope Flycatcher and Cordilleran Flycatcher cannot interbreed, their DNA and morphology indicate that they may be the same species. To confuse matters, a species may interbreed in one region but not in another. Take the North American populations of Red-Breasted and Yellow-Bellied Sapsuckers. In some parts of the United States, they breed and create hybrids, whereas in other areas where their ranges overlap, they do not. Ornithologists thus disagree whether the two birds are different species or one is a subspecies of the other, on the way to becoming a new species but not having yet arrived.[7]

If such disagreements exist today, imagine how challenging and confusing identification of new birds would have been in the late nineteenth century, when ornithologists had fewer tools and resources at their disposal. Few scientists wished to crawl around in the brush for weeks or months studying something when, they reasoned, it was much easier to blast it out of the sky and take it home to examine. Ornithologists used the physical traits of their specimens to reach conclusions about birds. But whereas the professionals tended to limit their time to the indoors, larger numbers of amateurs—many of them acute observers and note-takers—did the actual collecting in the field, amassing a large quantity of useful data about birds' ranges, habits, and habitats. Their observational fieldwork was especially important in supporting and enhancing the biological species concept.

Almost every new bird helped to fill in gaps between allied species and provided clues about geographic distribution. The first known specimen of a new bird has long been known in science as a type specimen. Although an illustration of a bird rather than an example of the bird itself sometimes served as a type specimen, modern use of the term implies an actual bird in the hand. Taxonomists used type specimens to describe and name the bird in print for the first time, thus claiming priority for the discovery. Although there was sometimes more than one type specimen for a species (as when two or more examples of the same new bird were killed simultaneously during an outing, and thus called co-types), the norm was for a single "ground-zero" bird to exist, usually in an institutional collection. Type specimens serve as the formal basis for comparison with later birds. It's a rather curious formal approach, because the first example of a newly discovered bird might actually be somewhat atypical—in size, coloration, plumage, or any number of other characteristics. Still, it has served as an effective method for "keeping order in nature's complex household," as one writer has noted, and it is a universally accepted notion in zoology today.[8]

Live examples of new birds occasionally found their way to museums, which allowed the describer to study the bird's habits. In a bizarre event in 1874, one living type specimen being held in Ridgway's office tried to eat another type specimen on loan to him from William Brewster. Ridgway had Brewster's skin of *Helminthophaga leucobronchialis* (now known as Brewster's Warbler)—at the time the only known specimen—in his office, sitting on his desk awaiting examination and formal scientific description. Meanwhile, an unidentified burrowing owl, new to science, had recently been captured alive off the coast of Georgia and given to the Smithsonian. The owl escaped from its cage during the night and tore the head off of the loaned skin. Ridgway, aghast at his carelessness, gave the damaged specimen to a taxidermist to reconstruct before returning it to Brewster.

One of the most striking aspects of this incident was Ridgway's response. In his remorse, and as an attempt to rectify what he considered a moral failure, Ridgway promptly killed the owl. He did so clearly not out of anger but out of a sense of righting the wrong: "Though I had been much amused by his oddities struck by his unusual intelligence and become greatly attached to him through his tameness and affectionate disposition, I gave him a dose of strychnine as soon as he could be captured," he explained to Brewster. His letting down a colleague bothered Ridgway far more than the killing of a bird new to science. Sharing the bad news with Brewster was the most painful duty he had ever had to perform, he claimed shortly afterward.[9]

Writing back, Brewster reassured his old friend: "There is no one bird in my whole collection that I would not have parted with sooner than that one and of course it is a very great loss to me but I assure you that I do not blame *you* in the least as I do not doubt that you took quite as much care of it as I should have done in your place. Furthermore I do not blame anyone not even *the owl*, as he poor devil, didn't know what a choice tidbit he was amusing himself with. I can consider it only an *accident,* a most *unfortunate* accident." Ridgway was greatly relieved by Brewster's response. In turn, a similar incident happened to Brewster with one of Ridgway's specimens, though not with such a rare bird: in 1882, a mouse got into Brewster's office and chewed the legs off three of Ridgway's skins. Ridgway's response has not survived, but it was no doubt as understanding as Brewster's had been eight years earlier.[10] No modern ornithologist would probably keep a live bird new to science in his or her own office, nor would the bird be killed for damaging something, especially as a result of human carelessness. This episode thus speaks to two differences of the era from today: the relative informality of caring for even rare birds, and a much deeper sense of honor.

By 1896, the Smithsonian's entire collection of type specimens of birds took up just a single case in the Bird Division. But they had an outsized significance, especially for ornithologists at other institutions. Museum workers often wanted to borrow these type specimens, for a variety of reasons that usually related to identifying other possible examples. Because birds could be lost or destroyed while on loan, however, the Smithsonian increasingly viewed loaning specimens as a risky practice. More stringent policies at the museum were partially informed by things gone bad elsewhere. Writing to a colleague in 1884, Ridgway noted, "It is a known fact that in the St. Petersburgh [Russia] Museum and perhaps some other, type specimens have been stolen or otherwise removed and other specimens substituted for them! While in other museums, of the first rank (e.g., that of the Philadelphia Academy) labels have been often transposed or lost so that in some cases it is impossible to be sure that one has the type in hand." In 1886, Ridgway asked Smithsonian secretary Spencer Fullerton Baird to restrict loans of type specimens to officers of public museums, rather than to individuals or private museums. This was ostensibly because officials were more responsible recipients than a clerical worker or someone else into whose hands the birds would land on arrival—and because public museums were larger, were supposedly better organized, and had a greater stake in being careful with items in their possession. The death of Smithsonian secretary Joseph Henry in 1878 and his replacement by Baird also influenced loan policies. Henry had believed that the collections should be scattered among specialists across the country, not kept in a single museum. Baird ended this practice when he became secretary, and the policy change at the Smithsonian in the 1880s was a logical conclusion of Baird's desires to keep specimens massed within the institution.[11]

The role of accountability as a marker of professionalization began to play a larger role in the 1880s. Beyond the changes wrought by Baird's tenure and his desires to centralize the institution's collections, the Smithsonian's move to regularize the loaning of type specimens was part of a larger movement toward government accountability. In 1881, the Smithsonian established more formal rules for the exchange and loan of study specimens generally, not just for type specimens. This didn't mean that people didn't break the rules. Ridgway's friends sometimes tried to circumvent Smithsonian policies on study collections. At Harvard, Brewster tried to get around the Smithsonian's restrictions in 1886 by asking Ridgway if the museum would loan him or his assistant a specimen for use at his home.[12] But although these personal pleas between curators sometimes worked, institutional restrictions became steadily more limiting and more formal.

This trend was taken up by other institutions, and Harvard followed suit not long after, prohibiting the loan of any specimens by the end of the decade. The Museum of Comparative Zoology's annual report for 1889–1890 noted that the practice of sending collections to specialists for study was ruinous to the specimens and that preparing each invoice involved considerable work on the part of the assistants, accompanied by the dangers of misplacing labels during packing and unpacking. They thus refused from that year forward to send any specimens out of the building, instead inviting specialists to see the birds at the museum. By mid-1893, the Smithsonian had passed a similar decree, forbidding type specimens from leaving the building. Even years later, however, missing specimens could turn up, having been loaned and lost before regulations were in place. In 1899, Richard Bowdler Sharpe of the British Museum wrote to Ridgway's assistant Charles Richmond, noting, "On cleaning out all the holes and corners of my previous halting places, I came upon a parcel which ought to have gone back to the Smithsonian years ago. It consists of a couple of Moorhens lent to me by Ridgway. Will you please tell him how sorry I am and it shall not occur again!"[13]

Even though the need for order became increasingly important with the rapid arrival of new species, issues of authority and legislation did not extend to the treatment of nearly extinct species, which were collected wantonly despite their endangered status. Ornithologists of the era were aware that some birds were about to disappear forever. Sending some specimens from Hawaii in 1899 to Charles Richmond at the Smithsonian, Henry Henshaw, who was an enthusiastic and sometimes indiscriminate shooter and collector and later the chief of the United States Biological Survey, remarked, "When you get the birds cherish them for it looks as the last hours of some of the species were drawing nigh." In another letter, Henshaw noted, "I have shot one *Drepanis pacifica* [the Hawaiian Mamo] but lost it in the thick forest. A native offers to bring me one for $50. They are on the very verge of extinction."[14] He was right, for the bird shortly blinked out of existence and is now known only in museum collections. For both collectors in the field and ornithologists in museums, there was no apparent contradiction in mourning the impending loss of a species and trying to collect a specimen. This was one aspect of birds that united amateurs and professionals rather than dividing them—neither group found any contradiction in collecting in the face of extinction.

However, it was also true that some birds that were considered rare in the nineteenth century have turned out to be relatively common today.

There are at least two possible reasons for this: either the species has made a steady increase in numbers or its habitat simply was not well known at the time. The Hawaiian Hawk, or I'o, was considered nearly extinct by the end of the nineteenth century. Henshaw, writing to Ridgway from the Big Island, noted, "The Buteo [Hawaiian Hawk] is doomed to sure and early extinction. Already it is uncommon and rare in localities where numerous a few years ago. Every planter shoots them."[15] More than a century later that species is alive and reasonably well, and although currently classified by the International Union for the Conservation of Nature as "near threatened," it remains solidly established on the island of Hawaii.

By 1896, while acting as a go-between for a private collector, Ridgway would note in a letter to the Natural History Museum at Tring, outside of London, that among the birds being offered, "you will notice the Ivory-billed Woodpecker and Carolina Paraquet (two birds now so nearly extinct that it is almost impossible to obtain them) among them."[16] Although these species would live on for some years, their scarcity throughout the last decade of the nineteenth century onward made acquisition of single examples by museums difficult, and getting substantial series of the birds was practically impossible.

Wrangling Collectors

We still know far more about the scientists than we do about the collectors who facilitated the work of science.[17] The Smithsonian and other museums regularly recruited an army of far-flung collectors to work on their behalf nationally and internationally. They tried—but were not always able—to recruit reputable collectors who could note locales accurately, skin and preserve the birds properly, and perform other tasks that would result in museum-quality specimens.

Working as a collector for the museums was a highly desirable assignment, despite the regular hardships involved. For many, collecting was far more than just a means of income. A San Francisco physician, seeking work with the Smithsonian as a bird collector, wrote to Ridgway's assistant Charles Richmond in 1899: "I am still in the medical profession (I see you took your degree too) but I don't like it a bit. My life is ornithology." Another collector wrote to Joel Allen at Harvard's Museum of Comparative Zoology, "I love the Birds and the work and study—and cannot bear to give it up, if there is any show whatever of making it a success."[18]

Henry W. Henshaw, 1904. Archives of the Museum of Comparative Zoology, Ernst Mayr Library, Harvard University.

These men were often on the front line of discovery, which certainly helped to keep their enthusiasm up. Institutions may have sometimes been a bit blasé about new birds, but collectors most definitely were not. "Some few of my specimens are still in the hands of specialists; but already over a dozen *new* species have been made out—one of them being actually a new *genus*," wrote Canadian collector J. H. Keen to a colleague.

"This, as you can imagine, imports great zest to one's collecting."[19] The pace of new discoveries was brisk. Scientists identified 479 new species of birds worldwide between 1876 and 1883, the years of the Nuttall Ornithological Club's *Bulletin*. Even more—some 670—were discovered and named between the publication of the first issue of the *Auk* in 1884 and the end of the century. These numbers do not include many dozens of species misdiagnosed in print as new and then proven to be a bird someone else had discovered earlier.

All this collecting, and searching, could be enormously difficult, and hardships abounded. Describing a collecting trip in West Virginia, Brewster complained to Ridgway, "The country was the hardest for collecting that I ever saw: all up on edge; nothing but steep hills." Outside the United States, things could go terribly wrong in the field, and humans, not terrain, were often the worst dangers. Ridgway's good friend George Kruck Cherrie, an experienced collector who did extensive work in Latin America for New York's American Museum of Natural History and had traveled with Theodore Roosevelt, was on a collecting trip in Ecuador when a recently dismissed employee shot him with a shotgun. Cherrie managed to kill his assailant by returning fire, but his arm was badly mangled by the attack. Somehow, he bandaged up his limb and made his way, on foot and by boat, to the city of Guayaquil, a hundred miles away. The American surgeon attending to his wounds decided not to amputate the arm because he was sure Cherrie would be dead by morning. After several operations and a six-month hospital stay, Cherrie and his arm both left the hospital intact.[20] Collecting was not a life for the faint of heart, as Ridgway had seen as a young man on the Fortieth Parallel Survey expedition in the western United States, when he had nearly drowned twice. And office life was not without its own hazards for bird men. British ornithologist Richard Bowdler Sharpe protested once to Ridgway that one of the large, heavy and slippery Smithsonian *Proceedings* volumes had slipped from his grasp on a high shelf. Putting up his hand to protect himself, he nearly dislocated his thumb, and his fingernail split in half. "Tonight Vol. 16 fell on my head and then into the grate, cracking both," he complained. "If my account of the Robin is faulty, you will know the reason why. I have a great mind to write this letter in blood from my thumb."[21]

But the most consistent risks were in the field. The Smithsonian sometimes went to bat for field collectors in other countries in exceptionally trying circumstances. William Lloyd, on a collecting expedition to Mexico with Frederick Du Cane Godman—one of the editors of the bird books in the fifty-eight-volume set *Biologia Centrali-Americana,* a gorgeously illustrated series of works on the fauna and flora of Mexico and Central America—was jailed in 1888 for killing a Mexican bandit in "defence of his life and property." Ridgway got wind of the event and recruited G. Brown Goode, the Smithsonian's assistant secretary, to help free Lloyd.[22] At other times, collectors caused problems because they seemed to be just plain lazy. In the 1930s, Ludlow Griscom of Harvard's Museum of Comparative Zoology wrote to Herbert Friedmann at the Smithsonian, griping about a Central American–based collector named Hasso von Wedel who had been working in Panama for the museum:

> He has been collecting with us for some time on a definite basis of employment throughout the year, and it was entirely improper of him to suggest selling specimens to you. Nevertheless we are more than glad for you to take him over if you wish. Wedel is more or less of a beach-comber, well over fifty years old, and inspires every white man with whom he comes in contact with strong contempt in about two months time. He makes very good skins, is not really able to remember one bird from another, but succeeds in the course of time in getting quite a variety of interesting material, largely on a hit or miss basis. He is very slow and does not average a thousand specimens a year. He is also lazy and entirely lacking in initiative, and cannot be persuaded to go more than a mile or two back into the woods from a settlement . . . As he always blows all the money he receives the moment he gets it, he is always strapped and down and out at the end of a month, and consequently when we would advance him no more money he must have lived a pretty miserable existence while he was collecting a thousand specimens of birds for us to clear up his debts. He really did some work all last summer and this fall for the first time in many years, and we are now quits. As a matter of fact he feels incensed at his treatment and would undoubtedly be glad to switch from us to you, if you care to do so! The basis on which we have been paying him is $1 per skin up to ten from each locality with a bonus of 50¢ for large birds and another bonus of 50¢ for all good or desirable ones . . . We have about all the birds from the tropical lowlands that we want, and if you wish to take him over we will fire Wedel with pleasure.[23]

Wedel did end up leaving the museum's employ and was hired by the Smithsonian, for which he did substantial work, including discovering several new species that ended up bearing his name, despite his apparently uncaring ways. Griscom's lament shows just how time-consuming and costly collectors could be—logistically, financially, and psychically. It demonstrated the high expectations of museums. It also points up the mercenary nature of some collectors. Logically enough, their loyalty was more to their own desires than to any one museum. Perhaps the great benefit for Wedel was the ability to live the very lifestyle that Griscom griped about, able to see and collect birds, unencumbered by messy scientific facts, a schedule, or recordkeeping.

Other collectors appeared to be more energetic. M. Abbott Frazar, collecting for the Museum of Comparative Zoology in the 1880s, was sent by Brewster to the Gulf of California with high hopes: "Frazar . . . will spend at least a year exploring the peninsula and collecting in my interests. I think he may be relied on to collect three to five thousand birds and at least half as many eggs in this time. He is very energetic and daring and is likely to ransack every nook and corner." These numbers were probably fairly realistic—Frazar's work included at least one shipment of a thousand birds to the museum—and provide numeric evidence of expectations by museums.[24]

Some species were of course very common and easy to obtain, even if individual collectors' practices were a bit suspect. A young collector in Chicago, using only his slingshot to procure birds, offered the Smithsonian twenty grosbeaks and complained that he could have killed hundreds more if not for "policemen and other obstacles." But the land area necessary to support large numbers of species was disappearing. Writing from Massachusetts in 1879, Brewster complained to Ridgway, "All my favorite haunts about Cambridge have been spoiled one by one, until there hardly remains a pica of cover where I dare to fire a gun, and indeed in most places there is nothing to fire at."[25]

The numerous collectors working around the world—and the enormous net quantity of birds returning to institutional doorsteps—taxed the infrastructure of the Smithsonian and other museums. The accompanying field notes and correspondence alone were daunting. A 1933 article in the *Auk* about Frazar's nineteenth-century collecting described his epistolary approach: "The letters were written on bills of lading, leaves torn from account books, or other scraps of paper, and not only did he use both sides for economy, but having filled a page in one direction, he would write completely across it at right angles."[26]

The Smithsonian's desire to expand its international collections did not mean that it neglected other parts of the United States. In particular, it tried to make use of many collectors on the California coast. Collectors, including the aforementioned Frazar, also fanned out across the Baja Peninsula in Mexico on behalf of the Smithsonian. The huge landmass out West made it difficult to find birds there, especially rare ones, but while the challenges were substantial, the rewards were often equally big for both collectors and institutions. The people collecting for the Smithsonian were typically in it for profit and included professional colleagues such as W. E. Bryant, the curator of the California Academy of Sciences. Bryant regularly went hunting for California Condors, which were nearly extinct in the 1880s. "I will say that I have already three applicants for the Cal. Condor . . . but have found reported *Condor* to shrink to Golden Eagles or Buzzards when I had arrived," Bryant noted wryly. He also kept a live juvenile condor on hand for several months.[27]

Working in the Far West could change people. Some collectors began with a built-in prejudice for the birds of their native locales, only to change their mind. Writing from Cambridge, Massachusetts, in 1874 several years before going west on the Wheeler Survey, which explored a portion of the United States west of the hundredth meridian, Henshaw suggested to Ridgway, "I tell you some of your Washington ornithologists ought to come up here collecting for a season instead of wasting your time out West." But once in California by the early 1880s, Henshaw, like Bryant, was smitten with the California Condor and was busy trying to track it down. No birds anywhere near that large existed in his beloved New England. Ridgway noted that according to Henshaw, the California Condor "appears to be now practically extinct in most parts of California, and may be entirely annihilated before many years, so if you haven't got it I would advise you to make every effort to secure one." Again, trying to secure an example of an almost-extinct species did not present a contradiction to any of the parties.[28]

Work for the Smithsonian was a far-flung affair. Collectors were often military men, and as such, they wrote to Ridgway from such places as Fort Custer, Montana; Fort Klamath, Oregon; Fort Walla Walla, Washington; Fort Huachuca and Fort Whipple, Arizona Territory; Fort Fred Steele, Wyoming Territory; Fort Garland, Texas; Fort Randall, Dakota Territory; and other military posts. They wrote from boats (the USS *Albatross* off Woods Hole, Massachusetts) and from other countries, including Costa Rica, Colombia, and other species-rich parts of Central and South America. They sometimes stayed at one another's houses (or at least often

Henry W. Henshaw, 1878. Archives of the Museum of Comparative Zoology, Ernst Mayr Library, Harvard University.

invited each other to do so). Many collectors moved around a great deal. A fellow native of Ridgway's tiny hometown of Mount Carmel wrote to Ridgway several days before departing for a new job in Tuscumbia, Alabama, explaining that he wished to collect for the Smithsonian there. "There must be rich and rare things to find in that fair region where the torrid hugs the temperate in a gentle embrace," he noted in florid prose.[29] All these efforts went toward building up the Smithsonian's collections and connections and helped to secure its authority as the nation's museum.

The Smithsonian sometimes used a carrot-and-stick approach with collectors. Ridgway responded to one request for free publications by noting that it was the policy of the institution to send printed works only to those who had contributed birds to the collections. The rejected correspondent then sent in more than a dozen specimens, in hopes of obtaining copies of unspecified Smithsonian bulletins and reports. At the same time, Ridgway would occasionally send along a publication—sometimes a fairly costly one—as an inducement and to gain favor with collectors. "I would suggest sending him at once Vol. IX [of the ornithologically related *Pacific Railroad Reports*] as a sort of encouragement, and ask him to send particular things," Ridgway wrote to Baird about a collector in a remote part of California. The government publication in question was large and expensive, and the cost of sending a big volume always had to be weighed against any benefit.[30]

A few things could skew an institution's willingness to receive specimens. If buying, cost was an issue. If being offered free specimens, scarcity or abundance was of course an issue. But for more common birds, size was also relevant. The American Museum of Natural History, for instance, generally preferred small birds to large birds for exchanges— probably because of its chronic storage limitations. Aside from their rarity, gigantic birds such as the California Condor proved difficult to manage in the field, much less in a museum setting. "Had the pleasure or the pain, as you will, of skinning two Vultures the other day, both fine adults," Henshaw wrote to Ridgway. Other huge species arrived intact at museums. The chief officer of a British ship captured and killed an enormous albatross with an eleven-foot wingspan. He promptly froze it in the ship's freezer and offered it up to the British Museum, which accepted it.[31]

Not all collectors were working for museums, of course. Some were

adding to their personal holdings of birds, and occasionally, over the decades, they would try to make off with an institution's goods. "Collectors
are utterly unscrupulous folk," grumbled a British natural history author
connected with Hope Museum in Oxford to a colleague in recounting
an attempted theft of beetles from his museum by one collector. "They
are all alike: 'collecting' seems to pervert the moral sense." Elliott Coues,
always ready to render an opinion on his colleagues' attitudes, noted that
there was no correlation between skill in collecting and a sense of ethical behavior. "One of the most vulgar, brutal and ignorant men I ever
knew was a sharp collector and an excellent taxidermist," he noted in
a book on collecting. However, numerous collectors followed codes of
ethics. For instance, if they determined that someone else was working for
an institution in a particular country, they would often bow out to avoid
the competition. Wilmot W. Brown Jr., an AOU associate member who
collected extensively for the likes of ornithologists Charles Cory, George
Cherrie, Charles Batchelder, Brewster, and Outram Bangs, canceled a
long-planned trip to Venezuela on hearing that the British Museum already had a collector posted there. "As I have had Venezuela in mind for
several years it cost me a severe effort to give it up. But as your collectors
preceded me on the *field* the only course left to me as a law abiding Ornithologist was to bow to the Ornithological law of 'priority,' and look elsewhere," Brown wrote.[32]

Specimens from private and public hands were intermixed into a
large river of specimens flowing in and out of and within countries. This
quantity of material was not limited to the United States and, in fact, was
probably even more extensive in Great Britain. The British Museum held
the world's largest collections by 1900. With 400,000 skins and birds, it
was more than double the size of the Smithsonian's holdings of approximately 185,000 items. The British Museum didn't take everything, but correspondents sent in a great deal. Examples abound in the archives of the
Natural History Museum in London, which holds historic runs of correspondence from collectors and potential donors to the British Museum.
One L. Du Bois in London wrote in 1897 offering "Emu eggs, perfect
specimens, from the unexplored interior of the Australian continent." The
letter was simply docketed "refused" three days later. People offered birds'
nests (sometimes accepted), bird eggs, skins, and "alcoholic" specimens,
with varying degrees of success. A Colonel Biddulph offered his collection
of bird skins, "Indian and mostly Himalayan, if the Trustees will accept

them. The collection contains more than 2,550 specimens representing some 680 species. The Himalayan 112 species and about 650 specimens I believe to be very good." These were taken, but with little fanfare. Sometimes, a sense of urgency arose amid international competition for particularly desirable specimens, such as birds of paradise. "With the Germans so active on our New Guinea frontier, it is essential that we should hold our own in seizure, and this Museum is most anxious to acquire specimens of the wonderful new forms you have described," Sharpe wrote to an Australian-based zoologist and collector, Charles Walter de Vis, in 1897. "We have got a male Loria at last, but there are a great many species still wanting." For the year 1897 alone, hundreds of letters to the British Museum deal only with donation or exchange of specimens from an almost bewilderingly wide range of private and institutional collections and collectors.[33] The business of deciding what to take and then processing the new materials occupied much of the staff's time.

Behind the bloodless paper-driven records of these institutional transactions and the specimens themselves, however, lay an inescapable fact: all these birds had to be killed by someone. Sometimes the killing seemed wanton, although it resulted in the specimens that would make up vitally long runs of birds in museums around the world. Henry Henshaw seemed especially enthusiastic about taking large numbers. In 1873, he wrote to Ridgway from Arizona while on the Wheeler Survey, "I took a sufficient number of specimens and then confined myself to shooting every bird about which there was the least doubt. I took no chances w[ith] the little chaps but slaughtered them remorselessly. I have about sixty specimens." He did uphold a notion of scientific utility in these deaths, however, observing, "It seems to me utterly impossible that among them should not be found one or two additional [that is, new] species." His somewhat indiscriminate shooting ways continued in later years. Writing to Ridgway from Washington Territory in 1880, Henshaw noted that he had encountered thousands of Red-Backed Sandpipers (now known as the Dunlin), and he had "shot two barrels into a phalanx the other day and picked up 69, and a considerable number got away wounded." A colleague writing to Ridgway in 1885 noted that "Henshaw was out Sunday I hear, and slew and skinned 50 feathered innocents."[34]

Others affirmed the life-giving nature of killing birds. "Glad to hear . . . that you did some shooting at Chatham," wrote Brewster to Allen. "Field work is a panacea for all ills that flesh is heir to." Still others confronted the paradox more directly. After shooting a warbler whose song had delighted him, Elliott Coues noted, its death was an inevitable tragedy—"for tragedy it is and I cannot, after picking up warm, bloody little birds for years, make anything else out of it or learn to look on with indifference."[35] Somehow, though, sentiments such as these seemed to remove the shooter from the act of shooting, as if he was somehow an observer watching a bird that had died by another hand. And indeed, the prevailing attitude among collectors as well as the institutions receiving birds was that the "hand" was that of science: the advancement of understanding through collecting and study.

People being paid by institutions to collect in the field did not always shoot their own birds, regularly employing people who were better shots or who had more extensive access to certain areas and knew the local terrain more intimately. These subcollectors worked for pay in the American West, Costa Rica, and a number of other locales, laboring for collectors from the Smithsonian and other American and British museums. On return of the birds to camp, the hired gun would measure, describe, and skin the birds and then turn them over to the collector of record. Collectors also worked under their own names, rather than outing themselves to their subcollectors as working for an institution, to avoid being charged unreasonable prices, which one Smithsonian collector noted would go up "a hundred fold" if the government's identity was known.[36]

Gathering statistics about specific birds was another form of collecting. For instance, Ridgway wrote a circular letter to numerous collectors across the country in 1891, requesting details for "a systematic investigation of the color-phase of the Screech Owl (*Megascops asio*), with the view of determining the relative abundance of the red and gray plumages in different parts of the country, the character of young produced by parents of known character as to plumage and other related questions." These queries led to some of the earliest studies of the evolutionary and chromatic phases of birds. In the case of the Screech Owl, Ridgway was asking on behalf of E. M. Hasbrouck, who used the data to publish an important paper on the topic in 1893 and whose data on Screech Owls was still being used as late as 1955.[37]

Other data collected in the field allowed Ridgway to incorporate details of ranges and frequency for individual species over a span of years. Some collectors trusted him enough that they sent him original field notes they would not readily show anyone else. Henry Coale, the founder

of the Ridgway Ornithological Club in Chicago, loaned Ridgway two journals, containing a dozen years' worth of written observations for Illinois
birds, obtained during a busy professional life: "Remember that since Oct
1872 I have worked 10 hours a day steady, with only 10 days vacation each
year, and sometimes less, that all my collecting has been done on these
few days, on Sundays or between daylight and 7 a.m. in Spring—Would
not care to have any one else inspect them—Am very glad that *my* notes
can be of *some* value in the cause of Ornithology."[38]

How quickly could collectors work? Shooting *and* skinning fifty birds
in a day was considered close to an upper practical limit, although maintaining that pace over numerous days, even during migratory seasons, was
difficult. "You will do very well if you *average* a dozen a day during the seasons," Coues observed. His own daily record was forty skins. Shooting a
lot of birds wasn't necessarily the solution to increasing numbers; often,
birds were too mutilated to serve as skins, or sometimes they were lost in
bodies of water or unapproachable underbrush. Not counting the time
in the field, the task of only skinning birds could be quite a bit quicker.
Coues recounted that an exceptionally good taxidermist could skin ten
birds an hour, although this seems optimistic. Coues himself said that he
once was able to skin eight birds in an hour, averaging seven and a half
minutes per skin. Four birds an hour was considered good work. It was
important to get the bird's shape right when creating a skin. "You must
pay close attention to the form and attitude of the bird," British naturalist Charles Waterton wrote about skinning birds in his widely read work
on the birds of South America. "In a word, you must possess Promethean
boldness, and bring down fire and animation, as it were, into your preserved specimen."[39] An artful eye and a familiarity with other birds of the
same species were as essential as careful, quick work.

The physical preparation of study skins was done by both amateurs
and professionals, usually in the field but also sometimes in the museum
when fresh specimens arrived. In most cases, the methods used have also
allowed specimens to survive into the twenty-first century, where they
have been used to answer a variety of research questions never imagined by their original skinners and preparers. The processes used to create study skins have not changed much over the past two centuries. The
first task was to open the bird at the breastbone, cutting down to the anus
and carefully working the skin free from the underlying flesh. The skinner
then cut the leg bones, usually at the knees, so that the lower legs and feet

would remain attached to the skin but the ribs, spine, and viscera could be removed. By the time a skin was completed, it consisted of the legs, the skull and beak, the wings and their bones, and as much of the skin and plumage as possible. The brain and viscera were removed, the body and skull stuffed with cotton batting to approximate the bulk and shape of the bird, and the incision on the breast sewn up. Well into the twentieth century, arsenic—in either powdered or soap form—was rubbed inside the skin to counteract the effects of deterioration resulting from any flesh, especially fat, still attached to the inside of the skin. The entire process of skinning was, and remains, a messy business. Getting squirted by bodily fluids of birds was a common experience, and careless workers could be cut by sharp beaks, talons, or even severed edges of bones within the bird, as well as by slips of their own scalpels.

The arsenic was especially problematic. Besides suffering its depressive effects, workers could experience other short-term physical ailments when preparing study skins. Arsenic could render one's hands nearly immobile. Ridgway wrote to Baird in 1883 (having his wife dictate the letter) that he recently had to stop preparing bird skins because his hands were so painful that he could no longer put his own clothes on. As a preventative against insects and to preserve the skins from deterioration, something called Maynard's Dermal Preservative was popular among ornithologists who swore by its efficacy in the mid-1880s not just in preserving birds but in protecting hands from arsenic poisoning. Ridgway and Henshaw both had high hopes for the expensive powder. Ridgway expressed excitement to Baird about the possibilities of the new preservative, having struggled with arsenic for years. The new preservative was claimed by Maynard to be totally harmless as well as an excellent preservative, and it was briefly very popular with the naturalists who used it. The preservative was not cheap, costing about twenty-five cents per pound, but Ridgway argued that if it was ordered in sufficient quantity, the price would drop. "At any rate," he concluded, "I should consider it really cheaper at double the price!" By the end of the century, however, the formulation had discolored plumages, was discredited, and disappeared from the bird preservation literature entirely. But the enthusiasm of Ridgway and others about the possibility of a new means of stabilizing bird skins illustrates how hazardous and undesirable arsenic was as a preservative in the field. Today, museum preparators use no chemicals when preparing bird skins. Even borax, which served as a natural soap and mild insect repellent for decades, has been eliminated because it has been implicated in accelerated color changes to plumage.[40]

Collectors typically charged between seventy-five cents and three dollars per prepared skin, depending on such factors as size, scarcity, and condition, although their work varied greatly in its thoroughness and quality. "I am not surprised at your being disgusted with the skins as the[y] are a pretty shabby lot," a collector wrote Ridgway in 1878, with whom he had exchanged some skins. "Trusting you will be lineant with me a new collector," he closed meekly. At the other end of the scale, collectors would go the extra mile to identify the sex of the specimens, sometimes a difficult job even within a museum setting, much less the field. "I will guarantee my skins to be first class in every respect—nice, clean skins. With great care given to the taking of sex and measurement," wrote a northern California collector to Ernst Hartert of the British Museum. Some collectors charged the same for any size, while others charged more for the greater work involved in preparing, storing, and shipping larger birds. A few rare large birds, such as California Condors, were worth as much as thirty dollars each. Some collectors sold their wares to museums in lots rather than as individual birds; others worked strictly on salary, regardless of the number of birds provided.[41]

These expenses added up. However, ornithologists working for their own institutions usually cost nothing beyond their regular salary. Ridgway's own collecting work in Costa Rica was especially cost-effective. The combination of his expertise with birds and his fieldwork meant that it practically rained multiple examples of rare species. Where his friend Henry Henshaw killed birds rather wantonly, Ridgway shot carefully and selectively. During his 1905 Costa Rica trip, despite having to borrow an unfamiliar gun, he was able to surgically collect no fewer than ten specimens of the hummingbird then known as *Arinia boucardi*, very endangered but still extant today as the Mangrove Hummingbird (*Amazilia boucardi*)—a particularly shy bird found only in the mangroves of Costa Rica and which until then had been known only by a single specimen in the Muséum National d'Histoire Naturelle in Paris, collected some thirty years earlier by Adolphe Boucard in Punta Arenas.[42]

But the wilderness was not the only place to find new creatures. A bird collector's careful scrutiny of the world around him could even yield new nonbird species, and sometimes they were *really* close at hand. The Reverend J. H. Keen, an enthusiastic amateur based in the Queen Charlotte Islands in British Columbia, wrote to British collector O. V. Aplin, "A mouse which I found frequenting my house like a common house mouse proved, on examination by an authority, to be a species new to science, and they have done me the honor of calling it after me." Keen's Mouse (*Peromyscus keeni*) lives on as a valid species, today primarily found in Canada. Furthermore, noted the sharp-eyed Keen in the same letter, "Among some bats which I captured here, too, was one sufficiently peculiar to constitute a new species, and will be described as such, if I can meet with another specimen." Such an encounter must have taken place: Keen's Bat (*Myotis keenii*), now found only in British Columbia, was described two months later.[43]

The Value of Series

Mice and bats aside, all of these monumental efforts at accumulation led to many long series of the same birds arriving in institutions. Multiple examples of the same bird have been amassed by collectors for centuries. Natural history collectors have been as obsessive-compulsive as most other collectors, and they often hoarded large runs of specimens as a fetishistic series of acquisitions. In the 1870s, duplicates were already considered useful for both study and bartering. The use of multiple species to identify variation within species was less critical than the ability to collect birds in different plumages, from immature to molting to adult, of both sexes, and—perhaps most important—as a medium of exchange. "How many birds of the same kind do you want?" asked Elliott Coues rhetorically in 1871. "All you can get." Just because someone had a bird in their collection didn't mean they would turn down another. "This is just as reasonable as to suppose that because you have got one dollar you would not like to have another dollar. Birdskins are capital; capital unemployed may be useless but can never be worthless."[44]

The Smithsonian and other museums worked hard to fill gaps in their collections by amassing full sets—family by family, genera by genera, and bird by bird. Duplicates were often traded for specimens institutions lacked or for better-quality examples of existing species. This could be challenging for institutions; space was tight in museums that had been built in an era when collecting was done on a smaller, more personal scale. And birds were not the only animals these institutions collected; they also counted extensive mammal collections in their holdings, including some stupendously large critters such as lions, bears, and even dolphins and whales. So bird quarters were often cramped for both workers and specimens, as an institution weighed its competing needs. But the value of long series of the same bird would prove vital for what they showed about how organisms could slowly change across a region, as they morphed over long periods of time from one bird into another. They were crucial for the use of individual institutions doing research but even more for the shared use of a network of institutions that constantly bartered, borrowed, and swapped specimens for comparative work. For instance, Ridgway, serving as an intermediary for a sale, described a collection being considered by the British Museum as a "fine series of specimens illustrating individual sexual and seasonal variations and the various stages of growth."[45] Variation in nature can be extremely confusing, and it had long been an object of intense scrutiny. Almost all organisms have some inherent variation. Humans, for instance, have differently colored hair, skin, and eyes; different heights; different voices. We are, however, a single species. So it was easy to confuse variation with speciation—and in fact, what may appear to be simply a variation is indeed sometimes speciation.

Identifying subspecies became a central task in the United States by the 1880s. What is a subspecies? It is—by at least one definition—a new species in the act of breaking off from an existing species for a variety of reasons. Ornithologists were the first scientists to implement subspecies into their taxonomic work, although the practice has now spread to all zoological disciplines. Bird classifiers are often known as either "splitters" or "lumpers." A "lumper" groups birds with similar characteristics into a single group, whereas a splitter assigns significance to finer gradations of color, shape, and other characteristics. Subspecies proponents tend to be splitters, recognizing that in any large group of birds, the reliable presence or absence of key characteristics, no matter how minor, is significant.

The road to fuller understanding of variation within species was filled with obstacles of various kinds. For instance, common birds could arouse suspicion as to any meaningful distinctions among numerous examples in their ranks, despite evidence to the contrary. This seemed to occur most frequently with the most common species—perhaps as prejudice against birds that often were considered pests. William Brewster, writing to Ridgway about the Common Grackle, noted that he hadn't paid much attention to the issue of any subspecific differences. Although the Museum of

Comparative Zoology held only fifteen examples of that extremely common bird, Brewster wrote, "I have—without any real grounds for so doing I must confess—regarded the difference between the supposed varieties with a suspicion amounting almost to disbelief but as I have just stated my convictions have never been very strong. I now have my series of specimens laid out before me and in the whole number I find only three that I have 'several tints on each feather' and the line of demarcation not well defined."[46] However, the Common Grackle is now known to exist as three separate races, or subspecies.

The United States was the founding crucible for what Mark Barrow has called "the subspecies research program." The modern concept of subspecies in ornithology did not emerge until Americans, notably Allen, Ridgway, and Coues, tested and retested assumptions about the variability of individual species by working over very long runs of birds. Even with hundreds of specimens in hand, determining just what constituted a subspecies was often a tortuous task. Because this feathery tribe working after the 1870s was the first group of people to try to identify many of them, the work was akin to writing a dictionary for a foreign language: terminology had to be established, boundaries circumscribed. Writing again about the grackles, Ridgway expressed frustration: "I do not see how any arbitrary line can be drawn which shall determine what proportion of a specimen exhibiting the peculiar impress of a region is necessary to constitute a geopolitical race, unless *a considerable majority* (say 75 per cent., or three-fourths) is considered sufficient."[47] In some ways, an attempt to identify a subspecies was a numbers game. It was easy enough to mistake a distinctive individual for a true species or subspecies—often a rookie move, but one that even experienced ornithologists could fall prey to if they were not sufficiently well versed in the characteristics of a particular bird. Time and time again, ornithologists in the nineteenth century identified birds new to science, only to discover that in fact the bird in question was a differently colored form or a hybrid, or had immature plumage, or simply displayed an unusual variation not found in other specimens. So percentages of birds that reliably displayed certain characteristics were an important verification that a type was worthy of designation as a subspecies.

Here, the effectiveness of collectors again came into play. One of the challenges of relying on collectors to capture birds is that for much of the nineteenth century, they were more likely to work in the centers of species' ranges, where they knew certain birds were likely to be found, rather than skirting the edges of birds' ranges and thus more clearly delineating those edges, which could most readily show where one species intergraded into another. "What was useless for personal collections was precisely what subspecies taxonomists required," one historian has noted, pointing out the hit-or-miss nature of collectors' work.[48]

But as the scale of field collecting grew, more and more specimens covering a wider area appeared. Collectors also tried to follow institutional instructions to collect the same birds in and around the known edges of their range. This led to a dramatic increase in collection sizes, because it was far easier to secure a hundred birds of the same species than to collect a hundred different species. It is difficult to draw generalizations about what duplicates were acceptable for a museum to donate or trade, because the value of every series of bird was slightly different, arranged along a spectrum. Keeping a hundred examples of the European Sparrow, for instance, would have seemed ludicrous to everyone because of its ready availability everywhere, along with scientists' near-universal dislike of the little birds, which they considered to be dirty pests that drove native species away. Slightly rarer birds, however, warranted slightly longer series, and birds newer to science warranted longer series, even if fairly common, so that a number of points could be teased from the runs of birds, including geographical distribution, plumage variations, and other finer differences.

In numerous cases where long series of common birds existed, specimens could provide extensive information on hybridization and speciation. Ridgway noted in 1899 that he had reviewed at least five hundred borrowed specimens of two allied birds: *Zonotrichia leucophrys* and one of its now-recognized subspecies, *Zonotrichia leucophrys gambeli* (the White-Crowned Sparrow and Gambel's White-Crowned Sparrow, respectively). Other institutions also borrowed common birds extensively. Before the AOU's Committee on Nomenclature met in New York at the American Museum of Natural History in 1889, Allen asked Ridgway to loan the Smithsonian's collection of Horned Larks to add to the five hundred examples Allen had already borrowed from other sources. Allen also pulled some three hundred larks from the museum's collections for the meeting. Although curators sometimes apologized for their profligate habits in borrowing so many birds ("I am afraid you will begin to think the city of Portland made up of mendicants," one Oregon curator wrote

Ridgway after a flurry of requests), the practice allowed for an invaluable extension of an institution's in-house series of birds.[49]

By the 1890s, the value of long series was firmly embedded in Ridgway's consciousness, and thus in the minds of those in the circle he influenced. As a result, ornithologists were less likely to give away birds as the century closed. In trying to help the bird curator at the Baron de Rothschild's famous British private museum in Tring, Ridgway noted that he would try to help fill in the museum's Galápagos birds, but he was not willing to emasculate the Smithsonian's duplicate collections to do so. "Large series of specimens of these inland birds are necessary to show the limits of individual as compared with local variation," he explained to curator Ernst Hartert in 1894. The British, considerably more resistant to the subspecies concept, were also thus resistant to keeping large series of the same species.[50]

The absence of sufficiently long series could prohibit ornithologists from acting on hunches that would much later prove correct. Ridgway occasionally noticed subspecific differences that he never published, but commented on in correspondence to colleagues. For example, he suspected the existence of two distinct forms of the Carolina Parakeet around 1900, a hunch that was finally validated in 1957 by the American Ornithologists' Union, well after the bird was extinct. Writing to Brewster, he noted, "You have brought to my mind a matter in which I was much interested several years ago—the unquestionable fact that there are two very distinct forms of the Parrakeet. We have a very good series of both, and the only reason why I did not publish a paper on the subject was that the matter of nomenclature bothered me, and, being unable, from lack of specimens from Louisiana, to settle the question I dropped the subject for the time and since then have been too busy to take it up again."[51]

The comparative morphological work of the era served as a foundation for much systematic work to follow. In 1881, William Brewster proposed that North American screech owls be grouped into three types. His three groups still stand, now as individual species: the Eastern Screech-Owl, the Western Screech-Owl, and the Whiskered Screech-Owl. "You will see that this material has a deep and far reaching significance," he concluded about the three groups. However, some problems just could not be worked out by morphological comparison, no matter how careful, because scientists labored without the benefits of mitochondrial DNA analysis. For example, Ridgway and Brewster corresponded at length about *Geothlypis trichas*—the Common Yellowthroat, first described and named by Linnaeus in 1766. "There is so much variation . . . in all sections of the country, in all these characters . . . that with what specimens I have examined I would not feel justified in separating the western birds as a race," Ridgway noted to Brewster in 1882.[52]

His caution was understandable in light of modern knowledge. In the case of the Common Yellowthroat, the record continues to be muddy today, with the authoritative guide *Birds of North America* noting, "Complex geographic variation and many poorly described (invalid) subspecies have contributed to confusion. Individual variation can be sufficient to obscure subspecific differences, which are further complicated by clinal variation where subspecies ranges meet . . . Consequently, much debate over ranges and descriptions of subspecies (particularly western races) has occurred."[53] No wonder Ridgway was uncertain about the bird's range!

Other obstacles to an accurate understanding of just what constituted a subspecies existed besides the lack of DNA analysis and collectors avoiding species borders. Labeling errors could also cause confusion or introduce error. Without accompanying geographic information, even long series of birds were of relatively little use for determining interspecies variation. Because knowing the location of their acquisition was critical to understanding their boundaries, and thus where they intergraded into other species, their paper labels were critical. These labels had been attached to birds' legs for centuries with a notation, but the text was often very casually written, and sometimes nonexistent. Labels got better as the decades passed and almost always noted at least the bird's name, in what general location it had been found ("Texas" or "Venezuela"), and the sex. Field notebooks, correspondence, and accession logs all helped provide metadata about bird specimens. Not everything important could fit on a label, though, and certain information was thus privileged, ignored, or abbreviated. By the last quarter of the nineteenth century, specific locations showed up more reliably on labels, but administrative problems were often an issue. The job of labeling newly arrived specimens usually fell to an assistant, who might work from loose, temporary labels or from a list of specimens that had arrived en masse in a box. This left room for considerable transcription errors, so that one bird might be misidentified as another. Location details would thus be attached to the wrong bird, among other errors. Mislabeling a specimen or excluding a label became a greater problem after Darwin, because it meant that range details might be wrong or even lost. "It is enough to make a sensitive ornithologist shiver to see a specimen without the indispensable appendage—a label," Coues remarked in his book *Field Ornithology*.[54]

Labels today contain considerably more detail than in the nineteenth century, although the size of the labels has not increased much. Ornithologists now typically record the name of the institution owning the specimen, the full species name, the sex, the exact location collected (including place name as well as latitude and longitude), date and time of collection, the molt status, amount of fat on the bird, the colors of soft parts such as eyes, the collector's name, a brief description of the habitat where found (for example, "second-growth oak-hickory of 20–25 feet"), measurements, weight, stomach contents, and sometimes even brief notes about behavior.[55] Getting all of these details onto a two-sided label that typically measures about twenty by ninety millimeters requires a very steady hand and shorthand codes. To lose these factoids now is even more problematic, although most institutions currently record them redundantly in a database. But lose the little tag, and any collection details recorded elsewhere might as well be tossed out, because they cannot then be tied to a specific bird.

Individuals and institutions had collected runs of birds for centuries, and had undertaken detailed systematic work to cluster similar birds together to understand their place in nature's tangled skein. They also undertook important anatomical work, making minute comparisons of bones, muscles, and other parts. Over the centuries, a torrential river of bird material—skins, alcoholic specimens, skeletons and loose bones, eggs, and nests—poured into museums from around the world and then rushed out again in trades, loans, and gifts. Collectors came from all walks of life, but by the 1870s, one hallmark of the professional ornithologist was that he was interested in using increasingly longer series of specimens for comparative purposes that illustrated aspects of Darwinian evolution through their comparative value. From the perspective of many collectors, the transactions involving study skins were mostly about income rather than the advancement of science. These collectors were somewhat accountable to museums for their work, but in many cases, they were far more accountable to their own desires and needs. In the Smithsonian's view, as well as that of other institutions, the advancement of scientific knowledge was paramount. Typifying that attitude was a letter from Ridgway to Charles Cory, the Chicago Field Museum's curator of birds, in 1882: "I believe that I told you when you were in Washington that we do not usualy so much consider the marketable value of specimens received and given in exchange as the paramount object of the benefit to science prompted by the exchange."[56]

The Smithsonian's collections in the National Museum grew rapidly from the 1870s onward because Joseph Henry, a physicist, had been adamantly opposed to amassing a large national collection and insisted throughout his tenure as secretary that specimens, even type specimens, be widely distributed to specialists. Baird recalled a number of collections back to the Smithsonian, maintained a formidable volume of correspondence with a large network of correspondents, insisted on keeping everything, and focused on institutional growth. Ridgway, being dedicated full-time to the Bird Division, could focus exclusively on expanding the collections. Another key effect of having long series of birds—which, by the end of the nineteenth century, were firmly connected to series at other museums by research trends and needs—was that accountability became vitally important. To be able to use the specimens of other institutions was essential, and if that was to happen, rules had to be put into place that would prevent chaos from descending on both lending and borrowing museums. The quantity of birds coming into and leaving institutions necessitated the development of policies necessary to regularize loans and prevent losses. Keeping specimens in order, in their home institution, properly identified, and available for research use also became more important as the significance of these series became more apparent. Priority for the discovery of new species was claimed by a formal approach that consisted of describing and publishing. In a related vein, type specimens—as the actual first bird of a new species—became increasingly subject to institutional rules about where they could and could not be used. As we shall see in the next chapter, newly formulated nomenclatural rules would dictate how these new species could be named.

Most important, study collections, when used in large runs of the same or similar species, allowed the drama of Darwinian evolution to be spread out for all to see. If a hundred copies of a bird were laid out shoulder to shoulder on a long bench and arranged in order of length, the upper and lower size, shape, and color limits of that species could be known. Careful examination of other parts of those birds might show that the longest 10 percent of the birds also had a slightly larger beak or a stouter beak. Those differences were important because a beak that was longer or stronger would allow a bird to exploit resources in a region that other birds of that species could not, thus conferring an advantage to those birds with longer beaks. Subtle differences in coloration or other characteristics could also convey an advantage in sexual selection. Eventually, these differences could become widespread, depending on where they occurred and the conditions under which those changes took place.[57]

Size mattered, too, and the American Ornithologists' Union formed a Subcommittee on Bird Measurements in the late 1880s to provide new standards for precision in measuring birds.[58] Upper and lower size limits, coloration, and other characteristics became increasingly important as they simultaneously became clearer through comparisons, and subspecies could often be decisively identified. Subspecies gave naturalists the ability to study how a species had been able to adapt to various habitats or environments, and began to give glimpses into the evolutionary history of a species. And subspecies, when accompanied by details of habitat and geography, provided confirmation of Darwinian natural selection. These factors contributed to the dramatic increase in the desirability of owning series. Long runs of the most beautiful birds had been eagerly sought for many reasons over the centuries, including aesthetic ones—simply for the lavishness of seeing such beauty aggregated in a group. By the last quarter of the nineteenth century, however, scientists were after a different kind of beauty: that of knowledge about the relationships between and among bird genera and families and across species and subspecies. Systematics—the study and classification of the relationship between life-forms and evolution—could not have provided supporting information for Darwinian claims without all these duplicate specimens. By the end of the century, scientific ornithologists had established a physical thesaurus from existing collections: they had created formal groupings of birds, with synonyms and related concepts laid out—and if individual terms were not entirely agreed upon, bird workers had at least established a basis for the arguments over the relative places and positions of birds in the natural world.

CHAPTER FIVE

Nomenclatural Struggles, Checklists, and Codes

No one appears to have suspected, in 1842, that the
Linnaean system was not the permanent heritage of
science, or that in a few years a theory of evolution was
to sap its very foundations, by radically changing men's
conceptions of those things to which names were to
be furnished.

American Ornithologists' Union's
Code of Nomenclature, 1886

D ISCUSSING NOMENCLATURE and taxonomy—the names
and relationships of things, especially in a scientific context—
doesn't really qualify as good cocktail party conversation. Broach
the subject, and civilians are quickly gripped by a look of panic. Their
eyes, if they haven't already rolled far back into their skulls, begin darting
furtively around the room in a desperate dance as they attempt to break
free of the conversational shackles in which they have been placed. However, the topic is vital for the study of the history of science and what it
has meant for the formation of modern biology. Biological classification
has always been untidy business. Humans have had a long and strong
desire to group, categorize, and name things as a way of applying order to
our understanding of the world. Ridgway knew what Darwin knew what
Linnaeus knew: it was possible to group living entities in ways that made
scientific sense, and that revealed a larger truth, be it religious or scientific. But trying to classify things with increasing accuracy created com-
plexities and confusion. Was what one person called something the same
entity as what another person called it? The buffalo, for instance, doesn't
seem easy to confuse with other animals. Yet the European buffalo went
by many names: buffle, urus, bubalus, catobeplas, theur, and the Scottish
bison, among others. Who had mentioned it first? Who knows, because

who knows what names were used for it? Even more significantly, more comprehensive understandings about the contents of the natural world meant that simplistic classification systems quickly fell apart. Humans didn't create order among plants and animals; they tried to discover an existing order. Existing categories were quickly split into more specific categories as scientific tools and knowledge advanced. As classification became more sophisticated over the centuries, exceptions and complicating factors arose at every turn.[1]

Ornithologists trying to keep up with nomenclatural change in the last quarter of the nineteenth century were like surfers straining to reach the crest of a constantly breaking wave. It was an impossible task, and because no consistent standards for naming birds, or for deciding on who had the authority to name, existed (or were just coming into existence), it was an era filled with contradiction, constant changes, endless synonymies, and, at times, passionate quarrels over very fine distinctions, the outcomes of which were essential to how birds were named and how evolution was quantified. Ridgway stood at the very top of the messy, challenging, and crucial taxonomic heap. Ernst Mayr described Ridgway as representing "the acme of descriptive taxonomy." During his lifetime Ridgway also described far more new genera, species, and subspecies of American birds than any other ornithologist.[2]

Evidence of this nomenclatural change appeared in a key document that a wider world could consult, argue over, and rely on as a form of authority in the names of birds. This document was the checklist: a published list of birds, their brief descriptions, and their ranges. Who had the authority to name birds? And more significantly, what was most at stake? As it turns out, it was the subspecies concept. In particular, how subspecies were named—using trinomial, or three-part, names—proved especially important for the advancement of science. This chapter covers two key elements: the checklists and the nomenclatural codes that determined how birds were to be named. As we will see, British and American practices differed significantly, for reasons that were proscribed by the respective sizes of their countries and the evidence of evolution that each country's geography provided.

If study collections were a kind of language, then nomenclature— well, nomenclature was akin to a dictionary. Ornithologists took the evidence they found in nature and tried to make language reflect it. One of the most important changes in science during the last quarter of the nineteenth century was the transition from common, much-beloved names to obscure, seemingly dry-as-dust two-part or even three-part Latinate names. Even common names could change dramatically. The well-known vernacular names of many North American birds were descriptive in plain English, and sometimes evocative—witness the Brotherly-Love Vireo (now the Philadelphia Vireo), the Bohemian Chatterer (now the Bohemian Waxwing), and the Fiery Redbird (now the much more plebian and somewhat inaccurately named Northern Cardinal, which is found year-round from southeastern Canada down to southern Mexico). If one considered birds outside of the United States, which constituted more than 90 percent of the planet's birdlife and were found in US museum collections, many more redolent names existed. Fortunately, a number of these scintillating common names survive, including the Echo Parakeet of Mauritius, named for the descending tone of its call, the Festive Coquette of northwestern South America, and the Purple-Bearded Bee-Eater of Indonesia.

Even gushing admiration appeared in names, as in the Lovely Fairy-wren, named by British ornithologist and artist John Gould in 1852 (and not to be outdone nomenclaturally by its cousins the Splendid Fairywren and the Superb Fairywren, all three endemic to Australia). Still other names hinged on emotional ties. "It is very trying to my mind to think of my sorrow, so I will talk ornithology at once," José Zeledón wrote to Ridgway in 1883 on the death of his beloved sister Antonia. He planned to name a recently discovered bird after her and called it *Carpodectes antoniae,* the Yellow-Billed Cotinga. Zeledón sent the bird to Ridgway to illustrate and describe, and Ridgway took responsibility for publishing the description and picture in the *Ibis* in 1883, thus claiming priority for the name.[3] Other scientific names were also evocative. The Latinate name for the Atlantic Puffin—*Fratercula arctica*—is rendered into English as "little brother of the Arctic." The Harlequin Duck—*Histrionicus histrionicus*—is well named, for its striking breeding plumage is reflected in its scientific name, which translates as "melodramatic melodramatic."

The system of binomial nomenclature—two-word names for birds, consisting of a genus and a species—was first proposed and codified by Carl von Linné, or Linnaeus. He laid out the idea for plants in 1753 in a work called *Species Plantarum,* and for animals in the *Systema Naturae* in 1758. In its original form, the *Systema Naturae* is nothing to write home about. It has a drab gray paper binding. Its text crawls along, its letters small insects on the page, unleavened by any illustrations. But it was a crucial moment in taxonomy because Linnaeus had finally provided a set of two-name standards for naming living things.

For most people, the scientific, binomial or trinomial names might as well have been Latin, which they ostensibly were. Or at least they were Latinized; it's difficult to argue that "jamesoni" or "wilsonia" or "macgillivraii" are actually true Latin words. It is also true that some young men, even as teens, who were eager for a professional life with birds, realized that if they used the scientific names of birds instead of the common ones in their correspondence with institutionalized ornithologists, they would be taken much more seriously. Ridgway, Merriam, Henshaw, and others did this as young supplicants for a life in science.

Like common names, scientific bird names came about for diverse reasons. In some cases, they were created primarily because of the difficulty in telling them apart from very similar species. In the flycatcher family, there was a species named *Empidonax perplexus* and an allied species with a name still in use, *Empidonax difficilis* (the Pacific Slope Flycatcher). In the course of working on the flycatcher family in volume 4 of *The Birds of North and Middle America,* his magnum opus on systematics, Ridgway noted that "of all the difficult and perplexing jobs I have had this case of *Empidonax* is the worst; and I feel it is with much *perplexity, difficult timidity,* and *trepidation* that I express an opinion concerning any member of the genus. Am now on *Myiarchus:* they too *are* cusses."[4] In fact, heaping difficulty upon difficulty, a subspecies of the bird had been identified by Baird, and named in 1858 as *Empidonax difficilis difficilis.* But it was on fine and often minute differences in appearance and morphology that the critical task of identifying and properly placing species and subspecies within the taxa of birds was based. These names often reflected these differences, and it was by such careful distinctions as those Ridgway studied that ornithology progressed, stepwise, into the twentieth century.

Bird men of the era often named both new species and new genera after their friends. For instance, millionaire bird collector Charles Cory provided the name for the now critically endangered Ridgway's Hawk in 1883. More than forty years later, and just a few years after Cory's death, Ridgway attempted to return the favor. Writing from his home in Illinois to his colleague Charles W. Richmond at the Smithsonian, Ridgway remarked that "since Cory was the first to describe the species, I would like to have the generic name commemorate that fact in some way. Cory was a fine fellow, and I would like to express my appreciation of him by naming a genus for him." But the genus *Coryi* did not survive, and this kind of tip of the nomenclatural hat was often a fruitless gesture as years passed, because bird names changed so frequently. However, many endure today, along with a number that Ridgway named after friends of his. These included Daniel Giraud Elliot (*Atthis ellioti*, the Wine-Throated Hummingbird, in 1878); Charles K. Worthen (Worthen's Sparrow, in 1884); Spencer Baird (*Vireo bairdi*, the Cozumel Vireo, 1885); José Zeledón (*Zeledonia coronata*, the Wrenthrush, in 1889); William Louis Abbott (Abbott's Booby, in 1893); George Cherrie (*Cypseloides cherriei*, the Spot-Fronted Swift) in 1893, and a number of others. As Ridgway described a flurry of new birds in the 1870s through the 1890s, they passed through a series of nomenclatural sieves in the decades that followed. Some were discovered to be synonymous with existing species; some proved to be subspecies rather than full species; and a minority proved to be completely new to science, even though many of their names changed in the subsequent years as nomenclatural reform brought older names into alignment with current naming practices.[5]

There was, and still is, no particular requirement—educational or institutional—to be allowed formally to name a new species. Getting the name published, however, was necessary. A competent printed description of the bird sufficient to identify its unique characteristics, typically in a fairly well known journal, along with the existence of a type specimen, was the accepted means of creating priority for a new species. Publishing something allowed for the news to be spread widely. Traditionally, the name given by the first person to describe a bird in the scientific literature was the name used from that point forward. However, this created big problems for later taxonomists trying to determine whether a particular bird had already been discovered, especially in an era before illustrations or photographs accompanied descriptions of new birds. An unwitting namer occasionally put the bird in the wrong genus, or used the wrong plural or singular form of the species in Latin, or otherwise erred in creating a reasonable and distinct name.[6] The issue of priority in names, along with other nomenclatural and systematics issues, created challenges that Ridgway finally addressed comprehensively in the early twentieth century in his massive work *The Birds of North and Middle America*. Checklists of birds, and prescriptive codes, showed the way forward.

Checklists and Codes

Checklists and nomenclatural codes went hand in hand in the last quarter of the nineteenth century. However, they are different beasts, and they each receive separate attention here. Each existed for a variety of biological sciences in the nineteenth and early twentieth centuries. The numerous checklists were often specific to particular kinds of plants and animals, whereas codes tended to be more overarching and applied to a wider spectrum. In ornithology, checklists detailed individual birds in specific regions. These lists included their full scientific and common names, their ranges, and an accompanying numeric identifier that users could refer to as a shorthand reference. They also usually included brief details on their synonymies (other usually obsolete but often still-used scientific names for the same bird). Nomenclatural codes, on the other hand, provided rules and recommendations for just how bird names were to be created and consistently applied. These codes, often published with checklists, as with the AOU's landmark 1886 work, dictated the approaches and practices naturalists should use in naming living things.

Books and lists that tried to inventory all the birds in North America were nothing new. Alexander Wilson's work was the first such attempt, early in the nineteenth century, followed by Audubon, and then by Charles Lucien Bonaparte, Thomas Brewer, and Spencer Baird—although Audubon created artwork that far outshone his sometimes questionable classification practices. Checklists provided authoritative, if occasionally contested, lists of zoological entities. The 1870s and 1880s also saw an explosion of checklists and survey results on a variety of taxonomic groups besides birds, including lists of beetles (*Check List of the Coleoptera of America*, by G. R. Crotch, published by the Naturalists' Agency in 1874), frogs and toads (the *Check-List of North American Batrachia and Reptilia*, written by paleontologist and comparative anatomist Edward Cope and published by the Smithsonian in 1875), and untold numbers of lists of plants, spiders, shells, rock formations, and more. On the bird side of things, several ornithologists—all whom would become or were already members of the AOU—issued an important series of checklists, starting in the early 1870s and continuing through the twentieth century. Their authority depended on a number of factors—most importantly, the reputations of the author and the publisher. Three key parties were involved in these efforts: Elliott Coues, who issued his own series of checklists between 1872 and 1895; Ridgway, who wrote checklists between 1873 and 1881; and the AOU itself.

The checklists vied with one another to be the most authoritative and most widely used, and their story is a vital one in understanding the ways that nomenclature became both more refined and more standardized. The first innovations in checklists appeared in 1873, when Ridgway published a list of the birds of Colorado in the pages of the *Bulletin of the Essex Institute,* a venerable museum-based journal in New England. Even though it was not a national checklist, Ridgway's itemization of the birds of Colorado, then still a territory, was an important use of trinomials.[7] Of the 243 different birds known to occur in Colorado under his heading "Eastern Species Found in Colorado," Ridgway described nearly one third of the birds as subspecies. No one had ever noted such a large percentage of the total number of birds in a region as subspecies, bringing the concept into considerably sharper focus. Some would later be found to be misidentifications or were supposed subspecies that would in fact prove to be entirely new species. But the extent of subspecies to be found in the Rockies was shocking. Another important aspect of this book was its field guide–like approach to describing species: it noted the habitat types in which each species or subspecies could be found: "Aspen woods below the pine region," "Rocky places in vicinity of water," "All wooded places," "Everywhere," "Meadows of valleys," and so forth. The listings also noted where individual species were sympatric (overlapping in distribution) with other species: "With *H. leucocephalus,*" "With *R. solitarius,*" and so on. All of these elements would later appear in modern field guides.

That same year Coues's own checklist appeared, published by the Boston firm of Estes and Lariat. It had appeared in identical form the previous year, but this time it was paired with a guide to field ornithology. The book's full title was *Field Ornithology; Comprising a Manual of Instruction for Procuring, Preparing and Preserving Birds and a Check List of North American Birds.* In the checklist, Coues described 635 species. One of his innovations was to list "varieties," as he and others called subspecies, with alphabetical notations (a, b, c, and so on). Each species was followed by the original describer's name, as was (and is) standard nomenclatural practice. After Coues's work, a six-year interval followed with no notable changes in the taxonomy or the naming of birds. But some things needed fixing. On the heels of Linnaeus and writing in the early 1790s, American naturalist William Bartram had attempted a descriptive regional catalog of the birds of the eastern United States. His list, which ran to 215 names, came into wide use.[8] However, when nomenclatural reform began in earnest in the 1870s, his names had become contested. Many of them appeared to be reworkings of names already given by Linnaeus and were often accompanied by descriptions so scant that identifying the birds with certainty was difficult or impossible. Many names were also cryptic in their origins. Given the importance of the *lex prioritatis*—the law of priority, which stated that the first name created for something should stand forever—and ornithologists' insistence on the sacrosanct nature of the first naming of a bird, this caused problems. Any of the names that had been accompanied by intelligent descriptions were usable—but what to do with Bartramian names of obscure origin or, worse, names that were trinomial? Bartram had a few of these, but with no explanations as to why he had given them three-part names. Coues, who wanted to resurrect Bartramian names, used a number of the problematic ones in his checklists. He proposed reviving them *en bloc,* proclaiming "Consistency is a jewel." Allen, however, took issue with this idea, claiming that there was nothing consistent at all about Bartramian uses of bird names, mostly because Bartram did not describe the range of key birds he described, nor did he often fully describe the bird.[9]

Ridgway and Allen agreed about the problems inherent in Bartramian names. However, no one took decisive action until 1880, when Ridgway's first national checklist was on the verge of publication. Allen, who had agitated against their use, wrote to Ridgway by way of advice, "[I] do not see that I feel disposed to depart at all from what I wrote on this subject four years ago." The compromise eventually reached by Ridgway, and later by the AOU in its own checklists, was to keep any Bartramian names that were accompanied by an intelligent diagnosis and discard all the others.[10]

In August 1880, Ridgway published his new checklist, "A Catalogue of the Birds of North America," which was published in the Smithsonian's *Proceedings.* For the first time, a checklist definitively dropped the "var." to note subspecies—a technique that had populated every list of birds that noted varieties since the term was first used in the eighteenth century. Ridgway issued the list in the form of a paper running to some eighty-four pages, listing 698 species, but promised a new version soon. Allen promptly wrote him about the proposed revision: "I notice that in your paper on "Revisions of Nomenclature" you refer (p. 1) to be a 'revised list following this paper.' Does this refer to a new check-list of N. Am. Birds? So many additions have been made, and so many changes in nomenclature, since the last check-list was published that a new one seemed indispensable. In reviewing your paper for the July Bulletin I allude to this fact, and would be glad to be able to state that you have in press, or in preparation, such a revised list. Can you authorize me to make such a statement? I certainly hope so, for a new list is so urgently called for that I fear some person whose ~~ambition~~ [struck out] desire for distinction may undertake it without having the proper qualifications for the work."[11] This seems to refer to Coues, for no one else appeared to have the same desires—and certainly not the ambition—to publish a new checklist of North American birds.

Ridgway did issue a new checklist in 1881, running to ninety-four pages.[12] It was mostly specific to the Smithsonian's holdings, but given the national museum's completeness in North American birds, it served as a generally useful checklist. He overhauled numerous names, many of which had been changed in recent journal articles. His work had a lasting effect on his fellow ornithologists. Not everyone was pleased. Morris Gibbs, a prominent Michigan physician who issued the first comprehensive checklist of his state's birds, wrote to Ridgway that year: "When will the everlasting war cease regarding nomenclature? I am sick and tired of it. At one time I had all of the names on the Smithsonian catalogue at my tongues end. Then when Coues came out I changed many as I thought it necessary for the good of all, and then out came your list with not only new names in scientific nomenclature, but also great changes in common names, so that I can not go through your list and tell what I have in my collection, or what I know by it." But Gibbs finally came around a couple of years later: "Allow me to say that I have wholly adopted your list, nomenclature and numbering. At first I was angry that you should have gone to work and changed the business so thoroughly but after looking over the Appendix with reasons for changing names, why species were eliminated &c I became reconciled and went through my work "Ornithology of Michigan" Mss a work of over 15 years notes, and changed the names of over 320 species, which took me two days leisure time."[13]

It was challenging to keep up with nomenclatural changes in the early 1880s. "It seems to be the fate of list-makers to get a lot of things just in time to be too late, and you and I are no exceptions to the rule!" Coues wrote Ridgway in the spring of 1881, on the appearance of Ridgway's list.

"My list is of course retarded by my sickness and absence, but is coming along." Others were immediately pleased with Ridgway's publication and noted its pioneering qualities. "A man who had no access to a good ornithological library hardly knew what to call many of our birds previous to the appearance of this paper," wrote Nathan Clifford Brown, curator of ornithology for the Portland (Maine) Society of Natural History and later a founding AOU member, about Ridgway's 1881 checklist.[14]

Meanwhile, after publishing his 1872 *Key* and his 1873 *Check List,* Coues issued forth a substantially revised edition of the *Check List* in 1882, under the same title, and partially in response to Ridgway's checklists. Coues disliked the competition. Writing to Allen, he noted of his 1882 book: "Did you ever get your copy of the Check List? There is a great commotion, I understand, among the small fry as to Coues vs. Ridgway. R. is of course working the *morale* of the Smithsonian 'official' list in his favor and my treatise has so much in it that his has not that it is about a stand-off between us. The List *ought* to be properly reviewed in the B.N.O.C. [the *Bulletin of the Nuttall Ornithological Club*] but of course I am not the one to do or say anything further in that matter. Brewster gave it a *very* handsome anticipatory notice, but that is not the required *review.*"[15]

Coues was distinctly worried about the standing of his checklist in the face of Ridgway's, and although he had no formal position at the Smithsonian, he felt that his, rather than Ridgway's, should be the institution's official checklist. The issue of authority arose as to which of list was the official one. Writing to Allen in 1882, he fretted again: "There is a general impression that R's list is in some way 'official' or 'Smithsonian.'" Ridgway's work was in fact an official Smithsonian publication, having been published in the museum's *Bulletin.* But Coues didn't see it that way. "The morale of this is very unjust to me, — as unjust as it would be to R. to let the philological erudition of my book tell against him. The point is too obtrusive to be passed over in silence. It must be raised. It is the simple truth, that R.'s list has no more 'Smithsonian authority' than mine. I of course do not admit that there can be any 'rivalry' in the case, or that my list will not become at once the standard authority. Still I am bound in self-justice to defend it against any *groundless* claim that another list may seem to have, or that may be wrongly attributed to it."[16]

The year 1884 brought forth a new edition of Coues's *Key to North*

Elliott Coues, ca. 1892. Smithsonian Institution, Division of Birds.

American Birds. As did his colleagues, Coues attempted to get reviews of his books placed favorably in important journals by his good friends, and his checklists were no exception. He also agitated for every advantage with his own publishers. "I have a queer pill rolled from our friend Wade [J. M. Wade, editor and publisher of the often-antagonistic *Ornithologist and Oologist*]. I hold the most extravagant praise of his, in a letter not to me, of the Checklist, 'grandest thing since Audubon' &c. &c., and am going to let it be used as his endorsement of the work, in my publishers advertisements!"[17] By June 1884, Coues was advertising his *Key* on the inside front cover and back cover of the *Ornithologist and Oologist*.

Coues also got an even bigger advertising boost for his *Check List* that year, because Estes and Lauriat, the Boston publishers and printers of his *Key,* were also the printers of the *Auk*—largely because Coues had orchestrated an arrangement between his book publisher and the AOU. Estes and Lauriat then reached an agreement with the *Auk* to run a full-page ad for the *Key* on the back cover of the first two issues. The complementary nature of this advertising was subsequently greatly amplified at the start of the journal's existence by a huge ruckus in print between Coues and Augustus Merriam, in the form of arguments about some of the finer points of ornithological nomenclature harking back to Coues's 1882 *Check List*: not over trinomials but instead concerning how Latin and Greek should be used to describe birds.

The spat began with a review by Merriam in the first issue of the *Auk* entitled "The Coues Lexicon of North American Birds."[18] He submitted it as an unsolicited manuscript to the editors of the *Auk* in November 1883, and in it, he scrutinized one particular aspect of Coues's 1882 *Check List*: Coues's use of Latin and Greek in the formation of scientific names of birds in his section entitled "Remarks on the Use of Names." Allen, the new editor of the as-yet-unnamed quarterly bulletin for the AOU, vetted the article and passed it along to Coues to read. Initially amused, Coues rapidly became enraged. Stung by this unexpected and rather late review (the *Check List* had been out for nearly two years), he nevertheless responded smoothly and sarcastically in print to what he called "pedantic hypercriticism" of his work.[19]

Having seen the article in draft form before it was printed, Coues was allowed to respond to it in the same issue in which it appeared. The practices of the *Auk* are illustrative of what was often a fluid and interactive set of dealings between editors and different authors. Being able to see critics' remarks before they reached print allowed them to respond simultaneously within the same issue, for the readership's benefit; two (or more) sides to an argument could be aired in a single issue. One net result was that the first issue of the *Auk* packed a wallop: a critical piece, and its sharp rejoinder, from two famous bird men, all in the inaugural number.

Coues's rejoinder was called "Ornithophilologicalities," a title designed to mock what he considered to be Merriam's pedantry. In it, Coues addressed many of Merriam's highly technical points about Latin and Greek grammar and took him to task for having nothing positive to say about the work. In private, Coues raged constantly about Merriam to his friends, but particularly to Allen. "Let him take care! Just so long as he boxes my ears, will I continue to spank his usual locality. Rely upon my keeping my temper—and the Lord help him if he loses his. I expect to teach him to keep his fingers out of my sugar-bowl."[20]

Coues next made a rare show of anger toward Allen:

> If anything remotely resembling the original article [by Merriam] should come to me in the proofs for the Auk, I should at first give the author an opportunity, courteously proferred, to withdraw it, for his own sake; if he did not avail himself of the chance, the ugliest possible row would be started in the Auk, and he should blush and tingle for the rest of his life. I am abundantly able to take care of myself, against any printed attack. It is only secret slander, treachery and disguised hostility that I do not manage well, because, being utterly free from deceit or guile myself, I never look for it in others, and hardly recognize it when I see it. Do you know what I should do if any body offered to attack you in a journal I controlled? I should tell them first, he must be a dunce to suppose I would permit it if I could help it. Secondly, that I would see him damned before I would lend myself to any such thing, right *or* wrong; that he could do it when and where and how he pleased, but without my assistance, and with my heartiest wishes for his discomfiture![21]

Merriam's original article was in fact only mildly hostile toward Coues personally and made some sound observations about Coues's use of language, as Coues at first noted on seeing a draft of the article before publication, before he knew who the author was. "There *are* some good points about the paper," Coues initially admitted to Allen on first reading the piece.[22] Its tone was certainly milder than much of the vitriol that had flown from Coues's own pen in recent years. However, Coues saw betrayal in everyone's actions in every sphere: by Ridgway for not agreeing with the suggested name for the AOU's new journal, a row simultaneously taking place with that over Merriam's article; by Merriam for his article; and even by Allen for agreeing to take the article in the first place.

Others on the amateur side responded to this dustup between two professionals bickering over the finer points of Latin and nomenclature. R. G. Hazard II of Rhode Island wrote in the July issue of the *Auk* that on reading the two articles, "the lay mind is filled with dismay." In his witty and pointed critique, Hazard noted that scientific ornithologists were in danger of missing the larger whole by their intense scrutiny of the highly specific. He also drew comparisons with past writers. "Not one of the later writers can compare with Audubon or Nuttall in the use of English, and more especially in a certain feeling for nature, a love of the natural for its own sweet sake, unless, indeed, I except John Burroughs." He closed with an admonition against museums and overly precise language and codification: "In spite of our advanced knowledge, our trinomials, our excessive subdivision, our flutterings from one name to its older synonym, and all the other abominations which the learning of our writers has forced upon them, they illustrate a decline in their art, and must bestir themselves to shake off the dust of museums and to draw fresh inspiration from a humbler devotion to nature, for herself."[23]

The 1882 *Key* and *Check List*, still in print two years later, gave Coues a financial boost from the controversy. "How queerly some things turn out, and what a sight of gratis advt. the publisher of the C.L. gets in this no. of Auk!" he exclaimed to Allen. It was an especially practical text owing to its completeness, and it went through four editions in his lifetime, as well as two editions posthumously. It was used everywhere, including in the most remote archipelago in the world—the Kingdom of Hawaii—by the end of the nineteenth century. No doubt it was worth its weight in gold to Ridgway's friend Henry Henshaw, who was by then a Hawaiian bird expert but who seems to have brought virtually no reference books to the islands when he moved there in 1894, as evidenced by his constant requests for information from colleagues on the mainland, including a plea to Ridgway's assistant Charles Richmond for a copy of the *Key* in 1900.[24]

Other activity continued in nomenclatural reform in the 1880s, spurred largely by the AOU, Coues, and Ridgway. Many authors issued small checklists of regional birds. Most of these list-makers—usually engaged in other professions—were upset at the changes for practical rather than philosophic reasons and felt the need to revise their own lists. When a new checklist came out, they would painstakingly update the names, along with any changes to taxa as reflected in Coues's or Ridgway's checklists. Then a new revised checklist would come out a year or two or three later, and they would feel the need to revise their own checklists again to conform with the new standards, whether set by Ridgway, Coues, or, after 1886, by the AOU.

The AOU Nomenclatural Committee

The American Ornithologists' Union was by far the most influential group in establishing the names of birds in the United States. No one was paying any sustained attention to bird nomenclature in North America during the late nineteenth century. European ornithologists remained relatively hidebound in their adherence to the 1842 Strickland Code, and the AOU committee members were eager to fill the gap and discuss the matter further.[25]

The first nomenclatural committee was officially titled the Committee on Classification and Nomenclature. It consisted of Ridgway, Coues, Allen, Henshaw, and Brewster, with Coues as chairman. Its initial goals were direct: to reduce Bartramian names and to trim down the number of genera. Most members were dedicated workers. Allen, for instance, did not miss a single meeting between the inaugural gathering in 1883 and the end of 1896. The group had to work out a few personality issues first. Several members felt that Elliott Coues's pushy temperament would have too much influence on the committee. Coues himself certainly thought that he and Allen were co-kings of the AOU generally. "Practically, you know, two persons decide to some extent at present what the A.O.U. wants to do," Coues wrote to Allen on the eve of the group's first gathering.[26] Other members pushed back, however, and democratic principles eventually prevailed among the five members. Once the group began to systematically work out names, they immediately ran into confounding problems related to previous namers of birds—chief among them William Bartram.

When the committee began its work in earnest in early 1884, Ridgway noted to Allen, "It seems to me that with these Bartramian names we should accept only those that are sufficiently characterized and at the same time perfectly recognizable; otherwise, I do not see where the matter is going to end!" The members were eager to begin the task of straightening out and regularizing bird names. Nine days after the inaugural meeting in New York in the fall of 1883, Allen wrote to Ridgway, "I feel that we shall all be willing to waive our preference on many [names], and all major points in the interest of harmony and unanimity, and that the outcome of our deliberations will go far toward giving stability to our ornithological nomenclature for a considerable period."[27]

The bird world, or at least the scientific men in it, were excited by the prospect of standardizing nomenclature. Augustus Merriam wrote to Allen in 1884 to say that "uniformity should be gained as soon as possible in the language of science, so that it shall become a universal language that may be understood of all men, no matter what their vernacular may be . . . the *future* is in your hands to shape . . . Over all new words the sharpest supervision should be kept." The issue of nomenclature was one of the key rationales for the existence of the AOU as a whole — specifically, gaining some kind of professional unanimity on scientific bird names. The subject of nomenclature was predictably volatile, because the naming of birds — primarily the issue of common versus scientific names — was directly tied to the role of the amateur versus that of the professional. In planning out the agenda for the inaugural meeting in September 1883, Allen had asked Ridgway about how pressing issues of nomenclature would be determined. Ridgway responded that a committee "composed exclusively of working ornithologists . . . known to possess intelligent views of nomenclature" be formed to begin working through key issues, rather than engaging the attendees as a whole. The larger group, he felt, included amateurs who had "the crudest possible information on nomenclature." However, other prominent AOU members like William Brewster felt strongly that amateurs should be included in any discussion of nomenclature. "Let all our leading ornithologists, amateurs as well as professionals, sink their personal prejudices and feuds and meet with the aim of helping to secure a broad, substantial, and wholly impersonal system [of nomenclature] and the thing is done," he wrote to Ridgway.[28]

The 1886 AOU Checklist

The 1886 publication of *The Code of Nomenclature and Check-List of North American Birds* was a landmark event in nomenclatural reform, and the checklist part of the volume stemmed largely from the confusions caused by Ridgway's and Coues's competing checklists, which differed on many points and led to a lack of uniformity. The volume began to take form in 1884 as the Nomenclatural Committee worked vigorously on

William Brewster, curator of birds at Harvard's Museum of Comparative Zoology, 1883. Archives of the Museum of Comparative Zoology, Ernst Mayr Library, Harvard University.

the code of nomenclature and checklist. At 392 pages, it was a substantial volume. The first 69 pages were given over to "Principles, Canons and Recommendations," with the bulk of the work consisting of a long and carefully ordered list of North American birds. All five committee members, along with Leonhard Stejneger and Clinton Hart Merriam, who got regular invitations to provide input, worked on the canons. Ridgway and Allen did the bulk of the labor on the checklist itself, despite the statement to the contrary at the front of the 1886 edition, which noted that the committee had shared the effort. (Ridgway, in fact, tried to rewrite the introduction and give Allen and himself primary credit, but Allen talked him out of it.)[29]

There were regular arguments, votes, agreements, and disagreements. Although the five-person committee had officially decided to split up checklist and code duties—with Ridgway, Brewster, and Henshaw being charged with determining the status of species and subspecies, and Allen and Coues tasked with formulating the canons of nomenclature and classification—in reality, all members could weigh in on either aspect.[30] The actual writing of the code may have been by Coues, as it matches his particular elegant but long-winded style. While much of the committee's work was done during regular meetings in person, the group agreed to allow written dissent or assent with any particular point as well. This meant that they maintained a lively correspondence about nomenclatural issues between their actual meetings. Lists of species under consideration would get passed from person to person and then returned, usually to committee chair Allen. In rapid-fire language, committee members moved quickly through many issues. "Yes," "Aye," Yes," Yes," and "Yes!" noted Ridgway, Coues, Henshaw, Allen, and Brewster, ruling on whether the Lesser Yellow-headed Vulture should be moved from the list of known North American species to the hypothetical list.

The meeting notes for the group's critical activities in the spring of 1885 as they attempted to finish up the work make for fascinating reading. A number of taxa of birds now firmly established were decided on that spring, usually on the strength of one member's suggestion and the four other members' subsequent vote. The word "Auklet," for instance, was first contrived by William Brewster on a Thursday afternoon in late April in Washington, DC, as a useful description of the small diving birds in the genus *Simorhynchus*. Until then, all auklike birds, regardless of size, had been just that. But some species, while related, were considerably smaller. "Auklet" seemed to fit the bill perfectly, Brewster thought, and the name became official. So the bird that in Audubon's time had been the Nobbed-Billed Auk became the Least Auklet, the Curl-Crested Auk became the Crested Auklet, and so forth. These were radical changes, for common names were much beloved. To change them was tantamount to renaming geographic features or the names of cities, and peoples' responses were predictable. "I do see a great deal of injustice in such changes as from 'Wilsons' to 'Common Tern'—we have certainly honored our Pioneers little enough," complained Joseph Wade of the *Oologist and Ornithologist* to Allen.[31] These name changes, however, were never done casually and had specific rationales. They regrouped birds into more appropriate categories, allowed for more scientifically descriptive and thus more accurate names, and corrected historical errors of fact. For instance, Wilson's Tern was named by the prince of Musignano in the first half of the nineteenth century, when he decided that the European Tern of continental Europe was different than the Common Tern in North America. He was wrong; they proved to be the same species; even Audubon had recognized the error, and said so. Since the name "Common Tern" already existed, Wilson's Tern disappeared as a moniker through the AOU's nomenclatural labors, for there was in fact no such bird.

Seemingly small points were a matter of considerable import for the committee. While modern practice in the names of birds is to use lower-case letters for all but the genus, this was not established when the committee began its work. Merriam wrote to agitate for the use of capital letters for specific scientific names—especially ones derived from proper names: "If you consider it worth while to so frame your rules of nomenclature etc. that the thing be adopted *without* a fight, would it not be well to advocate the use of capitals in proper names? . . . I am inclined to think that there will be a large majority vote in favor of caps. My own feeling is *very strong* in this direction. I am prepared to stand up for the 10th Linn. and see it through, but can't go small *bicknelli, parkmani* . . . , etc. without a hard struggle. I do not propose to make any argument in favor of caps. *now,* for I am perfectly willing to 'fight it out' when the proper time comes, and abide by the majority decision." The fear that chaos lurked right around the corner, or over one's shoulder, seemed to haunt committee members at every turn. To avoid dissent, Merriam suggested, the committee should "have the thing in such shape when presented to the *Union* that there will be no great kicking, for if a quarrel over any one point is commenced there is no telling where it will end."[32]

The checklist was of vital use to ornithologists in the late nineteenth century. It synthesized descriptive techniques from previous North American lists, eventually influenced European practice, and bore the imprimatur of a formal ornithology organization that included national experts. On publication of the AOU's checklist in late 1886, other scientists wrote in with their opinions. For some scientists, it went too far, but for others, it did not go far enough. A revealing letter from famous American entomologist Samuel Hubbard Scudder to Allen, a friend and confidant for many years, has survived within the Smithsonian's institutional archives, illustrating the reach of the new publication into other zoological disciplines. Scudder was an expert in entomological nomenclature, having published a key work in 1872 entitled "Canons of Systematic Nomenclature for the Higher Groups," in the *American Journal of Science and Arts,* as well as authoritative guides to North American butterflies. He had been explicitly mentioned in the introduction to the AOU's 1886 code, which noted that entomology was "by far the most extensive branch of zoology." Scudder found the AOU's new checklist filled with evidence of "skill and good judgement," and noted that it strived toward the key goals of stability, unity, and progress. He was even more of a splitter than most ornithologists and had agitated for the regular use of trinomials to describe insects as early as 1874. Many members of the general public felt that the finer details of subspecies were confusing and unnecessary, but Scudder, as a scientist, went the opposite direction, noting that insect species, at least, could be split into even more subdivisions than trinomially. "While I am *perfectly ready* to adopt trinomialism, I doubt its entire adequacy to express even our present knowledge," he wrote. His complaint also applied to common names as well as scientific ones. "Your categories of common names are not sufficiently detailed for Entomology. There are not enough wheels within wheels."[33] What was too much nomenclatural complexity for most members of the amateur community was not sufficiently complex for some scientists' descriptive and analytical purposes.

The nomenclatural committee's small size meant that the power to name birds in late nineteenth-century America was concentrated in relatively few hands. The group had considerable influence and autonomy in making determinations about names because their work was rarely contested outside of their committee. Even into the first decade of the twentieth century, nomenclatural matters remained of relatively little interest to most members. In 1906, someone—probably Allen—would note in a draft of a speech to the AOU Council that "probably at least one third of the members of this Council have had very little if any personal experience with the intricacies of nomenclature. While they are men of most excellent judgement in all matters of a general character that may come before them, and are all excellent ornithologists in their special lines, their scientific work has happened to be such which have not required an expert knowledge of matters involved in the revision of a code of nomenclature."[34] From its inception, in fact, neither the AOU Council nor the rank-and-file membership had concerned themselves with the names of birds in any involved way. This meant that the Nomenclatural Committee's preferences and prejudices carried uncommon weight, and its five members carried the nomenclatural day. In fact, a simple majority of just three members was sufficient to enshrine a new name in the AOU's checklists or to banish a previously established name to the waste-bin of history.

A number of issues were taken up in the course of compiling the 1886 checklist, mostly designed to rectify long-standing nomenclatural confusions not addressed by other guidelines. The Bartramian names were whittled down to about six by the committee, and then finally, down to just two (the Florida Jay and the Black Vulture). Decisions about what to do about introduced species and vagrants also took up a fair amount of discussion. The inclusion of some introduced species in a checklist was a no-brainer if they seemed well established and some time had passed. But what about more recent introductions—birds that had been released from cages, intentionally or accidentally, and had eventually established breeding populations in the wild? How long a time needed to pass before they were included? Introduced species could survive as a population, sometimes for months or longer, without becoming established, but might eventually die off due to a lack of long-term habitat or food needs or because their population was too small to survive the seasonal vagaries of drought or other adverse conditions. Should these be included in a checklist? This had been an issue at least since Ridgway's 1880 checklist. "If you intend to include *all* that have been properly turned loose here, but which have never gained a foothold, the list might be considerably extended," Allen wrote to Ridgway that year.[35]

In assessing the validity of new specimens for inclusion in checklists, ornithologists also had to be sure they hadn't encountered a recently released caged bird. Discussing the recent capture of a "Green Finch" in a swamp near Cambridge, Massachusetts, Joel Allen wrote Ridgway, "Any hardy cage bird seems able to live, for a time at least, in a state of freedom in its new home, and if captured would be hardly entitled to the same treatment in point of record as a true straggler that has found its way too us without human agency." But sometimes birds that were suspected of being cage birds appeared with no evidence of ever having been in captivity, leading to uncertainty as to whether it should appear on a checklist of birds of a region.[36]

Although the written record of committee members' discussions on the subject was usually brisk and brief, questions of the validity of North American species sometimes generated longer debates. Ruling on *Icterus icterus*, the Troupial, a bird now known to be found only in South America, three members voted no and two voted yes to remove it from the list. Apparently a single specimen had been shot from the top of a lightning rod in a residential area, leading to the possibility, at least, of the species' validity in the United States. William Brewster, one of the yes votes, noted, "In Charleston, Icterus spurius [the Orchard Oriole] sings regularly from the points of lightning rods as does also the Nonpareil and several other birds. Icterus icterus is found more or less regularly, and commonly in certain of the West Indian islands, if I am not mistaken!" To complicate matters, John James Audubon had apparently seen and described a similarly marked oriole, which is still named Audubon's Oriole (*Icterus graduacauda*), a rare visitor to southern Texas. Brewster continued, "I believe that it may occasionally stray into our Southern States and see no good reason for discrediting Audubon's very explicit statement." Audubon's opinions were enshrined in American lore and not to be discarded lightly. However, Henry Henshaw sensibly pointed out, "Only one spec[imen] captured and this shot from top of lightning rod near house. Doubt if the others 'seen' were this species. Lets' *hypothecate* it till some one gets some more from the trees where a wild Oriole ought to be." Augustus Merriam summed up the issue aptly. "In order to obtain analogical correctness, let it be understood that the new-comer has only a provisional acceptance," he declared to Allen. "Let him, like the Spartan babes, 'be exhibited after birth to public view, and if found deformed or weakly, be thrown out upon Mount Taygetus to perish.' Then let one that deserves to live take his place."[37]

Another key innovation of the checklist, and one still in use today in almost all bird field guides around the world, was the rearrangement of the families of birds. Before the 1886 checklist, the order in which birds appeared in lists went from the most evolved to the most primitive, with perching birds appearing first and waterbirds last. However, the AOU's Nomenclatural Committee stood this on its head, and listed the waterbirds first—ducks, fowl, and so forth—and ended with songbirds, who were felt to have gone through the most comprehensive evolutionary change. This arrangement caused considerable grousing from amateur quarters. In reviewing the checklist, Frederic H. Carpenter of the *Ornithologist and Oologist* noted that "The change in arrangement is very great, the birds which have hitherto been considered as belonging to the highest scale in the order of bird life, and therefore having a right to precedence, are placed at the foot of the list, and the family of Grebes placed at its head. Such an arrangement is hardly justifiable." Further, the reviewer remarked, "If this 'List' renders the true version of classified nomenclature, our ornithologists have groped long in the darkness of *unscientific lists*."[38] The members of the AOU's Nomenclatural Committee would probably have agreed exactly with that sentiment.

Allen, who served as the chief wrangler and central clearing house for the committee's work, passed the draft text repeatedly from member to member for their edits and emendations as it evolved. During the summer of 1885 every member read each entry at least once, while Coues— despite his silence as to the contents—proofread the checklist and code at least twice. Allen was thrilled to be done with the work as its publication neared. "Shall feel like jumping over the moon when this *op. mag.* is really through the press," Allen wrote to Ridgway in early 1886.[39]

As a complement to the 1886 and subsequent checklists issued by the AOU, the pages of the *Auk* also reflected a great deal of nomenclatural adjustment and change. The quarterly journal provided a way to get new names and species into print more frequently, and because several thousand people around the world received the *Auk* by 1900, it had a significant reach. The committee worked briskly through the rest of the nineteenth century, meeting quarterly themselves, but sometimes more often, in order to keep ahead of possible priority claims for names of new species in other publications. Supplements to the AOU checklist were issued starting in 1889, with a total of ten supplements appearing between 1889 and 1900, recognizing the fact that nomenclatural change and reform took place much more frequently than the intervals at which the full checklists were published. The AOU's checklists have continued up to the present day, appearing in six subsequent editions through the end of the twentieth century, accompanied by numerous supplements to keep the lists current between major revisions.

The 1886 AOU Code of Nomenclature

The 1886 Code of Nomenclature, born out of the original impulses for the founding of that group in 1883, was crucial for American ornithology, but a number of important precedents influenced the AOU's work. Other nomenclatural systems existed before its founding, the most important being the Strickland Code of Nomenclature, established in 1842 in England as an attempt to unify zoological nomenclature and modified by British ornithologist Philip Lutley Sclater and others in 1878. Henry Seebohm, a wealthy amateur naturalist and widely read author, praised the late Hugh Strickland—who had been an English ornithologist—in 1879 for "constructing a code of laws to save it [zoological nomenclature] from the hopeless confusion into which it was drifting." He noted, however, that British ornithologists had taken liberties with the vagueness inherent in a number of the Strickland Code's points. For instance, several of the most prominent British bird experts considered it acceptable to transfer an entire scientific name—both genus and species—from one bird to another. Henry E. Dresser, in his multivolume work *Birds of Europe* (1871–1881), transferred the name *Saxicola stapazina* from the Black-Throated Chat to the Eared Chat. In the same breath, *Sterna hirundo* became the Common Tern after being the Arctic Tern for some time. This incensed other British critics such as Seebohm, who wrote in the *Ibis,* "Ornithological nomenclature is once more disturbed by frivolous changes, and is rapidly drifting from the position of exact scientific accuracy to that of mere popular indefiniteness." He also accused two other countrymen of the same crimes: Alfred Newton in his various works on British birds, and Richard Bowdler Sharpe in his monumental twenty-seven-volume *Catalogue of the Birds of the British Museum* (1874–1898).[40]

In America, ornithologists carefully watched the nomenclatural work of other bird men and found much to be desired as well. "We all wish very much that you would send us a notice of the later Parts of Maynard's 'Bds of Florida,'" Allen wrote Ridgway in 1879, during the years of the Nuttall Ornithological Club's position as the only North American bird organization. "For personal reasons no one here cares to 'speak out' about certain matters that call for pointed criticisms, for instance his absurd classifications, and the newly-coined names of the groups, constructed in accordance with the rules of no known language, living or extinct. If you will send us just such a notice of these later parts, and such reflections on the general character of the work as you deem proper you will help redeem the fair name of American ornithology from the foul disgrace which ignorance and assumption now and then heap upon it." Charles Johnson Maynard (1845–1929) was a collector, taxidermist and writer who had done good fieldwork, including identifying the Ipswich Sparrow, a subspecies of the Savannah Sparrow, but his research often lacked the necessary classificatory and nomenclatural rigor for maximum accuracy. Allen was particularly expert in the subject of a big part of the book—the birds of Florida—and was galled by Maynard's approach, which, in attempting to classify birds, favored the structure of the bones of birds over such external characteristics as feather type and position.

Ridgway demurred, and the notice was ultimately written by Henry Henshaw, who strove for a more diplomatic tone in the published review: "In his classification Mr. Maynard has departed in many particulars from the beaten paths, the basis for most of his changes being anatomical. That he has labored diligently in this field of study is apparent, but we cannot but feel he has moved somewhat in the dark respecting what other workers have done. It not infrequently appears, too, as though his desire for originality were, in a great measure, responsible for the positions taken, and that in striving for this he often fails critically to examine all the considerations involved."[41]

There was a clear need for a coherent set of rules and regulations toward which to direct people like Maynard, and the 1886 AOU Code provided essential guidance not just for ornithologists but for naturalists in general. A variety of outside sources in turn informed their discussions, as when Allen wrote Ridgway to point to two obscure but key French works on nomenclature that had recently been published. One issue was just when to base the starting point—and thus the precedents for naming species—of binomial nomenclature. Both of the French-language sources urged using a point dating to the tenth edition of Linnaeus's *Systema Naturae*—considered by many, including the Nomenclatural Committee, to be a consistent starting place for the standardization of binomial nomenclature. Both French works also contained "radicalisms on other points, somewhat in line with our tendencies," Allen wrote. "It will be well, in the way of saving time, if you can make yourself familiar with them before we meet."[42]

Despite its desires to change key elements in nomenclature, the group's sense of history and precedent was strong. The committee—and biologists almost everywhere—hewed to the aforementioned *lex prioritatis*. This bedrock maxim stated that as a species was first named in print, so it stood for all eternity. Although the code presented stern guidelines about creating names, once a name had entered into the world via publication, it could not be changed. Canon 31 noted, "Neither generic nor specific names are to be rejected because of barbarous origin, for faulty construction, for inapplicability of meaning, or for erroneous signification." Beyond that, a great deal was specified, including how to treat names published simultaneously, how to construct and select names, how to create trinomial names, how to punctuate names, and much else. The results of following the code could be somewhat strange, if not contested. For instance, the code noted, in a section titled "Names of harsh and inelegant pronunciation" that as a general rule, zoologists should avoid introducing words of more than five syllables. But what was more than a five-syllable name in an Anglo-American argot could be less than five syllables if properly pronounced, so this rule could lead to unnecessary changes. For example, in 1887, Ridgway criticized Brewster's name "Ainophila chihuahuana," noting that the specific name exceeded five syllables.[43] But in fact, pronounced properly in Spanish, the specific name is only four syllables: "chi-wa-wa-na."

The 1886 code wielded international influence over the next two decades. Allen, in a written draft of a speech to the AOU Council in 1906 before beginning work on the AOU Checklist of 1910, noted that "as new codes of rules were elaborated by foreign national or international organizations of naturalists, the A.O.U. innovations were adopted, in part or in whole, until, in 1905, our [1886] code became the partially acknowledged basis of the Code of the International Zoological Congress. A detailed comparison of the two Codes shows that they are essentially the same. Not one of our rules has been rejected, and nothing new, of an essen-

Joel Allen, 1912. An indefatigable worker, Allen wrote some five thousand letters a year, many of them on behalf of the AOU. Archives of the Museum of Comparative Zoology, Ernst Mayr Library, Harvard University.

tial character, is added. Merely a few points are made more explicit. It has taken twenty years for the world at large to reach the A.O.U. plane of 1886."

By 1906, much had changed in the world of nomenclature. Allen, as head of the Nomenclatural Committee, understood that revising the 1886 code would have far-reaching consequences and urged restraint and a careful pace. He also urged involvement of other international bodies and officially invited to an upcoming meeting on nomenclature all of the relevant AOU officers as well as American members of the Nomenclature Committee of the International Zoological Congress, who could bring expert national and international knowledge to the discussions.[44]

The reach of the 1886 code was remarkable. Not just ornithologists but zoologists in general working in America tended to follow the tenets of the code for their own work. A number of the code's innovations were incorporated into the International Code of Zoological Nomenclature, which was adopted in preliminary form in Moscow in 1893 and printed in definitive form in 1905 in Paris. David Starr Jordan—the famous Stanford ichthyologist—was one notable example. He used the AOU's nomenclatural concepts in his own works, corresponded with Ridgway, and wrote a series of volumes on fishes whose scope and actual title matched Ridgway's own work on birds: *Fishes of North and Middle America.*[45]

The code's completion caused friction between self-described professionals and amateurs. Jordan, in writing a four-page review of the code for the July *Auk,* began with an unabashed dig at amateurs: "In spite of a good deal of amateur work, which, in one way or another, gets published, . . . American ornithology stands at the front of systematic science." The amateur contingent, which had followed the radical changes in both checklists and rules for naming birds with some alarm, was not pleased by this slam. Following on the heels of its critical June review of the *Code,* the *Ornithologist and Oologist* printed a letter from a correspondent who retorted that "the amateur is abused and nearly every evil laid at his door. Why is it thus? The question needs no answer. The selfish motive of the professional is too apparent. The amateur is called into service in reconciling migration and statistics, but no credit is given him; he collects stomachs of birds at request, and when fined for shooting birds by his county official, the [AOU's] scientific committee for bird protection pause long enough from their work at a heap of slaughtered victims to applaud him." The writer continued to note that "amateurs, by reason of profitable time spent in research, shall become the equal of any of the scientists."[46]

Important gender issues also came into play in the amateur-professional divide, because if amateurism was equated with conversational narratives, so too was much of women's writing about science equated with those narratives. The increasingly technical work of both the checklists and the AOU Code significantly masculinized the study of birds even further. One workable one-word definition of "professionalization" in late nineteenth-century ornithology might be "masculine," along with all the gender inequities that narrowly defined scope implies. In substantial ways, women had lost ground as contributors to the scientific dialogue from the eighteenth century. Older forms of conversation about science had suited women practitioners and distinctly established them as scientific authorities—and new directives about scientific practice from male-dominated groups like the AOU changed women's footing. Nevertheless, by Ridgway's time their role in ornithology and other sciences was critical. Even if they were not involved in formative technical elements such as checklists, they read them, used them, and were often in the audience at public lectures on scientific bird topics. Among other things, women played a key role in the retelling of science to a wider audience in print as well as in person, including to their children, many of whom were raised with an interest in science and its practice because of maternal influence.[47]

Trinomials

The enthusiasm in America to take up three-part bird names differed markedly from British reluctance to do so. Some historical background will illustrate the causes for these sharply differing attitudes. The first published work to mention trinomials as a valid means of distinguishing varieties of birds from "typical" species was C. F. Bruch's "Ornithologische Beiträge," in a 1828 article in a German journal named *Isis*, in which he noted that "would it not be a relief for the already overburdened memory to introduce a *triple nomenclature*, by leaving the old name to the typical form and characterizing the varieties with a third word?"[48]

Others pressed trinomialism into service, including its use in 1840 by a Swedish ornithologist named Carl Jacob Sundevall, based at the Natural History Museum in Stockholm, who apparently systematically treated subspecies as geographical varieties, giving a third name to the existing Linnaean name. German ornithologist Hermann Schlegel also made important contributions to classification work in 1844, working diligently to systematize the use of trinomials. But Linnaeus's binomial nomenclature had taken root deeply, and swaying large numbers of people to accept the variability of species was impossible at the time, at least as much for religious reasons as for classificatory ones. Linnaeus had believed in special creation, which held that God had immutably created all creatures and that they would continue to be immutable for all time. This made three-part names, and their implications of species transformation, unpalatable within what was then considered the scientific community.

In America, the first discernible application of trinomial nomenclature was by John Cassin, who used trinomials freely in his 1856 work *Illustrations of the Birds of California, Texas, Oregon, and British and Russian America,* noting a subspecific name and describing the geographic location for that subspecies. Cassin's book opened a new chapter in denoting subspecific differences. For the first time, it essentially ignored the life histories and behavior of birds, focusing instead on detailed synonymies. An important work followed in 1858, by Spencer Fullerton Baird, John Cassin, and George Lawrence, who also used some trinomials.

As early as 1871, Coues noted in a review of Joel Allen's work on geographic variation that "if we must continue to use a tool so blunt and unhandy as the binomial nomenclature, all cannot be expected to use it with equal skill and effect." Not everyone agreed; Daniel Elliot used very similar language to make the opposite case, warning that "a system which should be mainly trinomial will meet that fate, for it has proved to be at the best but an unwieldy instrument in the hands of even its warmest advocates." By the late 1870s many American professionals were making common use of trinomials. Allen wrote to Ridgway in 1879 to say, "Am glad you are agitating the subject of trinomial nomenclature. We must come to it sooner or later, and the earlier in fact [it] becomes generally recognized the better. In fact I think in this country we have pretty well adopted it already, but I think it important to present the matter formally. For my part I think nothing is better than a pure trinomial for subspecies, without 'var.' 'sub.sp.' or Greek letters, and that a modification recognizing this ought to be added to the standard rules of nomenclature, defining explicitly the significance, use and scope of a trinomial."[49]

But while American ornithologists charged ahead consistently in insisting on the use of trinomials, both the British and Germans essentially rejected them until well into the twentieth century. In 1891, long after Americans had adopted trinomials in both common and scientific publications, a German article on trinomials in the *Journal für Ornithologie* would note that "as a rule binomial description is completely sufficient." Trinomials were not in regular use in England and Germany even by the last five years of the century. Notable exceptions did exist. Ernst Hartert, a close friend of Sclater's, was pro-trinomial, and attempted to convince his colleagues in Germany of their importance and utility. Perhaps the smaller physical size of Germany contributed to a lack of broad understanding of intergradations. "We are still completely uncertain about the development of species, and therefore should not introduce our guesses into nomenclature," one German ornithologist noted at a congress in 1897. Other important Continental ornithologists rejected trinomials for purely theological reasons. Otto Kleinschmidt, a pastor and creationist expert on German birds, remarked that "from my point of view, species cannot be taken apart."[50]

Britain was even more insistent in its rejection of trinomials—a rejection that is at first made more puzzling by the stellar talent and meticulous work of its own bird men. Philip Lutley Sclater, the founder and first editor of the British journal *Ibis,* was no slouch, indeed, a formidable taxonomist with tremendous experience. He named 278 species of birds—more than any one person since Linnaeus—and his own name appeared in the scientific name of 19 species and in the common name of 5. These phenomenal numbers were a reflection of his amazing reach in British ornithology. Among other things, Sclater was present at the famous debate between Thomas Huxley and the Reverend Samuel Wilberforce in London on the evening of June 30, 1860, a scant seven months after the publication of Darwin's *On the Origin of Species.* Huxley, famously known as "Darwin's bulldog," was a supremely aggressive supporter of evolution on behalf of his much more retiring friend Darwin. Sclater, like many young naturalists of the time, worshipped Huxley.[51] However, although trionomialism supported the idea of evolution because of what trinomialism said about intergradations as one species changed into another, the British were among the last countries in the Western world to adopt trinomialism with enthusiasm.

Across the water, American ornithologists generally seemed more interested in talking than in listening to their British counterparts on the topic of trinomials. In 1884, Coues made a lengthy trip to England to try to persuade his colleagues there that they were wrong about the subject. He had been especially impatient at the glacial pace of British acceptance and, as usual, urged aggressive action. "I wish you would give the truly British stupidity of not seeing our Trinomial ideas a good rap in this case," he grumbled in 1883, on asking Allen to review a book by Seebohm. The issue was surely an ideological one on both sides rather than a scientific one. Ridgway also felt confident by the mid-1880s that British ornithologists would come around to American trinomial views. "We also know precisely the stand the B.O.U. takes on matters of nomenclature," he wrote Allen—"a stand, regarding certain points, which they are as little likely to yield as we are to our convictions. Let us go ahead with the work we have begun, and they will surely fall in line in good time."[52]

Coues's trip to England caused a major upset within the AOU's officer corps. His frustrations with the English were partly influenced by his own enthusiasm for trinomials, but he was also swayed by their flattering descriptions of him as a highly promising naturalist. The British Ornithologists' Union had elected him to foreign membership in 1872, as did other prestigious London science organizations. He wished to go as an official of the AOU and conduct official business—most urgently, he felt, to convince the British of their error in not accepting trinomialism. He wrote again to Allen, "It occurs to me that as I am going to England and shall see all the B.O.U. people, and discuss nomenclature no end, as a member officer and founder of the A.O.U. that probably it will be best for our interests for me to go in some formally 'official' way, as in some sense the representative of the AOU." Allen then wrote several of his AOU colleagues to ask about the suitability of giving Coues some kind of commendation or official letter of support. Such cultural exchanges were nothing new; just a few months earlier, the Smithsonian had sent its curator of mammals, Frederick True, to the British Museum to study British methods of museum administration, as well as to examine a variety of British collections of comparative anatomy. However, both Ridgway and William Brewster thought the idea was a terrible one. "I think we have already gone to far in giving 'sops' [minor concessions] to our English brethren, and I am also quite sure the whole thing is, to say the least, very *amusing* to them," Ridgway wrote Allen. "I cannot see how the interests of the A.O.U. would be advanced by the proposed measure—in fact, there seems to me not the slightest necessity nor even desirability of it. My views are not based in the least on any personal considerations, but wholly upon politic grounds." Nomenclatural Committee member Henry Henshaw also agreed with Ridgway on seeing Allen's request to Coues.[53]

But despite Ridgway's protestations that his views were purely political and not personal, others made it clear that Coues would be the worst possible choice. Committee member William Brewster gave one of the clearest surviving candid statements by a close colleague of Coues about just what it was that made him so disliked as well as untrustworthy in the eyes of many of his circle. The venerable and highly respected Brewster responded to Allen, "Having little, if any, faith left in Dr. Coues' good faith, tact, judgment, and discretion I cannot believe it either wise or safe to empower him to act as a formal emissary of the A.O.U. during his proposed visit to England." He continued to note that he would consider any other member of the council an acceptable representative, but not Coues. "I feel sure that Dr. C. would act in such a manner as to bring discredit on our Union if not to involve us in serious difficulties. He has absolutely no tact and his egotism is so excessive that 'a little bit of authority' makes him nearly unbearable which, as we know so well, he is apt to commit acts of the very rashest folly."[54] Oddly, Coues, who could almost preternaturally sniff out an antagonist, real or imagined, from the faintest whiff, seemed completely unaware of Brewster's animus and apparently liked and respected him. Coues could go as an individual, his fellow AOU officers concluded, but he would receive no official letter of support as a representative of the organization.

On hearing of his fellow Nomenclatural Committee members' unanimous rejection of formal AOU support of his trip, Coues was disappointed but defiant. "I am so used to having objections spring up to my best and furthest-reaching ideas, whenever the cooperation of others is required for their execution, that I cease to be surprised," he wrote Allen. "*You* were all right, and caught the idea exactly, but have been talked over and repressed, by counsel taken of fear and caution, when brilliant confidence would have been a sounder and wiser exploit." Coues concluded his letter by noting, "When will the A.O.U. learn to have absolute confidence in my ability to carry out some things as no one else could, and fall in with instead of antagonizing my projects for their benefit?"[55]

On his arrival in England, Coues pressed on with discussion of trinomials with the British, claiming triumphantly that he had turned them toward their use: "You will be pleased to know how everything is going our way. Present indications are, the full adoption of the 'American idea,' with P.L.S. [Philip Lutley Sclater] at the head of the movement—and the next thing in order, I suppose, will be as to who shall have the credit of initializing and leading the movement in England, among those who are quite ready and anxious to come to the fore of the matter!" Coues seemed convinced that Sclater would soon embrace trinomials, even though Sclater had never hinted at any support of them in his published writings. "You may be quite sure that P.L.S. will 'settle' it—for as *he* goes so go enough of the English naturalists to make the case for us here." Coues's optimism about British acceptance of trinomials seems misplaced, however, because attitudes changed very slowly. As Ernest Stresemann has noted, "No one in the British Isles dared to advocate trinomialism because the powerful P. L. Sclater, who had been accustomed for decades to set his imprimatur on English ornithological work, had a low opinion of all innovators."[56]

The *Ibis,* England's leading bird publication, did not even print articles about trinomials until 1912, nearly thirty years after the concept was first affirmed in the pages of the *Auk.* In contrast, by the early 1890s the *Auk* carried articles and debates on the evolution of birdsong, birds' wings, and feather coloration. It wasn't that the British didn't know their birds. Only a dedicated professional such as British ornithologist Christopher Alexander could write to his brother Horace in 1907: "Last night I dreamt I saw all the Hirundines" (the Hirundinidae is a family of flycatchers with a variety of genera and many dozens of species).[57] But dreaming alone couldn't make the language used to describe birds match the reality of their interrelations and the corresponding story of their evolution.

At least three factors accounted for the disparity between American and British attitudes. The first argument, made by Kristin Johnson and others, is that British ornithologists were highly skeptical of the mechanism of natural selection, whereas American ornithologists were less so. Some of Darwin's proposed concepts under the umbrella of evolution, such as the non-Lamarckian inheritance of traits, were widely accepted in England, whereas others, especially natural selection, were considered very questionable based on the evidence at hand. Still other elements, such as Darwin's feeble attempts to explain the mechanisms of inheritance, were more widely rejected. Trinomials implied incremental change, which implied Darwinian natural selection, and this was a mechanism most British naturalists roundly rejected until the discovery of genetic mechanisms in the first decade of the twentieth century. A number of British bird workers thus preferred non-Darwinian models to explain evolution. Mutation theory, for example, trumped selection theory as an explanation for evolutionary change. Lamarckian evolution also remained popular in Britain for considerably longer than it did in the United States.

A second reason for European ornithologists' rejection of trinomialism probably lay in their strong enthusiasm for the Strickland Code of Nomenclature, crafted in England in 1842 and substantially revised in 1865. Many considered the 1865 revision more harmful than good because of confusions it introduced. Above virtually all other principles, though, both versions of the Strickland Code stressed the law of priority in naming a living creature. As Ernst Stresemann notes, "Trinomials might perhaps have encountered less enduring and fierce opposition in England if their use had not involved serious consequences: the rejection of the Strickland Code of Nomenclature, to which British ornithologists had adhered since 1842."[58] Deeply rooted in European biological practice and rooted in Linnaean notions, the committee that drafted the 1842 Strickland Code had included legends of British science, including Charles Darwin, John Stevens Henslow, Richard Owen, and Leonard Jenyns (who had originally been selected as the naturalist for HMS *Beagle* in 1831 but who had been unable to go and was replaced by Darwin). Support for American trinomialism would have showed a lack of respect for the great minds who had crafted the Stricklandian Code and was nothing less than blasphemous.

While the British and Germans were ambivalent about the mechanisms of selection, a third distinguishing factor—geography—allowed it to be taken up with special vigor in the United States. The very fact of America's cultural and physical distance from England weighed in favor of American ornithologists' more ready acceptance of natural selection. For one thing, they lacked the Britons' more exalted view of Darwin and his circle. More important, though, the evidence in America for natural selection was far easier to see in its flora and fauna across a contiguous area. Thirty-five times larger than all of Great Britain, the US continent contained greater evidence of intergradations and species change across a spectacular range of terrain, climate, latitude, and longitude. Taken as a larger whole, North America—including Canada and Mexico—was, at ten million square miles in area, more than a hundred times larger than Great Britain. Writing to Richard Bowdler Sharpe in 1891, Allen patiently tried to make a case he had made frequently in the past. "We expect subspecies to intergrade; if we believed such forms did not intergrade we should treat them as *species* and designate them by binomials."[59]

Even after 1912, there was vigorous disagreement about the validity of trinomials within the pages of the *Ibis,* which called it "a much-debated question." But at least by then trinomials were being considered in England, rather than suffering from the outright rejection they had faced a decade or two earlier. The British Ornithologists' Union and its officers were finally aboard with trinomials by the start of World War I, but the issue caused great tension in the ranks, and some prominent naturalists were still against the idea. Writing to O. V. Aplin in 1918, A. Halle MacPherson asked, "Are you still a member of the BOU? I am on the point of resigning my membership. It seems to me that things have become past a joke. A few days ago when the 'Authorities' told us to call our Swift

'Apus Apus Apus' I merely smiled realizing that the authors of such an absurdity would be the laughing-stock of future generations." He couldn't have foreseen that *Apus apus apus* would remain an internationally recognized subspecies into the twenty-first century. His complaints seemed more about the increase in complexity that trinomials brought than anything else: "When year after year such species are manufacture[d] by the hundred and of every small variation and ornithological nomenclature is turned upside down and the whole literature of the subject thrown into such a state of chaos that future students will take years of study to extricate order, it became no laughing matter."[60]

Evidence of evolution—and scientists' greater desires to uncover it—led to the increasing use of trinomials and thus contributed to nomenclatural change. Checklists also provided a form of authority in print to thousands of naturalists. Coues, writing about William Bartram in 1875, noted that the distinguished namer of birds had been a Christian—a characteristic among naturalists that he observed "was a frequent element in the contemplation of Nature, among thoughtful and sincere men, before she had been much subjected to the scalpel and the microscope—those mighty props of the theory of evolution, which now threatens a revolution in revelation."[61] By the mid-1880s, checklists could be added to the list of "props." Even if they were not a divine set of texts, they helped to replace the language of religion with the language of science. American checklists were a form of authority and arrangement, corresponding to the order indicated by Darwinian principles.

As implied throughout this chapter, issues of gender also came strongly into play during this era, because the checklists were explicitly technical in their language, unmodified by sentiment, and thus considered "male" in a world where science's claim to authority has included notions of domination and subjugation. To valorize one thing—technical terminology, for instance—means that some other opposing thing is automatically reduced. So the wider authority that scientific terminology had, the less authority sentimental "feminine" writing could correspondingly claim.[62]

By the last quarter of the nineteenth century, American organizational systems were in the air. In the broader nomenclatural realm, a number of formal and powerful influences both within and external to the United States existed. The International Commission on Zoological

Nomenclature (ICZN) was established in Europe in 1895—a period when Ridgway and the ornithological community were working intensely to resolve conflicts over nomenclature. An International Code for Botanical Nomenclature predated the ICZN, and an American Code for Botanical Nomenclature also arose at the end of the nineteenth century. Melvil Dewey created his eponymous classification system in 1876. Designed to classify and categorize books by subject area, his proprietary combination of numbers and letters tried to organize all knowledge into ten top-level categories. The Dewey Decimal System quickly gained currency in libraries around the world, and it was followed in 1880 by another popular classification scheme for books. This latter system, called the Cutter Expansive Classification system, was the brainchild of Charles Ammi Cutter and used an entirely letter-based classification system. Bird workers, who relied heavily on printed works, would not have missed the significance of these attempts to organize and group commonly shared knowledge. Cutter's comment about his system could be transferred wholesale to ornithologists working on nomenclatural systems for birds: "No one, perhaps, can remember it all; it cannot be learned, even in part, very quickly; but those who use the library much will find that they become familiar in time unconsciously with all that they have much occasion to use," Cutter wrote in his 1882 work *How to Get Books.*

The federal government, too, played a significant role in naming systems. For example, the codification of geographic features in the United States also became official for the first time in the late nineteenth century. Confusions and disagreements about geographic names and how they were to be applied to places and features led President Benjamin Harrison to establish the US Board on Geographic Names in 1890. The board vetted (and still vets) a wide variety of types of domestic geographic names, laying out legal, procedural, and policy guidance for how features such as rivers, mountains, and so forth are named.

The mid-1880s brought intense and often acrimonious debate about the nature of language, the role of beauty, and nature. As R. G. Hazard, who had critiqued the Coues-Merriam imbroglio over the finer points of Latin in the first issue of the *Auk* noted in another letter to the journal, the era had brought with it an inappropriate preoccupation with the overly minute: "The tendency begotten of this precise controversial spirit, is to lose sight of the main object in pursuing the barren details. One who examines a landscape with a field-glass may be able to tell you that a man in a blue flannel shirt is rubbing down the farmer's horse in that distant farmyard, but, if fascinated by the power of the glass, he continues his examinations till the waning of the day, what is his knowledge of the details worth, compared to your own appreciation of the whole?" But the details proved exceedingly important in attempting to understand how the complex and intricate clockwork of evolution worked. In an unusual move, *Auk* editor Allen responded at length to Hazard's letter, focusing on just why ornithologists needed the names they used. "To do any piece of work we must have tools, and must also know how to use them. To mention objects, or their parts, we must have names for them, and in most cases the names have to be provided. The usual lay vocabulary is insufficient, and names must be invented, both for the objects and, to a large extent, for the parts, even if the object be merely a bird. The lay mind takes no note of the minuter structures and, therefore, has for them no designations. Yet they are the elements the scientific mind has most largely to deal with, and which afford the key to many a difficult problem."[63] Inventorying a bird's parts required a new level of precision, because identifying finer distinctions was essential to describing how similar birds were actually different. Evolutionary change was often minute.

Taxonomic and nomenclatural reform of course did not end at the close of the nineteenth century, but they did change form. In particular, the checklists of the 1870s and 1880s had no room to include details related to synonymies—tracing the various names used for the same bird over the decades. Brewster, writing Ridgway in 1897, remarked, "What a tangle we have since fallen into with the various problems of synonymy that the [Nomenclatural] Committee has had to deal with! I think the practice of deferring important questions about which we cannot agree and have not discussed at a full meeting a most wise and wholesome one."[64] Much of the basic groundwork, however, had been laid. In 1901, Ridgway published the first volume of his monumental *Birds of North and Middle America,* and in the process, he continued to transform the nature of systematics in the study of birds, untangling problems in synonymy and much more. This story occupies a substantial part of the next chapter.

CHAPTER SIX

Publications about Birds

It is hard work writing for the people. I dislike it, and we
cannot be too careful to write in a lively style and use as
few technicalities as possible. Thus it takes three times
as much room for a popular as a scientific article on
the same subject.

ALPHEUS S. PACKARD JR. to Joel Allen, December 6, 1866

PUBLICATIONS WERE important in the biological sciences
during the last quarter of the nineteenth century for two key reasons beyond the information they conveyed. First, they established
authority for their authors. This was not a simple task, because amateurs
and professionals had contested notions about the "right" way to discuss
birds, as well as whether the "advancement" of science was even a worth-
while goal. Thus, this was not just the territory of scientific publications.
More popular works, such as the *American Naturalist* and the *Ornithologist and Oologist,* also claimed a truth about birds: their authors and editors considered lyricism, anthropomorphism, and reference to a deity to
be a more accurate way to represent the natural world. Second, publication allowed describers of new species to claim priority in their naming
and discovery. This aspect of publication was, and still is, a mark of status,
and could help elevate someone who would otherwise have been considered an amateur into the professional realm. It provided the mantle of
prestige and place in publications, conferring a form of individual authority on its discoverer.

This chapter teases out the issue of just how technical works on ornithology, as well as more widely popular ones, advanced the study of birds,
and to whose benefit. Ridgway's own publications, because they span the
gamut from technical works to popular ones, are useful for analyzing differences in the professional and amateur domains. The ways that amateurs and professionals used language and rhetoric, and the places where

their writings were published, illustrate a great deal about what people felt constituted proper science within ornithology. Amateurs' attitudes ranged from timid to defiantly proud of their relative clarity and elegance. However, they often strove for standards as much as did established, paid professional bird workers. Their requests for supervision from professionals reveal what some of them thought "professional" writing should be like. Eugene Bicknell, the self-described amateur whose discoveries were discussed in chapter 4, while submitting an article for publication in the *Auk*, wrote Allen in 1884, "In reading these proofs I have felt that perhaps my treatment of some of the species was not sufficiently reserved and concise." Apparently "reserved and concise" were two key elements of successful bird publications in the professional journals. A lack of emotive language and a certain kind of precision were two key determiners of what most professionals, as well as some amateurs, considered to be successful language in publication.[1]

Bicknell's letters also reveal further detail about what he considered to be the differences between amateur prose and professional writing, and he puzzled through the distinctions. "My fault of using larger words when short ones would do, &c, I am myself fully conscious of. It is not an affectation—conscious affectation I detest—but longer words seem to come often more easily than short ones," he wrote Allen. "I have opposed the habit, but the goal of terse, clear Saxon is not easily won. My aim now, in going over my writing, is to cut away everything that is superfluous. But just here is the difficulty; a difficulty which training and experience alone can meet. After I see where one word for several will give terseness, but hardly an honest terseness, for the several words express a somewhat truer meaning. Then again words which add little or nothing to the meaning expressed, seem to be required to preserve the rhythm and easy flow of sentences, which is certainly an important factor in true literary style."[2]

As the French mathematician and philosopher Blaise Pascal famously said, "I would've written less if I had more time."[3] Scientific writing for publication involved much more than using lots of multisyllabic technical words. Writing science succinctly and well, as Bicknell had realized, took experience, technical chops, editorial discipline, and that difficult-to-learn quality, style. Some writers could write clearly in a technical vein, while others could not, and struggled. Personal ability thus had a substantial impact on scientific success in the writing and publishing domain.

Journal Articles and Priority

Periodicals were the dominant form of publication in the nineteenth century. Although they were not the only form of establishing priority of discovery, they were the most important means of doing so. As literacy and reading audiences expanded, and as book manufacture and distribution became more mechanized, a virtual torrent of titles became available to anyone willing to make the usually modest investment of a monthly or quarterly subscription. The quantity of these journals was extraordinary, even though many were very short-lived. Some 125,000 periodical and newspaper titles were issued in England alone in the nineteenth century. Statistics also exist for a tighter range of dates in America in the Progressive era. Between 1885 and 1905, for instance, some 7,500 new magazines (not counting newspapers or periodicals issued twice a year or less) were founded.[4] An annual subscription was cheaper than a book, and journals usually arrived in smaller, more digestible lengths that made them easier to browse and forward to friends. Readers could also interact with the journal and other readers by writing letters to the editor.

Journals and magazines were not the only things being generated in large numbers. In the bird realm, the number of new species identified in the last quarter of the nineteenth century was staggering compared to today. Some 1,200 new species were discovered and named in twenty-five years, as the number of known species climbed from about 6,500 in 1875 to nearly 7,700 by 1900. This meant that on average, a bird new to science was discovered every week for a quarter-century. Many of these proved to be invalid, because "splitters" of the era, who readily broke species apart into any number of subspecies, real and imagined, tended to create new species out of what were actually varieties that didn't even qualify as subspecies. But at the time they were considered new (at least by splitters), and these discoveries meant nothing to science unless word could be gotten out to a wider audience. Priority in publications was critical. When a new species was discovered, its announcement to the world was usually a straightforward matter: an author of some standing would provide a new name and publish a description of the new bird in a reasonably well-known journal, the bird would thus enter the record as a species new to science, and priority would be established.

Picking the right publication, however, was not always a no-brainer, and people disagreed on where a new bird should first be described. Ridgway sometimes successfully pressed his opinion on the Smithsonian's leaders, diverting articles about discoveries from a popular channel to a more scientific one. Writing in 1900 to his immediate supervisor Frederick William True, the Smithsonian's first curator of biology (a new, overarching position created in 1897), about the announcement of a new bird of paradise, Ridgway opined, "I do not favor first publication of this new Bird of Paradise in *The Osprey.*" The *Osprey,* founded that year, was a newly revived title with a fairly large national readership, and it described itself as a "an illustrated magazine of popular ornithology." Ridgway found it too unscientific for his liking and felt that the Smithsonian would be a better place to publish the first notice of the bird. "It is too good a thing to rob the 'Proceedings' of the National Museum of, for new Birds of Paradise do not turn up every day. Prior publication in The Osprey would remove any reason for re-publication in the 'Proceedings' . . . While an admirer of The Osprey and wishing it the success it deserves, I look upon it more as an exponent [of] popular ornithology than a proper vehicle for the publication of new species."[5] The article appeared in the *Proceedings* rather than the *Osprey,* and in Ridgway's eyes, priority was properly announced in a scientific journal rather than a popular one.

Getting one's licks in before competitors was important, although Ridgway claimed that it was less significant to him than getting information in front of a larger public in a timely manner—at least in 1877, when he penned a long note to Allen on the subject. "I believe it to be a duty which one naturalist owes his brethren, to give all the benefit of his knowledge as soon as possible (or rather, as soon as *convenient*), for life is too short to wait long for the publication of facts which each one is liable to 'stumble upon' sooner or later and then feel constrained to withhold them from the public from the fact that he may be aware some one else claims priority of discovery. And, should he be unhappily ignorant of such fact, incur the displeasure of his more dilatory co-laborer by unintentionally anticipating him in the publication of the matter."[6] In other words, publish promptly and when ready—rather than holding information back because one anticipates the possibility of being beaten to the punch in claiming priority. This was somewhat disingenuous on Ridgway's part, because in his opinion—as in the case of the new bird of paradise he had urged be published in the Smithsonian's *Proceedings* rather than the more popular *Osprey*—the scientific reputation of the journal probably trumped the need to get something into print quickly.

Priority issues not related to new birds also appeared in print. People occasionally claimed to be the first person to publicly announce a scientific concept, only to get smacked by their colleagues because someone had published previously on the same topic. Allen, Ridgway, and Coues had just such a brief but bitter clash in 1873. At issue was who had first raised the issue of climatic variation and its role in geographic variation. Birds that were farther south, in warmer and more humid climates, seemed to be consistently darker than their northern counterparts. Joel Allen played a significant role in scrutinizing this effect. It was a novel and important observation, or so many of his colleagues thought. In a long 1871 monograph in Harvard's *Bulletin of the Museum of Comparative Zoology*—the young curator's own home institution—Allen systematically laid out numerous examples of the relation between color and geographical distribution in birds.[7] Even Darwin made mention of it in *The Descent of Man* the next year: "Mr. Allen shews that with a large number of birds inhabiting the northern and southern United States, the specimens from the south are darker-coloured than those from the north; and this seems to be the direct result of the difference in temperature, light, &c., between the two regions."[8]

The trouble arose when Ridgway built on this issue in a two-part article of his own a year later without mentioning Allen's work—a slight that greatly annoyed Allen.[9] He complained to Ridgway about this lack of acknowledgment: "While you cannot claim to have been breaking new ground, you make no allusion to anything done previously on the subject, thereby giving the impression that you are presenting the subject for the first time. While it is satisfying to find ones opinions and generalizations so ably endorsed, it is not quite so agreeable to be at the same time wholly ignored."

After Allen's complaint, Ridgway wrote to apologize, but he implicated Coues as a potential troublemaker in the affair. "He wrote me long ago upon the subject, and in such a tone that I have never answered a letter from him since, for all his remarks concerning the paper were ungenerous, arrogant, egotistical and unthankful in the extreme." He was awaiting Coues's inevitable response to his article, he told Allen, so that he could "get even with him" and "show him up"—about as vindictive as Ridgway ever got in his private writings.

As Ridgway predicted, Coues promptly accused Ridgway of plagiarism in print. "He writes as if his views were both novel and original, which is not the case," Coues noted. After reading Coues's article, Ridgway erupted, in private and in print. "To be charged with literary theft must be unpleasant even when it is merited; to be charged with 'scientific plagiarism' without any provocation, is an accusation which cannot be borne in silence. In this case, the charge bears with it so much arrogance, that a simple defense against it is not sufficient," he wrote in a response in the *American Naturalist*. Ridgway agreed that Allen was indeed due all the credit due him but stated that "he is not the only one who has written upon the subject of climatic color-variation and geographical distribution." He noted that in writing his original article, he had assumed that the subject and its general principles "were so familiar that a preliminary review of its literature would be a superfluous addition to a paper already overburdened with references." Young Ridgway—just twenty-two years old at the time—noted that Allen's work itself was not the first on the subject and pointed to his mentor Spencer Fullerton Baird as one important antecedent in the case of regional color variations within species. Further, Ridgway noted, Baird was preceded by a German ornithologist, Constantin Gloger, who wrote an article in 1833 on the influence of temperature and climate on bird color.[10]

Coues then responded privately to Ridgway: "I see by the Sept. Nat. [*American Naturalist*] that you have 'gone for' me very savagely! That is all right, if you think my review justified it. I admit that the latter was 'more just than generous,' but I certainly did not suppose it contained anything that could give you *offense*. How you do pitch into me! It is a good strong reply, to which of course, I feel bound to make rejoinder. So look out for the Dec. Nat! I hope we shall not have another Marsh-Cope affair, and get put in an Appendix; for 'the way they heaved them fossils in their anger was a sin!' as Bret Harte says of the Society on the Stanislaus." Coues then listed off a lengthy list of things he had done in print for Ridgway by way of acknowledging aspects of his work. "In short, I make every acknowledgement, both in general terms and in specific instances, that I conceive to be due. What more can you desire? In the simple fact of my publishing before you, you have of course no grievance: for either of us is at liberty to print when we please. I trust you will see that the harshness of your article, especially in its implications of bad faith, is hardly warranted by the facts."

Further, noted Coues, the disagreement was not a personal one. "You mistook the tone of my review entirely: it was purely impersonal—I write all such articles as the Naturalist's reviewer, not in my own person; and write of the *cases,* not of the persons. I do not of course, know how the matter has affected you personally towards me; but for myself, I can say with entire truth, that my personal regard for you is not in the least altered by what you have done. I always separate personal and 'official' relations. You will find, in an early Nat., that I have 'gone for' you again; but do not suppose for a moment that it lessens, on my part, the friendly good will I have for you. Such controversies, unhappily are sometimes necessary to establish certain points of respective advantage and mutual respect."[11]

The squabble was patched up three months later, at least publicly, when the *American Naturalist* ran an editorial statement noting that "we are requested, by Dr. Coues and Mr. Ridgway, conjointly, to state that neither of these gentlemen desired to continue a controversy of no scientific consequence, and one which, furthermore, has lost its personal interest since a mutual misunderstanding in which it arose has been explained to their entire satisfaction."[12]

Ridgway's Publications

"While I have never met you or corresponded with you I feel—like all young ornithologists do—that they know you from your writings," a young Amos Butler of the Brookville, Indiana, Society of Natural History wrote to Ridgway in 1881. Ridgway—the quintessential office-bound ornithologist for most of his career—contributed to the literature on ornithology for sixty years, between March 1869 (when he published his first article at age nineteen in the *American Naturalist*) and February 1929 (in *Bird-Lore,* four months before he died at age seventy-eight)—a period marked by tremendous changes in attitudes about and approaches to publication.[13] Ridgway did not publish anything in only six of those years, and his output was prodigious: more than seventeen thousand pages in 553 articles and twenty-three monographs. These writings did not include works he edited, letters to editors, Smithsonian annual reports, or any of the hundreds of birds he illustrated for many different publications.

Ridgway's most substantial work was his monumental eleven-part *Birds of North and Middle America.* The first eight parts were published between 1901 and 1919 in his lifetime, and the last three were completed in 1941, 1946, and 1950 by others working from his notes. The magnum opus was popularly known and published as *Bulletin 50,* because it was a Smithsonian publication that was number fifty in sequence with other publications. Ridgway's Smithsonian colleague Charles W. Richmond and his successor Herbert Friedmann finally completed the last three parts.

Bulletin 50 is discussed later in this chapter. Two of Ridgway's books, however, fall into a rather different category—they addressed both popular and scientific needs inside and outside of ornithology—and are covered in the following chapter. In 1886 he published a book about color standards and nomenclature entitled *A Nomenclature of Colors for Naturalists*. In 1912, he self-published a more substantial successor to this work, *Color Standards and Color Nomenclature*. Consisting of 1,115 named color swatches, and printed with color-fast dies, this famous work served as a visual dictionary that gave naturalists, and many others in different disciplines, access to consistent visual information about color, keyed to a unique plate number, swatch, and name. The issue of standardization came into play with these dictionaries because they attempted to create an agreed-upon set of conventions for colors. However, the color work is most relevant here because a number of his other publications, especially *Bulletin 50*, used terms taken directly from his color dictionaries.

Ridgway evolved toward *Bulletin 50*; he did not spring into the world a full-blown systematist, describer, and organizer. He had to first cut his teeth on a number of other works. As detailed in chapter 3, a good amount of Ridgway's early writing was for the Nuttall Ornithological Club's *Bulletin*. He was usually requested by editor Joel Allen to write a piece for a particular issue, although he occasionally sent in unsolicited items for publication, especially as his cachet in the publishing realm grew. Allen instructed him to make these pieces "as short as possible without injuring their value."[14] This meant very brief notes, sometimes just three or four sentences, as well as occasional longer pieces. Ridgway's longest article for the *Bulletin* ran to nine pages, but it was the exception.

A variety of large and small publications constantly asked Ridgway to write articles. Requestors, almost always unsolicited, were sometimes downright pushy. Somewhat inversely to what one might expect, the newer and more parochial the journal, the more likely it was to make demands. One somewhat typical editor, the sole proprietor of the nascent Wisconsin-based *Western Oologist*, wrote to Ridgway in 1885 that his publication—"a new aspirant for favor"—needed not one but two articles of its own choosing (one on the English Sparrow and one on the naming of birds) and asked Ridgway in contemplating his fee "to have a kindly consideration for the pockets of those who publish the Western Oologist, and make them as low as you possibly can," and asking for "an *early* and favorable answer."[15] Ridgway doesn't appear to have taken the bait, since no works of his in that publication can be found.

Like Coues, Ridgway—who often did favors for the publishers and editors where he published his work—leveraged a number of relationships, and his reputation, to his great benefit. The writer-editor relationship could become a tangled nest at times, because it commingled friendship with professional activities, but without a doubt it led to Ridgway's improved access to publication. This happened with the Nuttall Club's *Bulletin,* with the *Auk,* and with *Forest and Stream,* which began in 1873. *Forest and Stream*'s editor, the auspiciously named George Bird Grinnell, corresponded with Ridgway regularly between 1875 and 1890 with a steady stream of identification questions, requests for information about certain bird books, and other queries that would have put him in Ridgway's debt. In turn, Ridgway published twenty short pieces in the magazine over the years.

Ridgway also curried favor by sending his Smithsonian-published works on birds to famous people. "Please give me copies of Bulletin 21 [a long Smithsonian monograph by Ridgway] for the Misses Audubons, and oblige," he wrote to Randolph Geare, the Smithsonian's chief of the Division of Correspondence and Documents, asking for publications to send to John James Audubon's two daughters.[16] The Smithsonian publications connection also meant that he could send out copies to colleagues in bulk; it was not unusual for him to have a dozen copies of significant works sent to others for their own use and dispersal as they saw fit— all of which increased the potential for his name to be seen in different spheres.[17]

Catalogs of natural history publications, which reached thousands of interested parties, were sent out by publishers of all sizes and extended Ridgway's reach. One Massachusetts publisher, the Naturalists' Bureau, issued a catalog called "The Naturalists' Quarterly," which was sent out to some 4,000 people. It was used by Edward Drinker Cope, Othniel Marsh, Baird, and other prominent paleontologists and naturalists. The Quarterly's editor, George A. Bates, solicited Ridgway with a request to include his publications, which he noted were in demand. Ridgway promptly used the service, which took a commission of 25 percent. He was nevertheless sometimes wary of such services, including the similarly named Naturalists' Agency, which took a more substantial 33 percent commission, but found the Naturalists' Quarterly to be a useful solicitor on his behalf for his many publications. The Mount Carmel *Register*, Ridgway's hometown newspaper, also helped by listing a variety of Ridgway's publications, as well as other government works, that could be obtained for free by writing him directly. This latter example probably is a reflection not so much of his savvy in getting his publications into more hands, but as a form of particular care and handling by his hometown of Mount Carmel. In any case, it was another shoulder to the wheel of publicity.[18]

Contributing money to a journal could help in getting articles published, despite the supposed editorial neutrality of staff. Authors sometimes also paid to have their article inserted in the next issue rather than waiting for future publication. In some cases, the more money the author offered, the longer the series of articles. Millionaire ornithologist Charles Cory paid for the printing of a book-length work on the birds of the West Indies, squeezed into periodical form, by subsidizing the cost of printing six lengthy articles in the *Auk* between 1886 and 1888.[19]

Despite these many advantages and tactics, publication was never guaranteed. Even Ridgway, despite his national name in bird circles by the 1880s, was sometimes rebuffed. He tried to place an article on an unknown subject in *Youth's Companion* in the spring of 1887, perhaps because his own son, Audie, was nine at the time. "Not being easily discouraged, I send another article for your consideration," he wrote to the editor. He didn't succeed the second time, either; no record exists of him publishing anything in that magazine. The same month, he sent a series of article manuscripts to *Cosmopolitan*, none of which apparently saw the light of day.[20]

Other things could spur rejection, even by publications to whose operation Ridgway was fairly essential. In August 1883, Brewster wrote to Allen asking him conjure up an excuse to reject a Ridgway paper for the *Bulletin*, on the birds of Illinois that also made substantial mention of the English Sparrow. "His paper seems to me to be really almost valueless. A mile from the center of Mt. Carmel takes in many stretches of primitive forest and he has 'done' that region to death already," he wrote Allen. "I have written him that I shall send it to you but that he must be prepared to have it declined as we have always refused to print anything relating to the Sparrow question not wishing to cumber the Bulletin with the discussion pro and con that would be sure to follow . . . The sparrow part of it gives us a convenient excuse if we wish to reject it." Ridgway quickly found a local Washington, DC, newspaper, the *Evening Star,* to publish a trio of brief pieces about the English Sparrow instead.[21]

Articles in bird publications thus served many needs. Perhaps most important, they provided for priority in publication of new species and new concepts. In doing so, their authors entered into relationships with other authors and publishers, and the benefits to the entire network of professionals were substantial. Amateurs, however, were often left by the side of the road by scientists. Regional articles, the staple of popular ornithology magazines, were discouraged in the pages of such technical journals as the *Auk,* the *Ibis,* and others. "Local" was often equated with "amateur," and unless they had lots of money like Cory, ornithologists who did not have ideas of national interest in mind were usually excluded. Writing about the *Ibis* in 1904, Arthur H. Evans of the British Ornithologists' Union wrote to O. V. Aplin that "It is quite true that the Ibis does little for the local ornithologist, I wish it could do more, but it was *hardly started with that object.*"[22]

Ridgway as Illustrator

Ridgway undertook artwork for many publications, a number of which remain undocumented because they were unsigned. His rate varied but was typically somewhere around fifty cents per figure for original artwork. His illustrations would then often be re-engraved or electroplated and hand-colored (or colored mechanically through chromolithography). Some pieces were joint productions by him and his brother John. In a number of instances, Robert would execute the drawings and John would color them. When artists split the work on bird illustrations with other artists, one would do the birds, and another the "accessories," as they were called: nests, foliage, eggs, landscape, or any other foreground or background elements, sometimes including lettering of the birds' names. Ridgway divvied up such tasks with his brother on a number of illustrations for many publications, which is why both men's signatures appear on many pieces of published bird art. At other times, they both drew and took turns at various aspects of the work. Robert was paid for many of these, although not for others, which would prove a considerable sore point for him, because most of the artwork he did was done on his own time. Paying bird artists for their work could be highly unstructured. The *Auk*, at least, didn't have a set rate for artwork for its pages, even a decade after its founding. Rather, it relied on artists to dictate the price, and then agreed (or didn't) on the fee. Writing Ridgway about possible artwork for an upcoming issue, Allen noted that "as to drawings for plates for 'The Auk,' we have no stated price,—in fact never fix the price, but pay what the artist demands."[23]

Ridgway also worked as a colorist, because his expertise in bird systematics and coloration would allow him to get subtle details correct. Sometimes he was paid well, as in the case of his early artwork in the popular 1874 multivolume *Birds of North America*, for which he received twenty-five dollars per image for the painstaking and expert work of hand-coloring multiple copies of thirty-four of the unnumbered colored plates, as well as doing some of the lithographic work on the numbered plates, printed chromolithographically. The set was widely circulated, with a thousand copies printed for a worldwide audience. The archival record implies that the sets colored by Ridgway numbered just fifty, with the rest being uncolored.[24]

Beyond his artwork, Ridgway contributed to the broad popularity of *Birds of North America* in several ways. He suggested to Baird that they illustrate a larger number of more familiar birds than Baird and Brewer had originally considered. Ridgway also felt that his own list of birds would make for more striking plates visually than those originally considered by Baird.[25] The special colored set was issued well into 1875, more than a year after the much longer print run of the uncolored set was issued.

A number of obstacles to creating presentable art arose. The heat and humidity in Washington was particularly damnable, because it could have a tangible effect on his artwork. For instance, in doing the drawings for the second volume of *Birds of North America*, Ridgway preferred not to work in the summer because the hot weather tended to soften the crayons he drew and colored with, making creation of sharp lines next to impossible. Executing a really good bird illustration was difficult, and although many tried, few could count on doing well at the task. "I myself have a respect now for the men who made even the worst alleged drawings in some of our bird books," one of Ridgway's correspondents wrote him after trying his own hand at feathery illustrations. For his part, young Ridgway thought highly of his own abilities. "With these done over, the series

Robert Ridgway's illustration of the Bald Eagle (*Haliaeetus leucocephalus*). Ridgway was an accomplished illustrator whose work appeared in dozens of popular and technical publications. Special Collections, Honnold/Mudd Library, Claremont University Consortium, Claremont, CA.

I am sure will be by far the handsomest and most accurate collection of plates of N. Am. birds ever published," he noted exultantly to Baird. This was wildly inaccurate in terms of the beauty of the illustrations—John James Audubon, Daniel Giraud Elliot, and a handful of other considerably more talented American artists had already seen to that, not to mention his own brother John, who surpassed him in ability—but the "accurate" part was probably right. Ridgway may have had the best grasp of bird coloration in the country, based on his access to long series of birds at the Smithsonian coupled with his understanding of the morphological characteristics that made birds look different depending on varied circumstances. In any case, four years later, he would sing a different tune about those illustrations, claiming that they were "not at all satisfactory to me, with my *present* knowledge of the subject," as he wrote to Brewster in 1879. Four years could mean a universe of difference in ability and perspective as Ridgway moved further into the technical realms he would inhabit more and more with each passing year.[26]

Ridgway was at the peak of his illustrative powers by the late 1870s, as his life among birds began to gain major traction both nationally and internationally. Although he was not a master in the same vein as earlier British artists H. C. Richter and Edward Lear, or others of his own era in the last quarter of the nineteenth century, he was nevertheless a skilled artist who worked diligently to advance his skills. When the publishers Little, Brown criticized his drawings during the creation of the *Birds of North America,* he pushed hard to improve them. He particularly scrutinized the illustrations of Joseph Wolf, the German illustrator whose lifelike works graced the pages of John Gould's *Birds of Great Britain.* "I am now paying more attention to back-grounds," he told Baird, "and am studying Wolf all the time."[27]

When Ridgway illustrated multiple images for a published work—either his own or for others—he typically worked on a number at once, which took him longer to produce individual images but allowed him to complete several plates at nearly the same time. He sometimes got his son, Audie, to run artwork back and forth between the Ridgway home and the Smithsonian. He also got a useful bounce for his other work on systematics while making bird art. When illustrating birds for another ornithologist's work, such as extensive paid work done for William Brewster on a book of the birds of Baja California, the two men could simultaneously send skins back and forth in order to sort out the systematics as well as have examples in hand to work out coloring and illustration issues.[28]

Other Books

The first monograph to which Ridgway contributed substantially was the aforementioned five-volume *Birds of North America,* published by Little, Brown. The first three volumes, on the land birds, were printed and produced in fits and starts, and completed in late 1875 (despite its stated publication date of 1874), and were followed a decade later, in 1884, by two volumes on the waterbirds. The work was of seminal importance for the

Robert Ridgway, early 1880s. Special Collections and Archives, Merrill-Cazier
Library, Utah State University, USU Caine MSS8, box 4, fol. 2.

field. Among other things, it presented the first detailed information on
the behavior of birds in Arctic breeding grounds and was a standard treatise on the life history of birds into the early twentieth century. By 1875
Ridgway had not yet developed the diagnostic expertise to have contributed substantially to such a wide-ranging text—the three land bird volumes ran to more than eighteen hundred pages—nor had he been asked
to do so. Thomas Brewer wrote most of the species descriptions for the
first part. The latter waterbird portion, however, was the more authoritative of the two parts, largely due to advances in understanding of ornithology over the intervening decade. By the time the last two volumes
were published, coauthor Brewer had been dead for four years, Baird was
no longer doing bird research, and Ridgway wrote the bulk of the work.
When it finally reached print in 1884, the *Auk* noted that "The publication
of the long-looked-for 'Water Birds of North America,' by Baird, Brewer,
and Ridgway, is *the* event of the year 1884 in the history of North American ornithology."[29]

Ridgway was paid well for his artistic labor on the land bird volumes but received nothing for the waterbirds. Despite appearing under the imprimatur of Little, Brown, the two final volumes were actually a publication of the relatively underfunded Harvard Museum of Comparative Zoology. The authors had found it difficult to procure someone to underwrite the cost of publishing the 1884 volumes until Alexander Agassiz—the wealthy son of the famous Harvard scientist Louis Agassiz—stepped into the breach and covered the expenses. This salvaged the effort while simultaneously giving Agassiz bragging rights for publication. No one, however, offered any payment to Ridgway for his role, and he was concerned. His one other long publication had been the Fortieth Parallel Survey report, and he had received no payment for that. Ridgway wrote Agassiz to show his good faith, offering to cancel a planned two-month trip to the West in 1883 if the printing of the book was about to commence. But there had been bad blood between Baird and Alexander's anti-Darwinian father Louis, and Agassiz never offered a penny to Ridgway, despite Ridgway's protests to Baird. This was probably influenced by Louis Agassiz and Baird's prickly association—one that was colored by numerous factors. The older Agassiz was a member of the Smithsonian's Board of Regents and a close friend of Baird's colleague and predecessor Joseph Henry. He was also a charming public speaker, a very successful fundraiser, and a renowned scientist. But one of Agassiz's key students left Harvard to work as a herpetological assistant for Baird—leaving in part because of Agassiz's antievolutionary views—and that move had angered Agassiz. The creation of a museum in the nation's capital also worried Agassiz, who felt that it would compete with his museum at Harvard. Agassiz would later try to blackball Baird from membership in the National Academy of Sciences, considering him a dilettante who, though a good descriptive zoologist, could not manage the theoretical or philosophical aspects of science. All of these elements no doubt leavened Alexander Agassiz's attitudes toward Baird's protégé Ridgway.[30]

Ridgway was more experienced by the time the second part was published, had contributed much more, and had been paid nothing. When someone from Little, Brown deigned to write to complain about some aspect of the book, Ridgway responded, "I think it decidedly unfair—if not somewhat remarkable—that you should ask any explanation from me."[31] He pointed out that he had undertaken the immense labor of preparing all the technical and descriptive manuscript material, as well as much other editorial work, and even covered the cost of postage for the proofs, all for no recompense. A third party, not associated with the publisher, had given him one colored copy and one uncolored unbound copy, but that was the extent of his compensation.

Issues of cost and remuneration were always part of these great publishing efforts. Ridgway's attempts to get payment for various kinds of work were often rebuffed, especially if he did not push hard for it. Many of the government's publications were already loss leaders, and decision-makers were loath to add to expenses any more than necessary. Writing in 1878 to Ridgway about a substantial thirty-three-page article about herons in a government publication, Coues (who was apparently editing the volume) noted, "I am trying to work your heron pictures through, but the Doctor [Baird] is opposed to pay me for them . . . The wood cut will be pretty expensive; and I think the illustrations will be thrown out altogether if H. [Joseph Henry, then the Secretary of the Smithsonian] has to pay for the drawing as well as expressing [that is, writing the text.] However, I will see what can be done. Do I understand you to present the manuscript, or do you expect pay for that too? . . . If you should charge at the rate of $1.00 a page for the text, solid long primer, you would see that it would make the whole article pretty expensive to pay for drawing and expressing also." Ridgway probably got paid nothing for either illustrations or text, since the government printed only two hundred copies of the work.[32]

Ridgway's work for Smithsonian publications did not emerge from a vacuum; it was influenced by Baird's sensibilities about visual presentations of museum topics, and he knew that Smithsonian works spoke to a larger audience of congressional supporters, expedition funders, and the public. He oversaw much of the Smithsonian's lithographic and illustrative work, both in its execution and in its printing, and became expert on many related facets. He was also acutely aware of the costs and constantly worked to economize on Smithsonian artwork.[33]

These experiences, however, did not stop him from continuing to work with other authors; his gripes about pay were primarily with the publishers. Several years after the waterbirds volumes were published, Ridgway and Stephen Alfred Forbes, a zoologist with the University of Illinois, coauthored a two-volume work, *Ornithology of Illinois*, on the birds of Ridgway's home state. Ridgway wrote the first volume, which detailed the birds themselves; Forbes wrote the second, which primarily dealt with "economic ornithology," or the financial effects of birds on the region's crops and other monetary impacts. Somehow Ridgway managed regularly to involve himself with people who had odd spikes of infamy earlier or later in their lives: as a rowdy lad of just fourteen, Forbes had briefly gained national notoriety while simultaneously embarrassing his family and fellow citizens. He had scolded Stephen Douglas from the audience while attending one of the Lincoln-Douglas debates, for what he perceived as an insult by Douglas to Lincoln. But as an adult Forbes was an excellent bird man, went on to medical school, became a member of the National Academy of Sciences, and served as president of the Ecological Society of America.

The Illinois bird volumes—of particular personal importance to Ridgway because of his extended childhood surveying of his home state's birds in his youth, and his emotional ties to his native region—had a long gestation. Forbes had asked Ridgway in 1883 to coauthor the work with him, and being a little savvier about money after his unpleasant financial experience with the waterbirds volumes, he contracted with Forbes for a handsome payment of three hundred dollars. "Only a fair remuneration could induce me to accept any additional engagement," he noted, and he got his asking price. The work was also delayed by a fire in the printing office, which destroyed the printer's copy as it was going to press. Fortunately, even though portions of the work had to be rewritten, Ridgway had retained some of the galley proofs, which allowed the work to be reconstructed. The authors recruited a corps of local Illinois volunteers for help, and by 1893, some forty birdwatchers were helping to work out the specific ranges of a number of species found in Illinois. The first volume appeared in 1889, the second in 1895.[34]

Ridgway's upbringing in southern Illinois also led to a love of the numerous plant species he grew up with, and for almost all of his adult life he maintained an interest in botany, particularly trees. He published at least thirteen articles on the subject, primarily in popular male-oriented journals such as *American Sportsman,* but also in several Smithsonian proceedings and in other venues. He also tried, unsuccessfully, to get the American Museum of Natural History in New York to mount an exhibit of his photographs of trees from Illinois.[35] Ridgway's feelings about plants and trees ran deep, but he was also defensive about the subject. This was probably because he was part of a larger cohort of men who did not recognize women as constituents in scientific study. In 1925, he wrote an article entitled "Is the Love of Trees and Flowers a Sign of Effeminacy?" for his local newspaper, the *Olney Times*. "Among many fallacies (which in plain English means 'fool notions,') which our system of education has utterly failed to eradicate," his piece began, "there is prevalent one to the effect that only old women, effeminate men, and 'cranks' are lovers of trees."

He went on to describe a visitor to his home, "a real man," who loved trees and flowers but was apparently ashamed to admit as much, and "casually remarked that he was 'just effeminate enough to love flowers'"— a view which Ridgway described as being that of "perhaps the majority of men." He then described "real 'he men'" who were devotees of trees and plants, including George Washington and Thomas Jefferson. He also listed a number of successful businessmen, journalists, and railroad magnates. In doing so, he denigrated those who did *not* like botanical life: "The truth is, that the majority of men who do not care for trees and flowers so much are those who value nothing except that which may be appraised in dollars and cents." The love of trees and flowers was not only a perfectly natural human trait, regardless of sex, station, or occupation, but was sometimes, he observed, a striking national characteristic, such as the love of cherry blossoms among the Japanese, "who surely cannot be considered an effeminate people."[36] In his argument, the love of trees and flowers was not only masculine but an almost patriotic imperative struck through with all of the era's manly implications of patriotism.

Ridgway's attitudes about gender are instructive, because in the fifty years straddling the turn of the nineteenth century, those working in science had much to say in print about women working in the field—most of it condescending. An 1887 article published in *Science* had served as the rallying cry of male botanists opposed to women entering the field, and Ridgway would have read this piece. "Is Botany a Suitable Study for Young Men?" the article title read, and in raising up males as exemplars of the study of plants, it simultaneously denigrated women. "An idea seems to exist in the minds of some young men that botany is not a manly study; that it is merely one of the ornamental branches, suitable enough for young ladies and effeminate youths, but not adapted for able-bodied and vigorous-brained young men who wish to make the best use of their powers," the author wrote. All sorts of intellectual skills could be gained through botany, he argued, including the tools of inductive reasoning; and it provided exposure to varieties of species and Darwinism. The work of botanists necessarily also included a great deal of walking, he continued, "and his frequent tramps harden his muscles, and strengthen his frame"—attributes that would not have been encouraged in women. Finally, botany was essential preparation for being a physician or pharmacist—almost entirely male pursuits at the time. "It is therefore preeminently a manly study," he concluded, and seemed to be saying that it was actually better suited to men than to women in both its specific and general utility. Ridgway echoed this attitude in his article nearly forty years later. Nevertheless, a number of women were writing short papers and notes in scientific journals at the end of the nineteenth century, as well as holding membership and officership in botanical clubs and societies, making herbarium specimens, and generally contributing substantially to botanical science.[37]

As a rule, the engine of Ridgway's work ethic pushed him forward in the face of an astonishing amount of labor, and although he would have been greatly pleased to be better compensated financially throughout his career, his work seemed mostly about putting knowledge forth to a larger world. In 1887, he wrote a 642-page work, *Manual of North American Birds,* designed to be a single-volume guide to the continent's avifauna, much more compact than the five-volume monster to which he had contributed. It was the first thing like a comprehensive field guide for North American birds. As Ridgway noted in the preface, the volume had been "reduced to the smallest compass, by the omission of anything that is not absolutely necessary for determining the character of any given specimen."[38] He solicited a Philadelphia publisher, Lippincott, whose first concern was, naturally enough, financial, and Ridgway was asked for some reassurance as to the book's likely success. His response provides a useful view of the ways authors persuaded publishers to get works into print. Writing Brewster, Ridgway stated that he would write letters to the Nuttall Ornithological Club, the American Ornithologists' Union, the Linnaean Society of New York, and the Ridgway Ornithological Club in Chicago—all groups with which he had enjoyed close relations—and would ask them each to address specific questions raised by the publisher about the potential of the *Manual.*[39]

The book ended up a decided success. Along with his writing and revision of the book, Ridgway also did 464 drawings to illustrate different genera. His volume had considerable legs, going into a second edition in 1896 and a reprint of that second edition in 1900. Allen, writing Ridgway in the summer of 1887, noted that he had spent many hours in looking it over and expressed his delight with it. "It strikes me as admirably planned and most carefully executed. Taking in, as it does, the birds of Mexico, etc. it will prove invaluable to all our more advanced ornithologists as well as to those less versed in the subject of which it treats. It is a great monument to your industry and ability, and as a work of reference will not soon be superceded." William Brewster described it as "a blessing to every working ornithologist." Montague Chamberlain of the AOU commented to Ridgway on its utility: "It is far away the best thing we have for the working naturalist—of course one has to fall back occasionally to 'B.B.&R.' [Brewer, Baird, and Ridgway's five-volume work on North American birds] but the Manual will be more used than any other work that has been issued."[40] However, a bigger work was on the horizon: *Bulletin 50, The Birds of North and Middle America*—Ridgway's multivolume masterwork.

Bulletin 50

The first part of Ridgway's massive magnum opus, officially titled *The Birds of North and Middle America,* appeared in the fall of 1901 as Part 1 of United States National Museum *Bulletin 50,* the number reflecting its place in the longer run of Smithsonian publications. Ridgway was officially directed to undertake the task by Assistant Secretary G. Brown Goode in 1894. The work was designed to be the most comprehensive yet on North American birds, including all of Central America. Ridgway's goal was to describe every species and subspecies in North America, and he decided that one critical aspect of the work would be to untangle the taxonomic Gordian knot of historic names for each family, genus, species, and subspecies of bird. What was the rationale for the Smithsonian's official support of the first descriptive, technical catalog of all the 684 thenrecognized genera found on the continent from Panama northward? According to Goode, it was because the institution had put exceptional efforts toward forming an exhaustive collection of the birds of the region ever since Baird's arrival. The descriptive work would be the other side of the coin, by making it available to a wider audience and thus realizing the collection's fullest potential.[41]

Ridgway began work on *Bulletin 50* at home in Brookland, DC, on the evening of October 23rd, 1894, a Tuesday night, about a month after Goode's directive. He worked mostly at home to avoid interruptions and disruption of his daily duties and routines. Goode, who oversaw the first years of the work, authorized Ridgway to take Smithsonian specimens home for his convenience. The job would substantially occupy much of his writing and publishing efforts until just before his death nearly thirty-five years later and would fundamentally change the character of his relationship with the Smithsonian. Taxonomic guides by the institution had been published before; the institution had issued an enormous number of other important works that served as catalogs of families or species of plants and animals or that specifically related to systematics and taxonomy. Both types of works were myriad and ranged from catalogs of minerals and their synonymies to taxonomic works on microorganisms to inventories of fishes in specific locations around the world. A wildly lush group of life-forms was inventoried, from fossil invertebrates to butterflies to dolphins. Some of these reports were based on the Smithsonian's collections; others were broader in their scope.[42] But by the time the last part was issued, *Bulletin* 50 would be the largest and most comprehensive systematic work published by the Smithsonian.

In the first few months, Ridgway worked at home in the evenings for two to three hours, making quick work of four families—grebes, loons, auks, and gulls and terns—by the end of the year, in just under twenty-six hours of labor. By now he was an extremely experienced systematist, with one of the world's great collections of birds under his nose or close by.[43] For comparative purposes, Ridgway steadily borrowed a stream of specimens as well as relying on the Smithsonian's own holdings. In the first months alone, he received material from the American Museum of Natural History, the Philadelphia Academy of Science, the Museum of Comparative Zoology at Harvard, and the Department of Agriculture's Division of Ornithology collections, branching out rapidly to other institutions' holdings as the years passed, including the Carnegie Museum, the State University of Iowa, William Brewster's private collection, and the Field Museum in Chicago.[44] In all, Ridgway borrowed more than four thousand specimens over the course of the eight parts in his highly comparative and synthetic work.

His assistant Charles W. Richmond soon assumed most of his daily office duties, leaving Ridgway free to devote almost all of his daylight hours to the task as 1895 progressed. He worked at home four days a week, where he had "perfect quiet," and at the Smithsonian two other days, usually laboring between eleven and twelve hours a day and sometimes longer. After the first part was well under way, Ridgway planned to press a stenographer into service, who would take down the description and remarks in shorthand and then type up the text for the printer. In a flurry, as 1895 closed Ridgway dealt with hawks, ospreys, and owls, along with the parrots, working out synonymies as one of his major tasks. In January 1896 he bragged to his friend William Brewster, "I am progressing finely with my work on North and Middle American birds; in fact, can safely say I am nearly *one half* through with it."[45] Little did he know how long a road remained.

The task was made formidable by its sheer scope. Ridgway was attempting to describe every species and subspecies of bird found on the North American continent, ranging from the Arctic to the Isthmus of Panama. He also included the birds of the West Indies and the other islands of the Caribbean (excluding Trinidad and Tobago), as well as the birds of the Galápagos archipelago. His list included accidental and casual visitors as well as introduced and naturalized species, which swelled the ranks of birds needing work. In August 1896 he reported to Goode that since beginning he had made a first pass through a total of 72 families, 559 genera, and 2,022 species, leaving 25 families, 142 genera, and 636 species yet to be taken in hand.[46] This did not mean, however, that he had properly written up a manuscript on those families—only that he had made a solid outline of just what birds existed in North and Middle America and in what families they belonged. Getting his ducks in a row required him first to understand the lay of the water, air, and land.

Ridgway was especially pleased with one element: the use of outlines of generic characters to be used as artwork in the publication. Baird had begun these as a convention in an earlier book, *Review of American Birds* (1864–1872), mostly in the latter third of the work, and Ridgway had elaborated on them considerably. "These generic outline illustrations are of the utmost utility, very often proving superior to the best possible diagnosis as a means of identification," he noted proudly. He had used them in his 1887 book on North American birds as well. "They are distinctively 'Smithsonian' in their character, the few irregular attempts at similar illustrations, in the 'Catalogue of Birds in the British Museum' and elsewhere, being crude in the extreme," he remarked to Goode.[47] The engraved drawings, which appeared at the end of each part, ranged from tiny images fewer than two inches square to ones that filled the page.

By the summer of 1897, work on the families of crows, flycatchers, manakins, ovenbirds, thrushes, mockingbirds, warblers, nuthatches, creepers, and others had flown by. Each species' entry began with its name, followed by detailed measurements—the bird's overall length in study skin form, as well as the lengths of wings, tails, exposed culmen, tarsus, and specific toes—critical identifiers for separating very closely allied species, and all averaged from the multiple examples he examined for every bird. These were provided for adults, for males and females if they differed, and for immatures. The entries also noted locality and range details, exhaustive synonymies, plumage and coloration, and other physical details. The highly taxonomic focus meant that some elements were excluded, such as the birds' life histories. Also absent were descriptions of the birds' eggs and nests, probably because the Smithsonian's collections of those items were not as comprehensive as their skin collections, and tracking down those details for thousands of different species via other institutions would have added tremendously to the effort.

By the summer of 1900, the strain of working so intensely on *Bulletin* 50 caught up with Ridgway. He found himself unable to continue, having worked to exhaustion as he neared the press deadline for the first part. Ridgway's wife, Evelyn, wrote to Frederick True, chief curator for the Smithsonian, in early July, to inform True that "Mr. Ridgway now realizes that he has been attempting the impossible, and has broken down under the effort to get the work done by the 15th. He has been under the doctor's care for the last two days, and is still unable to sit up."[48] It is difficult to tell whether he had actually been affected by the undiagnosed arsenic poisoning detailed in chapter 3, perhaps from examining and measuring so many specimens prepared with arsenic, but in any event, Ridgway was prostrated for days. He rallied as the summer progressed, and in early September 1900, he wrote to Richard Rathbun, the assistant secretary of the Smithsonian, triumphantly noting that he had completed the first part, with the manuscript weighing in at 1,774 pages. It consisted entirely of the family of Fringillidae, or finches, numbering 389 species and subspecies.

Ridgway had many suggestions for his bosses as to the size and configuration of *Bulletin* 50. For instance, he thought that none of the parts should exceed 600 pages to maximize their utility. He imagined a somewhat larger format for the book, on the order of a small folio similar to a work he greatly admired, Herbert Dresser's *A History of the Birds of Europe*. The finished product, however, was in a smaller format and longer in length. Part 1 weighed in at 715 pages of typeset text, plus twenty plates of illustrations at the end of the book, and it ended up as the second-smallest of the eight parts done in his lifetime. With the exception of Part 6, which ran to fewer than 600 pages, every other part was at least 840 pages long.[49] Ridgway was giddy at having the first one out of his hands and in the printer's that fall and looked forward to working on the second one.

However, on February 22, 1901, while the manuscript was in press, tragedy struck. Ridgway's only child, Audubon Whelock Ridgway, died at the age of twenty-three, of pneumonia contracted while out skating in Chicago. Known as "Audie" to friends and family, he had started a promising career as an assistant in the ornithology department at the Field Museum in Chicago, where he had worked for just three months before his death. His father's correspondence over the years shows occasional evidence that Audie suffered from pulmonary difficulties most of his life, which may have contributed to his ultimate fate. Ridgway's assistant Charles W. Richmond penned a paragraph noting his passing in the *Auk*. "He was very popular among the young folks of his acquaintance, hundreds of whom have in his death met with a personal bereavement," he concluded. Audie had been a gentle young man, averse to shooting birds. He was also a proficient photographer. Ridgway was on leave from the Smithsonian for several months, recuperating from his *Bulletin 50* labors, and was in Florida in the field at the time. Audie's mother was at his side as he died, and she accompanied his body back to Chicago.[50]

Ridgway's sorrow over his son's death was mingled with frustration at the Smithsonian and his inability to be closer to nature. His wife's increasing depression also worried him greatly. These feelings were encapsulated in a letter to William Brewster at Harvard, written two months after his son's death. Although he never directly expressed his sorrow, it seeps through the pages. "I have been so constantly distracted by cares and worriments as to feel unequal to anything beyond attention to the most ordinary and urgent matters," he wrote. "Of course it has been impossible for my wife to help me in this, for she is still completely prostrated, and instead of getting better grows steadily worse. In fact I am now, as a last resort, making arrangements to send her to a hospital or sanitarium, where she can be under constant medical supervision. She is utterly crushed, with all the light and hope, apparently, gone out of her life. I have serious doubts whether she will ever recover from the terrible shock which she sustained. As for myself I have been kept up mainly because I was obliged to, though many a time I have felt like giving up, the strain has been so great."

At the same time he endured the grief of his son's death, he also felt embittered by institutional obstacles blocking his way. He pointed to the Smithsonian's logistical inability to support his efforts on *Bulletin 50*: "I little realized, when I began this work more than six years ago, the enormous labor involved in it, and still less the most serious disadvantages under which it would be done," he complained to Brewster. "These disadvantages, instead of growing less, increase. The simple fact is the National Museum is not in any way, so far as facilities are concerned, equipped for an undertaking of a work of this character, and I am heartily sick and tired of trying to make up for these shortcomings by devoting *all* my time and energies to it. On the other hand, to relinquish it at this stage would, I fully realize, be a very serious thing. It would practically mean the sacrifice of my life's work, and, besides, I would feel, more or less strongly, that I had shirked a duty which I was in honor bound to perform . . . I feel so discouraged at times that it is hard for me to keep silent."

In addition, he confessed that he was bone-tired of technical ornithology and pined for the outdoors—not the harsh outdoors of collecting expeditions but the gentle outdoors of country life, with a house nearby. Just fifty years old, he sounded like an old man. "It is almost too late in life for me to hope that the *intense longing* to spend at least a few peaceful years in closer communion with nature—as far away as possible from the to me wholly uncongenial associations of city life—feeling that I am free to work when I feel in the humor and to enjoy life in my own way, may ever be realized; but this one thing is what I have for many years looked forward to more than anything else affecting myself personally, and now that my son is dead, and my wife *almost* taken from me, I feel that all courage to continue the struggle of past years has gone from me." Ridgway rarely got things off his chest in writing, but his old friend William Brewster—who was generously inclined, nonjudgmental, and honest in his dealings—was a safe place to vent. All three of Brewster's siblings had died in early childhood, and he would have been an exceptionally sympathetic ear to Ridgway's woes and sorrow. He closed his letter to Brewster apologetically: "In spite of myself I have been very personal in this letter, but I hope that under the circumstances you will forgive me."[51]

Ridgway was discouraged, and for good reason. "There is a limit . . .

to the duration of ones life but also ones ability to work," he told his boss Richard Rathbun, Goode's successor. His sense of fatigue and discouragement was almost tangible: "Even my unceasing energy during the past six and a half years together with my best future endeavors may not suffice to complete this enormous undertaking," he wrote. "Sometimes I seriously question whether it is really worth while for me to continue it under all the circumstances; not from doubt as to the results being worth the sacrifice, but whether I would be justified so far as my duty to myself is concerned in keeping up for the remainder of my existence a continuous drudgery which, however much it may benefit science or add to the sum of human knowledge, can only give me as recompense unceasing monotonous labor and the loss of many things which would add much to my pleasure and physical good." He also observed that he had given the project more than twice as many hours as he was being paid for by the Smithsonian.[52]

But the calendar marched on, Ridgway continued, and the first part appeared on October 24, 1901. One of the most important features was Ridgway's exacting treatment of synonymies: earlier names for the same bird. Over the decades, authors and printers had made numerous blunders in presenting bird names for the first time: typically, mistakes that appeared while the manuscript was being prepared for publication. Sometimes obvious spelling errors crept in, or authors used wonky Latinate terms, which solidified into the almost biblical authority given to the first appearance in print of a new species. Ridgway attempted to trace the synonymies of these birds and, in so doing, hewed to this concept of first name authority, condemning those who attempted to correct published bird names: "The correction of an author's orthographical errors is a pernicious practice, though much in vogue; 'science is not literature,' neither has it any concern with what an author should have done or meant to do, but only with what he actually did," he noted. The ferocious labor involved in Ridgway's corrections and clarifications caught reviewers' eyes. "The bibliographical references alone almost stagger us with their suggestion of the work involved . . . We trust that the author's life may be spared to complete what has been undertaken," noted a typical commentary in the *Wilson Bulletin*.[53]

Some aspects of the initial part were puzzling. Why, for instance, did Ridgway begin with the finches, a "higher" order of birds, rather than the seabirds, which were typically the lowest order? The AOU Checklist, after all, had set an important precedent in starting with the low orders and working upward. The reason proved to be a practical one. In the preface, Ridgway explained that the Smithsonian's facilities were inadequate for properly arranging the larger birds (seabirds typically being the largest birds, including as they do albatrosses, pelicans, herons, shearwaters, and seagulls and other inconveniently outsized examples). So he started with the smaller ones, as they were most readily available and best organized. Ridgway's description of these less-than-ideal conditions brought a supportive shout-out from the editors of the *American Naturalist* in their review of the first part: "Such a statement . . . may well make us feel ashamed," wrote the reviewer. "Is this country too poor to provide facilities for such a man as Mr. Ridgway, who returns to it a thousandfold the small means it has placed at his disposal? Are we so blind that we cannot see that scientific knowledge is more than the equivalent of money . . . but is in itself far more nearly an end of national existence?"[54]

In the opening of the first chapter of Part 1, Ridgway also penned a passage that has been widely quoted (usually with several minor errors) and is useful for several reasons, not least for what it shows about his attitudes toward professionals and amateurs at the turn of the century. He began by noting that "there are essentially two different kinds of ornithology: *systematic* or *scientific,* and *popular.*" He continued:

The former deals with the structure and classification of birds, their synonymies and technical descriptions. The latter treats of their habits, songs, nesting, and other facts pertaining to their life histories. Although apparently distinct from one another, these two branches of ornithology are in reality closely related and to a degree interdependent. The systematist who does not possess an intimate knowledge of the habits of birds, their mode of nidification [nest-making], the character of their nests, eggs, and young, is poorly equipped for the work he has in hand, while the popular writer who is ignorant of scientific ornithology is placed at an equal disadvantage—his writings may entertain, but are far more apt to mislead, through erroneous statements, than educate. Popular ornithology is the more entertaining, with its savor of the wildwood, green fields, the riverside and seashore, bird songs, and the many fascinating things connected with out-of-door Nature. But systematic ornithology, being a component part of biology—the science of life—is the more instructive and therefore more important. Each advance in this serious study reveals just so much more of the hidden mysteries of creation, and adds proportionately to the sum of human knowledge.[55]

Ridgway attempted to do several key things with this passage. He wanted to leaven the often-vocal criticism of the scientific approach in the popular press by noting that the popular approach was also useful to the scientist, and vice-versa. He also hinted at the literary appeal—the "savor"—of popular writing about birds as a means of mitigating criticisms. His final sentence, about "the hidden mysteries of creation," was not meant to be religiously tinged but rather a statement about how science could tease out the mysteries of evolution.

Others, however, took issue with his differentiation between the scientific and the popular. Joel Allen, for instance, thought that Ridgway had made a hasty generalization. "An important field of research has not only been disregarded but, by inference at least ruled out as not only not scientific but as not falling within the author's definition of biology. This is, in a broad sense, the life-histories, including the relation of the animal to its environment, and the many problems of evolution that depend for their solution upon the study of the living creature."[56] Allen was right. The dawn of the twentieth century brought new tools—methodological constructs, statistics, genetics, and other elements that quantified the living bird in ways not previously possible. The study of birds' life histories was no longer the domain of eighteenth-century naturalists who studied birds in the field and described them primarily as evidence of the glory of God, and it was the next logical step in the evolution of the study of birds. Ridgway, in fact, ironically helped to put the nail in the coffin of systematics as the key means to study birds as a profession, because his work on *Bulletin 50* was so definitive that it would remain unsurpassed for decades, leaving ornithologists to turn to other venues after Ridgway's domination of avian systematics.

Others complained, too. Although praising the first part on many fronts, the editors of the California-based *Condor* also attacked Ridgway's distinction between "systematic" and "popular." Joseph Grinnell, the authoritative California ornithologist, noted that "this does not seem to be to be a fair discrimination. One is led to believe at once that 'popular' ornithology as here understood is unscientific; and that systematic ornithology alone is scientific!" Grinnell called it a "queer idea" and, like Allen, criticized the work for not covering any details of the food, habits, nests, or anything else that would help in the study of the relation of the species to its environment. Environmental ornithology was of great scientific importance, he pointed out, "perhaps in line with the discovery of the methods of the origin of species." He also took Ridgway's implications further. Were students of economic ornithology, or migration, necessarily "popular"?[57]

Even the *American Naturalist*, normally a reliably sympathetic Ridgway fan, objected to what it viewed as a dichotomy. "The present writer has to confess that this statement quite took his breath away. The study of living birds, then, is 'popular,' but the study of their mortal remains, stuffed with cotton . . . this is 'scientific.' It is hardly possible that Mr. Ridgway could have intended his words to be taken literally." Freud, who had brought the clinical study of mental processes into mainstream discourse, was highly regarded in science and provided a useful if oblique foil: "One might as well say that the study of alcoholic brains is scientific (because the literature of that subject is so), but that the study of mental processes is popular, or non-scientific, for the reason that we have a mass of trivial literature on psychology," the *Naturalist* concluded.[58]

Despite the monumental nature of his work, Ridgway was slowly but surely falling out of step with wider trends in biology, and it began to show in the critiques he faced regarding his attitudes toward the living bird. As the century turned, biologists had begun to recognize that elements long considered "popular"—behavior, habitat, nesting and breeding practices, and other aspects of bird life—were becoming increasingly essential for accurate evolutionary classification. This is not to say that Ridgway ignored the living bird in other writings, as a number of his articles based on his own fieldwork attest. But by the publication of the first part of *Bulletin 50*, Ridgway's focus on a suite of physical characteristics seemed more typical of Linnaeus than Darwin, and his critics were right on the mark.

From his perspective, Ridgway had other concerns more subtly related to the issue of popular versus scientific ideas in the preface. Evolution was an essential aspect of systematics, and he had little patience for anyone who believed otherwise. "Accepting evolution as an established fact—and it is difficult to understand how anyone who has studied the subject seriously can by any possibility believe otherwise—there are no 'hard and fast lines,' no gaps, or 'missing links' in the chain of existing animal forms except as they are caused by the extinction of certain intermediate types," he noted in the introduction of Part 1.[59] This was a powerful idea. Darwin's notion of modification with descent served as an engine for the details of species and subspecies that Ridgway so carefully described, and he laid the idea out explicitly. "The question of species and subspecies and their nice discrimination is not the trivial matter that some who claim a broader view of biological science affect to believe. It is the very foundation of more advanced scientific work; and without secure foundation no architect, however skillful, can rear a structure that will endure." This was nontheistic science writ large, and a huge chasm separated it from the writings of a half-century before, when species were considered immutable and the "architect" of nature was God.

One of the problems, though, with trying to accurately reflect what was really a continuum of species caught in the middle of evolutionary change, on the way to becoming a subspecies and thus to a new species, was trying to reflect that continuum without causing nomenclatural chaos. The editors of the *Wilson Bulletin,* reviewing the first part, pointed to the woes of being a splitter: "The increase of trinomials over previous lists and catalogues is partly due to the extension of the application of trinomialism to each member of a group . . . This practice is entirely logical, but we cannot but deplore its adoption into a system which is already too cumbersome . . . Furthermore, the adoption of an ultra scientific system which can be used only by the expert in color values and careful measurements of many specimens, is divorcing the scientific from the practical." Somehow, "scientific" did not equal "practical"—a common complaint of many observers of the era. "We no longer have use for a science which is wholly for scientists," the article concluded.[60]

Ridgway also fretted about the "depressing effect" of one of the few highly public reviews of the book, in the pages of the *Chicago Tribune.* The venerable newspaper had published, highly visible on the second page above the fold, a brief and sarcastic piece entitled "Sesquipedalian Ornithology." The *Tribune* blasted Ridgway for what seemed like an incomprehensible book, and quoted his description of a small hummingbird to make its point about the obtuseness of his language: "Aegithagnthous, anomologonatous birds, with short (non-functional) colic caena and nude oil gland; first toe (hallux) directed backwards . . . first toe with its flex or [flexor] muscle (flexor hallucis longus) independent of the flexor performs digitorum; hinder plantar tendon free from the front plantar; ambiens and fermoro-caudal muscles absent." The paper called the volume a "truly appalling work," and its severest criticism was reserved for what it considered inexcusably obtuse language. "The width of the channel between scientists and laymen is well illustrated by the latest publication of the Smithsonian Institution," the review observed. "The lay reader can hardly get a glimmer of sense from any of the seven hundred-odd pages." Some colleagues, however, found those technical aspects to be highly aesthetic, in an intellectual sense. What was incomprehensible to many amateurs and casual observers was a glorious work to his fellow technical ornithologists. Brewster wrote to say that "it strikes me as one of the most beautiful as well as dignified things of the kind that I have ever read."[61] This kind of technical writing about nature distilled tensions between nature and the constructed, technological world into a particularly dense conundrum. It was language not about man-made things but about nature, granted — yet it was not the unique and authentic nature experience Americans longed for, either in writing or in person, during the rise of technology and growth of urban centers of the turn of the century. The nature study movement spanning the Progressive era urged Americans to get out of doors, to have a vital firsthand experience.[62]

Despite the *Tribune's* critique, though, technical details were important, just as they were in medicine. No one in or out of the medical field would complain about a rigorously technical medical work detailing the human body's workings. Nor would geologists' writings about strata, rock types, and other technicalities necessarily raise eyebrows. Why, then, did a broad public balk at the same technical descriptions of birds? A clue is revealed in the *Tribune's* passage above. "The lay reader can hardly get a glimmer of sense," it noted, and this was crucial: there were countless people who had long experience in past decades in grasping and appreciating a clear, literary language about birds. But in Ridgway's view, people writing about birds in literary fashion were distinct from the scientists: useful supporters, and perhaps lyrical writers, but people who should leave the details to the professionals.

If *Bulletin 50* was the face of scientific ornithology at the turn of the century, what did the most widely read popular ornithology look like? The bird writings of William Henry Hudson, a literary contemporary of Ridgway's, provide a useful counterpoint. Hudson's books on natural history sold 1.3 million copies by 1933, making him among the most popular of all bird writers. Some of his bird books were praised by scientific ornithologists, including his 1895 *British Birds,* coauthored with Philip Lutley Sclater of the British Museum. But Hudson was also sometimes taken to task for a lack of accuracy and was looked down upon by systematic orni-

thologists for his lack of scientific thoroughness. "Hudson was very helpful at first in putting me on the trail of the commoner species but it is most
aggravating that it is so incomplete," wrote legendary British ornithologist
Wilfrid Alexander to his brother Horace in 1921. "In addition his accounts
of the habitats of the birds are very much exaggerated and sometimes
misleading. For instance he says the Gallito or Little Cock (Rhinocrypta
lanceolata) practically never flies, yet though I have only seen them a few
times I have frequently seen them fly up into the bushes when disturbed.
No doubt they are chiefly ground birds but to emphasize this he makes
a decidedly misleading statement. And this is only one of many similar
cases."[63] For serious ornithologists, errors of fact—especially easily re-
futed errors—were a deadly sin.

Some popular writers about birds were dismissive of any systematic
approach. "Studied the birds? No, I have played with them, camped with
them, gone berrying with them, summered and wintered with them, and
my knowledge of them has filtered into my mind almost unconsciously,"
nature writer John Burroughs claimed. Burroughs, who had accompanied
Ridgway and others to Alaska on the Harriman Expedition of 1899, emphasized the living bird, not the museum specimen, and did so anthro-
pomorphically: "There is a sympathy between us that quickly leads to
knowledge. If I looked upon it as something to be measured and weighed
and tabulated . . . my ornithology would turn to ashes in my hands." He
added an even more explicitly and unapologetically emotive element to
his study of birds: "To interpret nature is not to improve upon her . . . it is
to have an emotional intercourse with her, and reproduce her tinged with
the colors of the spirit."

Despite his literary aims, however, Burroughs claimed to be deeply
opposed to anything he felt was inaccurate regarding the natural world.
He played a key role in an early twentieth-century American debate,
known as the nature fakers controversy, highlighting the conflict between
science and sentiment in popular nature writing. For him, descriptive
accuracy was of vital importance, even if it had literary leanings. Writing
to a fellow British naturalist in 1908, he remarked, "I am interested in all
honest endeavors to get at the truth about the lives of the wild creatures
around us." He also found the need for a middle ground between the technical and the idealistic: "Our nature writers are either all very dull, or else
very romantic and unreliable," his letter concluded.[64]

A hallmark of some forms of amateur writing about birds was not

just their occasional appeal to a higher power but their references to children—a population that secular scientists had no particular reason to
mention. "Children especially delight in them," opined a popular adult
book on hummingbirds. "This need not surprise us, for musical sounds
and graceful forms, bright colours and lively motions, cannot fail to please
a childish heart; God meant that they should do so; and this belief, which
I would have my readers cherish, gives us an exalted idea of His goodness,
as well as his Wisdom." Sometimes these same works would explicitly call
on science, despite the lack of scientific evidence about supernatural elements. "What is a bird? It seems to us that we have not yet answered
this question properly—scientifically," the same author opined, and proceeded to detail the vertebrate, feathered, oviparous nature of birds.[65]

Bulletin 50, then, was everything that literary writings on birds were
not. Dry and impenetrable to a lay audience, it was nevertheless a tower-
ing monument to nomenclatural and systematic rigor. Ridgway could
write clearly and compellingly, and his writing and editing work showed
up in numerous popular publications over the years. But for him, part
of being a professional was to use rhetoric completely stripped of emo-
tion and filled with technical details that specialists would immediately
understand. He was certainly sensitive to language, and he admired the
work of other writers. In particular, he thought that William Brewster
had an outstanding style. Writing to his friend in 1912, he confessed that
"there is no one who, to my mind, writes so well about birds as you do. By
this I mean not only excellence from a literary point of view but, what is
of greater importance, truthfulness of description without the least exag-
geration, and evidence of a familiarity with and love for the words."[66]

Brewster responded to this praise with mixed insecurity and pleasure.
"I certainly try to make them good and true to life and when I am engaged
in the work of composition I enjoy it hugely and sometimes also feel that
I am doing fairly well but everything that I write seems stale and hack-
neyed and commonplace and unprofitable after it gets into print . . . It is
delightfully comforting and reassuring to have such praise in your recent
letter come to me from one whom I know to be absolutely sincere."[67] Such
was the blurry boundary between professional and amateur at times in
the late nineteenth century. Writing with scientific precision was an expected goal, but to display literary excellence was a high ideal, even if professional ornithologists at times scorned value-laden and emotive writing.

The story of the remaining parts issued during Ridgway's lifetime is primarily one of hard, unrelenting labor. The stultifying number of measurements of specimens for all of the parts were mostly done by the time he finished the first part. The exception was with new specimens from other institutions that came in as he worked up the subsequent parts. Clearly feeling better in the spring of 1903, Ridgway wrote Brewster to thank him for a series of specimens received. "I can truly say that never before have I been so well equipped for the work, and if I dont reach sound conclusions it surely will not be because I haven't specimens enough!"[68] Much of Ridgway's work in the subsequent parts consisted of working through reams of scientific literature to ensure that his synonymies were accurate, which meant that they were necessarily exhaustive, given how many names any one bird might have been given in previous decades or centuries. But measuring was just one of his tasks, and he had to examine many, many more birds in order to write accurate physiological and taxonomic descriptions of them.

As each part was published, reviewers scrutinized it. Some of Ridgway's distribution details about birds came under criticism, most often by the painstaking Joseph Grinnell, writing in the western-centric *Condor*. But even as he pointed out errors as to what species were and weren't known to occur in California, Grinnell concluded his review of the second part by remarking that "it is not exaggerating to aver that Mr. Ridgway is accomplishing the largest and most useful piece of systematic bird work ever carried out by one man." Part 3 was published in 1904, and Ridgway's poor financial circumstances made him feel obliged, in part, to continue. He could not afford to retire or move back to Illinois. He wrote Brewster: "I have quite lost hope of being able to make a change and have decided (or, perhaps I should say it has been decided for me by circumstances), to stick to my task until I 'peg out' [die]."[69]

The relatively long intervals between publication of each part added to the labor necessary to stay current. New species were still being discovered regularly, especially in tropical America, and any new birds to be included in the work would need to be noted elsewhere first in order to claim priority in discovery and naming. Ridgway thus published a number of separate papers in journals describing new species as the work progressed, providing diagnoses and descriptions of at least fifty new species or subspecies of birds from around the world. He continued to receive and measure new specimens as he wrote and described, and he liked to have access to at least ten examples of any one species so that he could average their measurements.[70]

Despite the publication of Part 4 in 1907, coming fairly closely on the heels of Part 3, discouragement dogged Ridgway, who felt unappreciated as well as just plain tired. The published volume, he complained to Brewster, was "work which in its final form bears little proportion to the amount of arduous labor involved in its preparation, is adequately appreciated by comparatively very few, and to the majority either utterly incomprehensible or considered 'pure foolishness.'" Despite Ridgway's laments, many of the reviews were highly laudatory—if not of the content, then of the industry involved. "One of the most exhaustive works ever undertaken by a government man of science is about half completed," wrote the *Washington Herald* in 1907 after the appearance of Part 4. "The enormity of the task is well shown by the size of a single volume." Bird journals like the *Condor* agreed upon the appearance of the doorstop-sized, thousand-page Part 4, calling the work "the most thoro of all the systematic treatises on American birds ever published."[71]

Despite his protestations to the contrary, not every waking moment was given over to *Bulletin 50*. Ridgway took subscriptions to a number of popular periodicals, including *Country Life, Country Calendar,* and the local paper. Life was not constant work. But the drudgery of Washington life began to overtake him again. He had worked at home for much of the first several years of the twentieth century, but by 1906 he was back to commuting daily to the Smithsonian from Brookland Station, a trip of some miles by trolley. Ridgway's work on *Bulletin 50* was also regularly interrupted by his other duties. F. W. True, the head curator of the Department of Biology and Ridgway's boss by the first decade of the twentieth century, asked him several times to stop his book work for several months in order to accomplish a variety of pressing in-house tasks, including organizing the surplus mounted bird collection.[72]

He and his wife yearned for a life closer to nature and continued to try to sell their spacious Brookland home in a Washington suburb, on a half-acre of land, so that they could move to southern Illinois. Their house was too large for the two of them, filled with the ghost of their dead son, and too far from the wilderness. His family and colleagues also questioned his desires to move. "When we have talked of the contemplated change we have at once been placed on the defensive," Ridgway complained to

Brewster. "I know that you can sympathize with my love for the country . . . unfortunately most people cannot even understand such feelings. Many (members of my own family included) look upon my 'notions' of this sort as a mild sort of insanity." Perhaps forgetting his awe and excitement on arriving in the city as a young man, Ridgway also claimed that he had never liked Washington or its surrounding area. "Existence here for so many years would have been impossible to me but for the saving congeniality of my work," he wrote. "I know that to dislike Washington is to many people 'rank heresy.'"[73] Though he had once considered Washington to be the center of the universe, Ridgway now wished to relocate back to Illinois, no doubt for the relative ease and quiet of rural living.

Ridgway also struggled mightily with depression for years after the sad death of his son. Although in good physical shape, he suddenly broke down in 1907 and could not continue working. He wrote Brewster: "The long and trying illness of my wife following the death of our son bore heavily on me in many ways but it was absolutely necessary for me to keep up; after her recovery the reaction came and although I have fought it to the very utmost I have been losing ground steadily and of late rapidly." Finally consulting a doctor, he was told that a prolonged abstinence from mental labor was absolutely necessary. The doctor brought a specialist to see him. "The trouble is purely a *mental* one, the result, as before stated, of hard work and worry, continued for several years," he concluded.[74]

At the time, the Smithsonian was small enough that personnel issues could be addressed directly to the institution's leaders. With these medical affidavits in hand, Ridgway persuaded both doctors to call on the Smithsonian secretary and assistant secretary to explain his difficulties. After that visit, he was granted an indefinite furlough. "Very reluctantly, and only on my assurance that the remaining work is purely mechanical, they have allowed me until the middle of June to complete the color work," he wrote to Brewster, referring to the less-taxing color nomenclature dictionary that he would self-publish several years later. "Then, I must go to a Sanitarium for at least two months—a thing which, naturally, I hate to do." He ultimately talked the doctors out of the sanitarium visit, which he said he didn't need, at least not half so much as a good long rest in the country. But the event essentially marked the end of his formal time in Washington. Off for most of the rest of 1907, working intermittently from home in DC but largely resting, he would drop by occasionally, but mostly for appearances. As he put it, those visits were "more to 'show up' than anything else, owing to crowded state of the gallery, bad light, noise, and unfavorable conditions generally."[75] Diffident under the best of conditions, Ridgway's shyness became almost crippling when he was under stress.

The tug of a more rural life in Illinois persisted, and the same year the Ridgways sold their large home, opting for a smaller rental in Washington while planning a new house in the town of Olney, where they had bought property several years earlier not far from Mount Carmel. The Ridgways were comfortable in their new city place. By that year they had a telephone, convenient access to rail lines within Washington, and other amenities, and they regularly visited with relatives in the city. They left Washington rarely, and then only to travel back to Olney. Some aspects of Robert's depression may have been situational. In late 1907, he wrote to Brewster that a recent trip to Olney had done him much good and that he'd felt as well as he ever had, working very hard every day all day long. "But unfortunately the moment my face was turned towards Washington the old feelings of discouragement and depression returned, and I am now little if any better than before I went away," he lamented.[76]

Living in the bustle of the city in a little rental in the Columbia Heights neighborhood, Ridgway missed the privacy of the spacious house, with its large porch and attractive grounds, but was still very happy to have sold it. Evvie described home life to her brother-in-law John as very quiet, though livened with occasional visits from friends, which seemed to suit them both. In her rare letters, her Quaker heart, as well as echoes of her sorrow over her son's death, comes through: "I am *so thankful* (as I look back now) that I never crossed Robt. in his desire to help and aid his parents, brothers and sisters, when necessary. A union of those who love, in the service of those who suffer, can lighten many a sad heart. Its our sacred duty to do it, when possible," she wrote to John.[77]

The days marched along, into weeks, into years. In early June 1913, the Ridgways finally made their long-anticipated move to Illinois, relocating into a newly built house. Part 6 of *Bulletin* 50 appeared in the spring of 1914, and a fresh flow of congratulations from Ridgway's friends ensued: "Volume VI is actually issued," Frank Chapman marveled. "Accept my heartiest congratulations on the appearance of this addition to the greatest systematic work on birds which has ever appeared."[78] The work was referenced and used constantly by ornithologists from the publication of the first part onward, and not just for the utility of its work. Waxing a bit rhapsodic, William Brewster wrote Ridgway in 1918 to describe his set of *Bulletin 50*, bound in fine red morocco, stating that they "are facing me, within arm's reach, now. To glance at them every now and then is always pleasing and also helpful for they cannot fail to remind me of what *may* be achieved by patient and untiring industry, although *not* by anything less than that."[79]

At least during his early seventies, Ridgway awoke each morning at 4:00, which he described as his "usual getting up time." But even a surplus of hours in the day was not sufficiently invigorating. He wrestled with just how and when to retire. On the one hand, he felt deeply committed to his work on *Bulletin 50*. On the other hand, he felt he was no longer up to the sheer physical labor. His eyesight was failing, a recent attack of influenza had laid him low, and his health was in decline generally. He recognized he was not going to complete the work in the time he had remaining, and for the first time, he began to suggest others to help with parts of the task. He suggested his trusted Smithsonian associates Charles W. Richmond, Joseph Riley, and future secretary Alexander Wetmore, as well as colleagues Edward Nelson and Henry Henshaw. He also earlier considered his friend Henry Oberholser to take up the quail family, although he had lost faith in Oberholser's judgment about subspecies at some point along the way.[80] None of this came to pass. Ridgway published Part 7 in 1916, and 8 in 1919, and it would be the final one completed in his lifetime. However, he pressed on with the work, making substantial contributions to what would be published considerably after his death.

Even as his health failed, Ridgway's star continued to rise. In August 1919, he was awarded the Daniel Giraud Elliot Medal and an accompanying honorarium by the National Academy of Sciences for his publications work. He did not attend the ceremony, although he was proud of the award and the accompanying medal and often showed it to friends.[81]

During much of her husband's work on the later parts of *Bulletin 50*, Evelyn Ridgway was in frail health. Her regular doctor for thirty-five years was Julia M. Green, a Washington-area homeopathic practitioner who would take Evelyn in under her own roof for months at a time for care. By 1925, she was living away from her husband, who remarked to his brother John: "Poor little woman; I feel very sorry for her, for, while she is fully prepared to go when her time comes and has no fear of death, she is anxious to live for my sake. I see her every Sunday, also occasionally during the week."[82]

By 1923, Ridgway's life was increasingly difficult. He tried to retain a

Robert Ridgway, ca. 1925. Special Collections and Archives, Merrill-Cazier
Library, Utah State University, USU Caine MSS8, box 4, fol. 2.

sense of focus. Although his brother John wanted him to move to California, Ridgway expressed no interest in moving, or even visiting. "I have
not the slightest desire to see it—least of all the centers of highly artificial
'civilization,' like Los Angeles and San Diego. Neither does a sight of the
Pacific Ocean appeal to me in the least, for I have seen all that I want of
it, having cruised for nearly 10,000 miles over its surface. There are only
three places on the Pacific Coast that I would care to see; the redwoods,
the big trees, and Yosemite Valley; but I would not think of making the
trip across to see all three of them." In truth, Ridgway's work on *Bulletin*
50 gave him his primary purpose to continue working and living. "Do you
know, I am now in my 74th year, and cannot expect to live many more.
The remainder of my life's work is all laid out—carefully planned—and
there will be far too little time in which to do it, and it must be done right
here."[83]

Ridgway had finished Part 8 in 1919 and would complete no more in his lifetime. Still, he continued pushing hard to work out anatomical characters of orders and suborders for the projected ninth and tenth parts. He exclaimed to Richmond in 1925 about the work that "I have never tried to solve a cross-word puzzle, but imagine it would be no more difficult than the puzzle that I have to work out!" His brother John, meanwhile, continued to pester him about moving to California so that Robert could be closer to his brother and nieces. Ridgway refused to even consider it. "It is a very risky thing to transplant too old a tree," he warned, and insisted he remain in Illinois, where he could maintain his routines—which he now clung to like a life preserver—to continue his work on the great questions of just where specific birds belonged in the vast organizational schema that constituted avian life. He knew that his task was essentially unfinishable. Without a doubt he would have agreed with a comment by one of his contemporaries in England in the early 1920s, John Latham, writing to a friend: "To make a complete ornithology is impossible—we can only arrive at a certain point, and our Successors must do the rest."[84]

His work on *Bulletin 50* also affected his views about subspecies somewhat, and he found himself recanting some of his earlier ideas about them. Writing to Alexander Wetmore, he noted that "there has been a somewhat recent recrudescence of the 'fad' in which, during 'the seventies,' ornithologists (including myself) indulged too far . . . in reducing real species to subspecific rank." Although he had long favored subspecific descriptions, he came to think of them as a necessary evil, finally penning a piece titled "A Plea for Caution in Use of Trinomials." From the vantage point of the 1920s, Ridgway realized that, as he put it, "when much additional material, from more numerous geographic areas, had accumulated and been carefully studied it was found that many forms must be reinstated as species, and so a healthy reaction took place." A recent trend back toward trinomials had begun, "the modern professors of the cult being even worse offenders than the original culprits," for reasons he extolled in the article. Trinomials should, he concluded, only be used when present-day intergradation of birds could be clearly proven.[85] A long series of subspecies named in the 1870s and the 1880s had proven, with the passing of decades, to be spurious because they were later shown—partly in the course of Ridgway's own determined work tracking down synonymies and species descriptions in the course of *Bulletin 50*—to be invalid.

In the early summer of 1927, his increasingly limited world fell apart. Evvie—in very frail health for years—died, most likely of a heart attack.

She had written out her will eight months earlier, with detailed instructions: "If possible have a Universalist minister to say a few words and a prayer. Have everything simple as possible. *Strictly private*. Do not put me on *public exhibition*. Services from your study. Only *intimate* friends to be with you. *No flowers*, only a *sprig* of *evergreen* and few sprays of berries, picked and arranged by you. Remember *I shall be near you for a while*. Think of me as being with those who have gone before, and released from this torture. Keep on with your work as long as you can. God will take care of you. Such love as I have for you will never die." Her passing was swift. "So sudden and unexpected was Mrs. Ridgway's death that I have not yet recovered from the shock—indeed probably never can . . . I was not prepared, for not more than half an hour elapsed between the first paroxym and the end," Ridgway wrote to his friend Edward Nelson. Her body was cremated, per her wishes, but she had not specified where they were to be scattered. So, "on the morning of June 5th—the first bright day we had had for a long time—I took them to Bird Haven and scattered them near where the little cottage used to stand, for I could think of no more appropriate place than the spot where she and I together spent many of our happiest and most care free days, and where my own remains are to be interred," he told Nelson.[86]

He struggled on. By mid-1927 Ridgway was essentially done with *Bulletin 50* and continued to look for someone to take over the work. Writing to Nelson, he noted that "for Part IX I have contemplated the work down to and including genera; but for part X (the concluding volume) I shall not be able to go beyond families, for my capacity for work is very uniformly less than it was a few years ago, and the groups yet to be treated consist mainly of birds too large, or too numerous in forms, to be handled here." Out of space in his modest home, far from the Smithsonian, and having left the largest, most physically demanding birds for the end of the work, Ridgway was running out of time. His health was also failing, most notably his eyesight. Reading for even brief periods of time caused his eyes to water so profusely that he was unable to continue.[87]

By 1928, the Smithsonian's assistant secretary Alexander Wetmore could call the eight completed parts of *Bulletin 50* as "recognized as standard throughout the world." Although he tried to retire in 1928, Wetmore talked him out of it, and effective July of that year, his salary was raised to $4,800—comparable to just over $60,000 today.[88] Ridgway was finally making a reasonable salary, which pleased him after years of feeling slighted financially by the Smithsonian.

After his death in 1929, Ridgway's notes on the unfinished portions of the work were assiduously collected by Charles W. Richmond, his former assistant. Richmond made some efforts to continue the work before setting it aside. Herbert Friedmann, who had been appointed curator of the Division of Birds in 1929 on Ridgway's death, began working with Ridgway's notes in 1933 in an effort to complete the project. After long delays, Part 9 of *The Birds of North and Middle America* was published in 1941, followed by Part 10 in 1946, and, finally, Part 11 in 1950. The publication thus spanned just under a half-century between publication of the first and eleventh parts.

Friedmann made it clear in the preface to Part 11 that although he had generally hewed to Ridgway's plan for parts 9 and 10, he made substantial changes of his own to the final part, deleting genera that were no longer considered valid, and generally rewriting the text. "As in the two previous parts of this series," Friedmann pointed out, "the author has felt himself responsible for the entire contents and has not considered himself as an editor of an unpublished work." The *Condor* agreed, noting that the last three parts were really Friedmann's work, built on Ridgway's plan. By the middle of the twentieth century, systematics seemed to have reached a logical conclusion. A colleague wrote Friedmann in 1947 to congratulate him on the publication of the tenth part, noting, "I know that this is not exciting work especially when it is somebody else's dead horse. It appears to be an excellent piece of work, and I know you are glad to get it off your hands."[89] The Smithsonian Institution Archives also holds unpublished drafts and notes for Parts 12 and 13, which were projected but never published.

Widely read authors like William Henry Hudson and John Burroughs forged an emotive bond for their readers, and this made them far more popular than technical ornithological writers. Amateurs often rejected more scientific approaches, while scientists considered much literary writing to be amateurish in the scientific domain. Popular writings referred to nonscientific elements and often called on a higher deity who had no particular relevance—at least in the minds of the scientists of the nineteenth century—in scientific inquiry. Even when not religiously tinged, literary writings were frequently highly anecdotal and filled with snippets of dialogue and verse. They also often pressed lyricism and anthropomorphism into service in the cause of writing about birds. Scientific writing, on the other hand, expressed specific ideas in terser and more accurate language.

The Progressive era also saw the arrival of complex but fundamental tensions between a competing fascination with, and fear of, technical science. Countless people found joy in nature—a deep personal bond that provided a more meaningful existence. At the same time, an increasingly industrial society and selfish individualism became more highly valued, squeezing nature to the margins at times. The problem, however, was not with science itself, as Kevin Armitage has noted. The nature study movement of the 1890s through the 1920s did strive for modernity while shunning narrowly technical approaches. "The scientific worldview entered the daily lives of people as never before through the union of science and technical rationality in the workplace," Armitage notes. Authority systems and regularized timekeeping—including standardization of time zones in the United States, reliable schedules for an explosively growing national railroad network, and other arguably technical elements—were an increasing part of everyday life. Avoiding these systems probably seemed undesirable to most people for the material benefits they provided. But they caused a sort of queasiness in the public mind. As Robert Wiebe noted in his classic study on turn-of-the-century industrial America, people "groped for some personal connection with the broader environment, some way of mediating between their everyday lives and its impersonal setting."[90] I contend that the technical study of ornithology actually helped bridge this gap. Even if descriptions of birds were obfuscatingly technical, they were still about something in the living world, however far removed they seemed to be from the living bird.

One important idea that is implied throughout this chapter, but which I make explicit here, is the idea that the evolution of scientific language and grammar, distributed to an audience via print, makes the evolution of scientific thought possible. That is, it is impossible to advance science without advancing the linguistic tools scientists use to discuss their work. This involves creating and using words that carry resonant meaning in a single term rather than several. "Anomalogonatous," a word found in the *Chicago Tribune*'s passage dismissing Ridgway's writing, describes birds that lack something called the ambiens muscle (a muscle that passes obliquely over the knee joint and assists in the control of the bird's toes). Not all birds have this muscle, and most parrots in particular lack it. Its absence or presence determines how they walk, how they perch, and how they fly. One word, no matter how technical, can be pregnant with meaning that would otherwise require many words to describe. It is useful because it is simultaneously compact and descriptively rich. Relatively obscure words and phrases in science, medicine, and technology conveyed (and still convey) a great deal of meaning. These forms of words, in one writer's words, "became a powerful means to condense information, convert events into objects, and minimize negotiation of ideas, thereby giving scientific writing a privileged position."[91] Science is not literature, as Ridgway noted in the introduction to *Bulletin 50*, and a *Rosa banksiae lutea* by any other name would not smell nearly as sweet.

CHAPTER SEVEN

Standardizing the Colors of Birds: Ridgway's Color Dictionaries

One of the most remarkable books of the whole world.

LOS ANGELES TIMES, reviewing Ridgway's 1912
color standards book

IF YOU DISCOUNT his relationship with the birds, Robert Ridgway had one long and passionate affair—condoned by his wife and supported by his employer. It was an affair with color, and it led to the book for which he is most widely known: his privately published 1912 work *Color Standards and Color Nomenclature*. Ridgway had a sensitivity for color descriptions of birds as young as age fourteen, when he first wrote to Spencer Fullerton Baird. "Bill pea-green; iris yellow; toes yellowish gray; claws dark-horn," he wrote in describing a bird he wanted help identifying, and proceeded to describe another twenty birds with preternatural clarity and precision.[1] His experiences making his own colors as a boy, and his interests in using color with precision to identify birds, informed his desires to create a work of standard colors that could act as a universal language.

Classification meant standardization, in large measure: to make sense of the apparent messiness of nature through placing it into a Darwinian hierarchy meant that its location was then fixed, to some degree, as was that of its neighbors. Ridgway's work in the late nineteenth century, and into the twentieth, generally focused on the value of standards for professionals. Although he was interested in the orderly classification of birds according to their presumed natural relationships, Ridgway's work also included other important aspects of color classification. His first concerted efforts on the subject appeared in 1886, when he published *A Nomenclature of Colors for Naturalists*. The first book as well as its 1912 successor were part of his ongoing efforts to create a set of color swatches with accompanying standard names, organized by a particular color system, and thus offering a language that all biologists could agree on for describing birds and other life-forms. Because of their wide applications, these color works had a significant impact on a variety of tasks, including the description of colors of birds, fungi, mollusks, mushrooms and many other entities in biological studies. They were also used widely for nonliving things such as rocks and postage stamps.

Although his works about birds, especially *Bulletin 50*, were renowned among scientists, the 1912 book was the mostly broadly used, cited by anyone desiring a standardized palette of colors. If you were a naturalist in, say, Belgium, and you wanted to correspond with another naturalist in Kansas, Ridgway's book gave you a common language to talk about colors. Instead of saying that a particular flower or bird or rock was a very, very light blue, you could describe it as Pale Glaucous Blue, as shown on row F of plate 34.

Many of the color dictionaries that predated Ridgway's 1912 work used soaring, beautiful, and sometimes strange language to describe colors, and often in multiple languages, which extended their use to an international audience. One of the most colorful, a 1905 two-volume work by the Société Française des Chrysanthémistes, went to great lengths to describe the colors shown on the unbound plates that made up the work. Havana Brown, for instance, was described as "the ordinary color of choice cigars." One rich shade of red was described as matching the blood of a freshly killed rabbit. Snow White was "the color of snow observed in diffuse light." Pale Ecru was "the most ordinary color of human skin, excepting the face and hands." Apparently fair-skinned Frenchmen working in chrysanthemum fields had highly variable coloration, and one wonders what all of the nonwhite French flower workers would have thought of Pale Ecru. Because the book was ostensibly aimed at flower enthusiasts, plant descriptions were prominent. Holly Green was described as "the ordinary color of the foliage of the common holly, viewed from 1–2 meters away, and without considering reflections." Other colors seemed somewhat arbitrary in their descriptions. Old Silver was "the most ordinary color of pieces of silver that have been in circulation for more than fifty years." And even though the work was meant for international consumption via its translations of the color names into three other languages, its soul remained French. Sky Blue, for example, was described as "the color reminiscent of pure sky, in summer (in the climate of Paris)."

Ridgway's *Color Standards and Color Nomenclature* tried to be more neutral and to reach a wider audience. It had several key innovations, including a state-of-the-art attempt to make the colors as fade-resistant as possible so that they would remain reliable indicators far into the future. Earlier works of color samples typically relied on chromolithography, which was neither pure in its appearance nor reliable in its lack of fading. Ridgway's book was simple to use, highly portable, sturdy, comprehensive with more than a thousand named colors, and did not rely on a knowledge of English because each color swatch could be identified by plate, row, and column number. He also printed an unusually large number of copies—five thousand—which allowed the book to get into wide circulation around the world. And because the book was privately published, the cost was reasonable enough that all of the copies were apparently sold.

The issue of fading was especially important as an obstacle to scientific accuracy in describing species. The books ornithologists used constantly in their work included color illustrations—but these were unreliable. As Joel Allen noted in a letter to Ridgway in 1888, "Hand-colored plates are apt to vary in different copies of the same edition. I have met with many striking instances of this." To determine the light phase of a particular species from the dark phase, or one species or subspecies from another, would be a critical identification task. But when ornithologists in far-flung locations could not compare actual specimens, they often had to rely on printed and hand-colored works. Still another reason for Ridgway's interest in color dictionaries—and in the permanence of their colors—was because some bird skins and parts faded over time. One of his own preservation techniques to prevent fading, at least of the bill, was to coat the inside of a bird's upper and lower mandibles with a thin layer of varnish. This blocked the air flow he felt was responsible for decomposition of the small amount of remaining blood left in the bill, and which was the main cause of destroying the color seen in life or in fresh specimens. Birds also changed colors after their death for other reasons, including the use of solvents such as turpentine in the course of their cleaning or preservation.[2] So establishing baselines for bird coloration was important.

Ridgway's focus on the subject did not emerge from thin air. His interest in bird nomenclature certainly influenced his desires to standardize color names into dictionary-like form. As ornithological science strove toward greater precision and standardization in the early 1880s, the issue of color nomenclature also went hand in hand with standardization of the names of birds. At precisely the same time he was undertaking research

on color systems and working to solve a variety of problems related to producing a book of standard color names and examples, he was deeply involved in the American Ornithologists' Union's newly formed Nomenclature Committee, attempting to regularize the names of birds. Ridgway's
frustrations with his nomenclatural work were many—he was fighting
with Elliott Coues, who had created a competing list of bird names in the
mid-1880s, and he was becoming increasingly caught up in the public outcry over whether birds should be called by their scientific names—or at
least to what degree such naming should saturate the wider culture outside of the scientific community. The color books were thus a productive
outlet for Ridgway's desires to forge an authoritative and agreed-upon set
of descriptive standards. Importantly, he had the color nomenclature field
almost entirely to himself—at least among his American brethren.

Another important factor also drove the work: Ridgway's interest in
the relationship between color and the geographical distribution of birds.
Attempting to correlate colors of birds from different regions was a highly
subjective task without tools to quantify those colors. A number of books
were issued forth about bird colors at the end of the nineteenth century, but what they lacked was an agreed-upon vocabulary. Several ornithologists earlier in the nineteenth century had realized, almost simultaneously, that the colors of birds changed depending on their location. To
accurately trace changes in geographical distribution of birds, Ridgway
reasoned, he needed to be able to precisely note color distinctions. Because ornithology relied on a wide network of collectors, naturalists, observers, and correspondents, Ridgway keenly felt the need for a standardized vocabulary to describe those colors. It was also true that very little
was known at the time of the causes of dichromatism, or the occurrence
of two different kinds of coloring in the same species. As Ridgway noted
to Allen in 1884, "We evidently know very little as yet of the true nature
and meaning of so-called 'dichromatism' in birds, and I anticipate interesting and important developments when our knowledge is extended."[3]
So the reach of color, and its importance to science, was a long one, and
of great interest to many.

The First Color Book

Ridgway's first color book, published in 1886 by Little, Brown, was a
relatively minor undertaking and is scarce today. The book was entitled
A Nomenclature of Colors for Naturalists, and Compendium of Useful

Knowledge for Ornithologists. Ridgway hoped to reach professional naturalists as well as amateurs. "Popular and even technical natural history
demands a nomenclature which shall fix a standard for the numerous
hues, tints, and shades which are currently adopted, and now form part of
the language of natural history," he wrote in the opening section. Ridgway
also surveyed the numerous works on color that had appeared since the
second half of the century. The popular nomenclature of colors, he noted,
was in an almost chaotic confusion through the coinage of a multitude
of new names, many of them synonymous, and still more of them vague
or variable in their meaning. He would, he explained, bring order to color
nomenclature by providing sensible names, attached to colors that had
been produced by careful combinations with other colors.

Ridgway began work on the book in relative secrecy sometime in 1881,
and by that summer he had made sufficient progress on it to show his
friend William Brewster.[4] By the fall of 1882, he considered it far enough
along to start hunting up a publisher and asked Brewster for a letter of
endorsement of the work. Ridgway hoped to use a Washington publisher
in order to keep a close eye on it, as he worried about the cost. Reproduc-
ing colors in printed form was not cheap, even with the highly mechanized aspects of most book production by the early 1880s, and publishers'
margins were often slim for books with modest press runs. The book also
involved expensive manual labor because it included hand-painted color
swatches.

Ridgway had a very keen eye toward sufficient compensation for the
work, especially after some of the frustrations of recent years in receiving
no payment for other books over which he had labored. "The work was
conceived and begun and will be finished with the bettering of my financial condition as the *main* object in view," he wrote Brewster. "At the same
time, I have used my very best endeavors to make the book worthy of the
want it is intended to fill." Brewster agreed to write a half-dozen letters for
Ridgway's uses. Joel Allen also supported the work and in fact served as
its midwife. He knew someone on the staff of Little, Brown and pitched
the book to him in the fall of 1884.[5] Ridgway then provided the plates and
manuscript to Allen, who in turn showed them to John Bartlett, the firm's
president. Bartlett, a veteran of the publishing trade by the 1880s, had also
been the author of *Bartlett's Quotations*, which was itself a kind of dictionary, so Allen probably found a sympathetic ear.

Little, Brown issued the book in early 1886. The final volume had

129 pages, with 17 plates—10 consisting of hand-colored chips and 7 of printed outlines—illustrating 186 colors, along with outlines of diagnostic features of birds and feathers. The book was a bit rough in places, despite its production by a commercial publishing house. Colors were not terribly consistent, and the brush marks from the hand painting are obvious on many chips. The legend for plate II also describes the wrong colors. The work was nevertheless a substantial contribution. The first part— "Nomenclature of Colors"—discussed the principles of color mixing, while the second—"Ornithologist's Compendium"—offered up a glossary of bird terminology, a table of English-to-metric measurements, and the color swatches. It was issued by the publisher in at least two colors—a snappy olive green cloth binding and a yellow-brown cloth binding.

In the text, Ridgway began by noting the difficulties that had arisen through the coinage of different color names in recent years, remarking that there were too many differently named colors to be useful. He also decried a new class of names currently in vogue—squishy ones invented at the caprice of the dyer or fabric manufacturer, such as "Zulu," "Crushed Strawberry," "Baby Blue," Woodbine-berry," and "Night-Green." He considered such names as "Ashes of Roses" and "Elephant's Breath" to be even more nonsensical. There was, he concluded, a complete lack of standard classification of color names, thus rendering existing terms "peculiarly unavailable for the purposes of science, where absolute fixity of the nomenclature is even more necessary than its simplification."[6]

Ridgway suggested a simple and consistent classification system for colors, in two categories. The first would consist of visible colors in the solar spectrum—red, orange, yellow, green, blue, indigo, and violet— which could then be broken into colors that were primary (not produced by mixture) and secondary (mixtures of two primary colors). The second category consisted of "impure" (that is, nonsolar spectrum colors), consisting of a variety of tints, shades, and "subdued colors." To further the book's international value, Ridgway also provided translations of his own names into German, French, Spanish, Italian, Norwegian, Danish, and Latin.

Immediate reviews were largely favorable. Allen, no stranger to the utility of color descriptions, noted in the *Auk* that Ridgway's task had "required much research, a nice display of judgement, and other qualifications which only experience and skill as a colorist, combined with critical knowledge of the requirements of descriptive ornithology, could give." Fresh from his own work on the 1886 AOU checklist and nomenclatural code, Allen also raised the issue of standards. The book did little to create standardized terminology, he noted, but this was no fault of Ridgway's, only a reflection of the lack of agreement on color nomenclature. Ridgway's other bird colleagues also applauded the book. Among other things, it could easily be used by anyone from the rank novice to the experienced ornithologist. Writing to Ridgway a year after publication, AOU official Montague Chamberlain wrote to say that "to amateurs the book will be especially welcome."[7] Ridgway mailed a stack of the books to various friends in February 1887.

But despite glowing reviews and kind comments from colleagues, Ridgway clearly considered the job unfinished. As he sent out copies, including ones sent to reviewers, he often asked the recipient for suggestions for a possible subsequent edition. The responses ranged from expanding the work to include greater utility to all naturalists by including more colors to adding such aspects as illustrations of bird markings in its scope. The hardcover book sold quickly, at a price of four dollars. Ridgway sent a dozen copies to William Eagle Clarke, the British coeditor of the *Naturalist*; they were quickly sold, and Clarke wrote asking for six more.[8]

Not all published opinions of the book were stellar, but as they trickled in over the next several years, Ridgway was able to correct several misunderstandings, as well as bring interested readers up-to-date on his most recent efforts following the book's publication. In late 1889, a letter from C. R. Orcutt of San Diego to the editor of *Garden and Forest* lamented the lack of a "complete nomenclature of colors for field use" and said that because Ridgway's book was "primarily designed for orinthologists [sic], only those colors that obtain in our native birds are shown, and many colors and shades of common occurrence among our flowers and in the lower ranks of life" were wholly unrepresented. Orcutt was no slouch naturalist; he was Charles Russell Orcutt, a pioneering naturalist in the Southwest who made major contributions in botany and malacology.

Ridgway responded in the next issue that a work showing every color in nature would be impractical. Perhaps half of the colors in the book, he added, weren't found in any North American bird, while virtually all of the colors of flowers were included. He also noted that since publication of the book, he had increased the number of colors to 411, based on his system.[9] Orcutt's critique was at least partly on the mark, as the book was clearly designed for studying birds; it included a glossary of bird terms, and all of the non-color plates related to bird or feather anatomy. But

Ridgway absorbed this critique and others, and continued to make improvements to the work to broaden its usefulness and accuracy.

The Second Color Book

Color Standards and Color Nomenclature, published twenty-six years later, proved to be another story altogether. It was a much bigger deal, and it caused great personal difficulties for Ridgway as well as proving to be a standard color identification guide for decades after his death. Although superseded by other systems, "Ridgway colors" are still cited and used today by mycologists, philatelists, food colorists, and many other workers in fields of study where precise color definitions are important. Ridgway published the book himself in 1912, and almost everything about its production and financing proved difficult. It also involved developing a new set of tools for describing and quantifying color.

The work began as a potential Smithsonian publication, and the preliminary discussions were held in 1898. Ridgway and Secretary Samuel Langley—best known today for his pioneering work on early human flight—must have first talked about the idea, given Ridgway's book on color a dozen years earlier. Langley, writing to Richard Rathbun, the newly appointed assistant secretary, described the planned opus as "a work upon the standards of color in Nature, suitably illustrated . . . with a very large number (perhaps 500 or more) of exact reproductions of tints which may be found in Nature, presented in some unchangeable material." Langley, however, was trained as a physicist and mathematician and had overseen the publication of many technical analytical works for the Smithsonian. He thus had a particular view of how the work should proceed. "I should certainly counsel the intimate union of this work with that of a physicist," he wrote to Rathbun. It would run to two volumes and would involve extensive research and testing on the institution's part. Langley speculated that the book as envisioned would perhaps be the most expensive the Smithsonian had ever published.[10]

Ridgway, however, had other ideas. He specifically wanted to create color standards to describe birds. Langley was a bit dismissive of Ridgway's background. Writing to the famous color author Ogden Rood, Langley wrote about Ridgway, "his scheme, as shown me, consists of about two hundred small tablets of colors from violet to red. These are not, as a rule, simple spectral colors nor is their composite given in spectral

Robert Ridgway at his Smithsonian desk, August 1898.
Smithsonian Institution Archives, RU 7440, box 99, fol. 67.

terms; neither do they seem to be arranged on any physical principles. The
gentleman in question has no knowledge of the spectroscope."[11] Langley
noted how he had instructed Ridgway to read up on his Rood. Langley
wished to represent the colors by their numeric wavelengths as well as by
showing a color sample, while Ridgway wanted to simply show the colors
and name them, much as he had done in his first color book. This was
ironic, because in some ways, it showed a divide between professional and
amateur from Langley's perspective: as a physicist, Langley was the professional, as he saw it, and Ridgway an amateur because of his interest in
and love of colors, accompanied by a total lack of understanding of their
optico-physical properties.

    As with the first book, the fading of printed colors over time was
of concern to all parties. Writing Ridgway in late 1898, Langley—who
was quite the micromanager in many aspects of his Smithsonian life—
reiterated: "It would seem to be a desirable thing to have not merely a scientific classification of color, but a method by means of which the color
can be accurately reproduced when the tints have faded. These tints
should be practically reproductions, if not of the pure colors of the solar
spectrum, then of combinations of spectral colors." Langley suggested
Ridgway make a preliminary analysis of the tints in question, using a mir-
ror and a spectroscope, and offered to show him the ropes to produce
the desired numbers. "The pigments will fade. The solar spectrum will
not. Your book, I hope my dear Professor Ridgway, will last as long as the
spectrum itself," Langley concluded in his letter.[12] Ridgway gamely went

along with the physics aspect for several years after it was first suggested, although he knew almost nothing of the subject, beyond the basics of the key colors of the spectrum, which he had used in his first book.

Ridgway answered several days later, taking Langley up on his offer. In doing so he noted one central reason for the color work's importance: *Bulletin 50*. "Were it not for the intimate connection between the two works—the color scheme as an aid to the preparation of the *opus magnus*—I should not have taken up the former at all," he wrote.[13] Although he seems to have been more personally fond of the color project than *Bulletin 50*, his labors with color were also a means to an end for Ridgway. He used his own color terms extensively in *Bulletin 50*, and in fact, after 1913, perhaps 80 percent of the thousands of color descriptions he used throughout each volume conform to the nomenclature he laid out in the 1912 color book. Ridgway practiced what he preached, and his use of those color terms in *Bulletin 50* have helped generations of ornithologists understand just what Ridgway was trying to quantify.

Although a hiatus of some twenty-six years passed between publication of the first and second books, Ridgway never stopped working actively on color matters, attempting the entire time to work up a more comprehensive palette. His primary difficulty in finding a publisher during that time was not a technical one (although those issues proved formidable). Rather, it was a marketing issue. Ridgway queried publisher after publisher, only to be told that demand for the work would be limited. Until late 1898, it had been Ridgway's intention to have an outside publisher issue the work, but after failing repeatedly to raise interest through that channel, he asked Langley about whether the Smithsonian might be interested in publishing it. Langley declined but noted that his own personal interest in the project was so great that he was pleased to offer Ridgway as much assistance as he could.[14]

Ridgway estimated that a press run of two thousand copies would cost about $3,000, or $1.50 per copy, with the expenses to be borne by the Smithsonian. Langley cautiously concurred, requesting only that any profits from the book be used to pay back the printing expenses. As a way of figuring out what the publication might cost, Rathbun also asked *Contributions to Knowledge* editor A. H. Clark to supply him with costs from previous works, so that Langley could figure out where the color work might fit in the strata of expenses and priorities for Smithsonian publications. Clark listed thirteen titles published over the past decade, ranging in subject matter from venomous snakes to solar coronas to the study of birds, and noted that they had entailed just over $16,000 in expenses for the institution. By far the most expensive was a two-volume work on birds, by Charles Bendire, the Smithsonian's curator of bird eggs, and the same man Coues had tried to drum out of the AOU's corps of founding members. Bendire's work had cost nearly $6,000 to produce and served as a useful financial metric.[15]

As the century drew to a close, Smithsonian administrators asked color experts for their opinions about a color standards book. Milton Bradley, who later would gain greatest fame as a maker of board games in the twentieth century, was one such authority. He had worked up a color wheel for use in elementary schools, as well as a somewhat more complex version, and sent copies to the Smithsonian for their evaluation in considering a book on color standards. "For every day experiments and practical results I prefer the High School wheel, but for *class* exposition in Universities the more elaborate machine is novel and unique," he observed. The Smithsonian ordered the high school wheel, apparently not even bothering to examine the more intricate version.[16]

Langley continued to offer Ridgway various forms of technological and financial assistance to move the work ahead. However, Langley cared little for the textually descriptive aspects of color standards, and most of his offers of help related only to the aspect Ridgway found least interesting: spectroscopic analysis and the assigning of numerically expressed wavelengths to describe colors.

Meanwhile, the steady stream of scientists and visitors passing through the Smithsonian heard about Ridgway's color work and occasionally chimed in with a letter of support asking the Smithsonian to publish the work. "Among other most valuable things which teached me very much," wrote Dr. A. B. Meyer, director of the Royal Zoological Museum in Dresden, "I saw Mr. Ridgway's colour schemes which interest me immensely." He urged the institution to publish the work as a service to working zoologists and botanists who describe species.[17]

Rathbun thought well of Ridgway and his expertise in the subject. "No one with whom I am acquainted has given so much practical attention to colors in nature," he wrote Langley. "He is an artist, an indefatigable worker, and in this line he would take particular pride in producing something that was creditable, especially since he has come to realize that the Institution is back of the undertaking." Despite his technical bent, Langley did understand that the work had artistic appeal and merit as well as scientific utility, and on Ridgway's behalf he asked a number of artists what would make the most stable, nonperishable pigments, including the possibility of using vitrified pigments. Writing to the versatile American painter and stained-glass artist John La Farge, he asked for suggestions on nonperishable pigments but also explained that the work would include numbers describing the colors' wavelengths. He explained that he was seeking a color scheme that would meet the approval of both artists and scientific men. This question for an artist such as La Farge, he remarked, was "perhaps as near to a question for the artist as we can ever get without stepping over the boundary line of pure science, and I venture to hope that you will consent to give me a hint from the artist's point of view, as to how to obtain a result which the physicist is apt so imperfectly to realize." For Langley, there was a boundary between art and pure science, and although the two could never meet, they could at least gaze at each other across a professional, if not ideological, fence.[18]

Smithsonian officers also asked other artists what approach might be best for producing an authoritative work on the nomenclature and classification of color and pigments. Robert Coleman Child, a Washington-area artist, responded to a Smithsonian physicist at length, suggesting colors' names in a variety of languages (fairly standard practice for color dictionaries); the chemical composition of the colors; notes about the purity of colors and their attendant impurities and possible details as to their manufacture; the results of a series of researches on the permanence and behavior of common pigments under the action of light and the atmosphere, including the behavior of binding mediums and varnishes; the physiology of color vision; and considerably more. There were no shortages of advice, both solicited and unsolicited.[19]

In wrangling the project, Langley seemed especially fond of committees. Perhaps this was an artifact of the way that many aspects of the physical sciences were managed, or maybe it was simply a way to ensure institutional oversight and a sharing of responsibility for the work. Langley first suggested a three-person committee, consisting of Ridgway, Dr. Cyrus Adler (the Smithsonian's librarian), and Dr. Theodore Gill (an ichthyologist who was interested in taxonomy and birds as well as fishes). Rathbun agreed that a committee was essential and expanded it to include himself, as well as William Henry Holmes, an experienced artist and colorist, who served as chairman. Holmes was also chair of the de-

partment of anthropology and director of the gallery of art, so he was a
major Smithsonian administrator as well. The committee was not to be in
charge of the work (which would have been awkward, with Ridgway as a
member) but merely advisory. Ridgway was considered the editor, not the
author; a number of other experts on various subjects would be commis-
sioned to write essays. The working title for the book was "The Constants
of Nature in Color." To ease Ridgway's labors, Langley also approved a
hundred dollars for the purchase of copies of other color works as references (to be bought only after checking the nearby Library of Congress's
holdings on the subject).[20]

Langley's interest in the project was somewhat unusual, because he
continued to offer financial and logistical support to Ridgway even after
it was clearly not going to be a Smithsonian publication. He probably felt
that there was a chance that Ridgway would eventually turn to the Smithsonian for publication when the work was done. Because the institution
had offered to cover the cost of printing, it retained some say in the development of the project. In any case, a wait-and-see attitude prevailed.
Discussions about the profitability of the book thus continued among the
interested parties. Until the end of the decade, there was much talk but
relatively little labor expended on the color book project. Work began in
earnest, however, after Ridgway had been assured that the Smithsonian
would support the efforts financially, at least to keep its options open as
a possible publisher. He and Rathbun agreed that he would take the first
half of 1899 to work on *Bulletin 50,* and then renew his work on the color
book. He began buying color samples and watercolors of many hues in
the fall of that year, obtaining them from a local art store in the form of
half-pans of oils and watercolors. He soon switched exclusively to watercolors, as he found them more satisfactory for his testing and mixing purposes than oil.[21]

As Ridgway attempted to meet the Smithsonian's desire to exactly
match spectral colors, he found that it was extraordinarily difficult to do
so, because he had to mix the tints by eye to match an existing color—a
slow and necessarily imprecise process requiring endless trial and error.
In the bird illustration world, if a tint was close to the color desired, it was
sufficient; artists worked from memory and from study skins to match
bird colors, and because plumage colors could change a bit between life
and death, and from bird to bird, there was some wiggle room. But match-

ing a particular tint exactly to another existing printed one was next to impossible.

Another problem arose, this one more political and logistical than technical. With so many hands in the project, Ridgway felt that he was losing control of the undertaking to his committee. "It looks very much as if my own identity as originator and author of the work were in danger of being lost—a contingency which I cannot, of course, view with equanimity," Ridgway wrote to Rathbun in the fall of 1900. "Indeed, I can see no reason whatever why the proposed work should be anything but my own, so far as authorship is concerned, especially since up to the present moment no material or practicable modification of my original plan has been suggested by anyone, excepting that of the Secretary regarding measurement of the standards by their wave lengths, which most excellent idea had not occurred to me, and is besides entirely practicable."

This letter from Ridgway implies that besides pushing for the use of numeric wavelengths, the committee also apparently tried to move it out of the bird domain to address broader color issues. "Any color scheme and nomenclature which shall be adequate to the needs of the ornithologist will obviously meet the requirements of the botanist, the mineralogist, the meteorologist, or worker in any other branch of Science," he observed to Rathbun. He pointed out that color variation in birds was probably greater than in any other single part of the natural world. "Certainly the range and variety of color reaches its extreme development among birds; consequently, the ornithologist requires the most elaborate treatment of the subject that could be devised," he noted. He stubbornly held to his guns regarding the specific practical use of the dictionary for bird uses, and the number of 200 colors remained steady throughout this discussion.[22] However, Ridgway would later budge, and the book as it finally appeared in 1912 would be considerably more general than his earlier book on colors for naturalists. It would also include 1,115 named colors, not 200—albeit without any numeric spectral identifiers, only a color swatch, a name, and a plate number for each page, and several small marks to delineate shades that were intermediate between the spectral colors.

But the Smithsonian leadership's desires for the book did not seem to be a deal-breaker, and Ridgway continued with the project. There was to be a chapter called "The Chemistry of Pigments" and another called "The Aesthetics of Color," neither of which appeared in the final work.

In early 1901, committee chairman Holmes also asked Ridgway to travel to New York and other large cities to consult with other artists about aspects of the book. Langley wanted the work to be suited not only to the naturalist (and the artist and the decorator) but also to color makers and manufacturers. Ridgway would, it was hoped, visit painters, decorators, color makers, and manufacturers of colored goods, "in the hope that the practical business sense of these people may result in suggestions which will be important for the work."[23]

This suggestion, however, raised Ridgway's ire. "I can scarcely conceive of anything that would be more difficult or distasteful to me or for properly doing which I am more unfitted," he wrote Holmes, insisting that he felt so strongly about it that he would rather drop the project entirely than have to travel.[24] Ridgway's shyness—always lurking in the background—was almost completely debilitating by this point in his life, and travel to a big city to talk to strangers about a nonbird subject was impossible for him to contemplate. One can almost imagine Holmes and other committee members rolling their eyes at his strenuous objection to a perfectly reasonable request. But Ridgway's frustration was also understandable; he was being asked to move further and further from his birds as a subject of investigation. The committee gave Ridgway the deference he had earned over the decades, and the request was quietly dropped.

By the start of the twentieth century, the Smithsonian's original vision for the work was much more complex, and in some ways much more ambitious, than the version eventually produced by Ridgway. The committee suggested a two-volume work, with one part consisting of text and the other of the plates, the latter arranged in a loose portfolio for convenience as a number of other earlier color dictionaries had also done. It suggested that two thousand copies would be printed, and the total cost to the Smithsonian would be around six thousand dollars, on a par with Bendire's costly two-volume work.[25]

Langley sent off a copy of the committee's recommendations to color expert Ogden Rood, who said the plan appeared to him to be "quite good." Langley also wrote to Ridgway, and the retained draft of his letter is telling, for it includes a number of elements he later thought better of and lined out. He suggested Rood as a coeditor with Ridgway of the work. He also thought that the artist John La Farge should write the introductory text, noting that "there has been no great essay written on color, that I know of, in the last hundred years." He then crossed out that sentence, no doubt thinking that Ridgway might take offense at the slight to his own essay on the subject in his 1886 work on color standards.[26]

Responding a few days later in early 1901—just eleven days before the death of his son, Audubon—Ridgway noted that the plan seemed fine, except that he felt Milton Bradley's criticisms of Rood's work might mean that he and Rood would not see eye to eye editorially on color matters. Ridgway had been deeply impressed with the apparent soundness of Bradley's articles criticizing Rood's work but wasn't familiar enough with Rood's efforts to know for sure how it would play for the two of them to work together. "Professor Rood might easily convince me that he is right and Mr. Bradley wrong," Ridgway pointed out. He then dropped a bombshell: he was leaving Washington for a long rest, and someone else would have to assume editorship of the work. With a flourish of his pen, Ridgway brought his relationship with the Smithsonian version of the color standards work to a halt.[27]

Three months later, Ridgway—back at work but fundamentally changed from the grief of losing his only child—wrote a long letter to Langley. He could, he noted, handle the inclusion of spectral numbers for colors; he had agreed to the idea years earlier. But the proposed expansion of scope of the color book to consider colors outside of the natural world was too much for him. "When, some two or three years ago, I offered my work on the nomenclature of colors to the Smithsonian Institution I had in mind, and well advanced toward completion, a very simple scheme consisting of little more than the color-standards themselves, with their names and symbols and a small amount of concomitant text, designed specially for the use of naturalists. Of course it was intended that the standards should be fixed by spectrum analyses and that the colors should be classified according to scientific principles. As you are well aware, the Secretary's ideal is far different, involving a greatly extended scope and increasing proportionately the difficulties of preparation." He would be, he felt, incompetent to handle this version of a book on color standards. Besides, he wanted to return to his work on *Bulletin 50*, which he felt compelled to complete. One week later, Langley formally released Ridgway from the color book project. Holmes and Rathbun considered asking the elderly Rood to serve as editor but felt that his age might make him less willing to take up the work. A few days into March, Langley and Rathbun decided to drop the project altogether, with or without Ridgway. The Smithsonian's involvement in the color standards work was finished.[28]

Ridgway still badly wanted to move forward on resolving the book's various challenges, and he continued working on the problem on his own. Writing in early 1903 to Rathbun, he told of some progress. He had made elaborate experiments with something called the Maxwell disk, which he had not previously used. The Maxwell in question was the great James Clerk Maxwell, who had pioneered spectral analysis of colors in the 1860s. Different Maxwell disks could have sectors consisting of any number of different colors, and spinning the wheel rapidly would make the colors appear to be blended into a single color. This allowed colors to be mixed in a way that Ridgway found invaluable. He was also able to use dyes and colors of his own making. "The experiments which I have recently made are of greater value for the reason that I have not only used the Bradley "standard" disks but also others of my own preparation, in the coloring of which I have used aniline dyes instead of the artists' pigments usually employed," he wrote Rathbun by way of a newsy update, two years after the Smithsonian's involvement had ended.[29]

Left to his own devices, Ridgway quickly abandoned the idea of using numbers to describe the specific spectra of individual colors. He was probably not philosophically opposed to their use because his work required discussing the color spectra by way of providing background to the subject, but it was not his area of greatest comfort, and it involved expensive technical apparatus he was not familiar with.

The years 1902 through 1905 were largely taken up with his work on *Bulletin* 50 and other Smithsonian duties. Then, in the fall of 1906 Ridgway took a three-month furlough from work in order to focus on the color book, although he ended up splitting his time with his labors on *Bulletin* 50. As he continued his efforts on his own in 1906 and 1907, Ridgway attempted to make his new system of colors track with his 1886 color book. He found the task extremely frustrating. "I am surely 'up a stump' on the color work, and dont know how to get down! Had a notion this morning to put the whole business in the furnace," he wrote to his friend Edward William Nelson in 1907. "The thing has so nearly driven me crazy that I haven't enough 'gray matter' left to last me to the end. The great trouble now is in matching the colors in the old 'Nomenclature' . . . If I could *ignore all previous nomenclature* the thing would be simple enough. It may seem strange, but with nearly 1300 colors to select from there are between 50 and 60 colors in the old book that cannot be even *approximately matched!* Now what am I going to do? . . . I must confess that I have really

Richard Rathbun at his desk, ca. 1900. A capable and even-handed administrator, Rathbun managed many of the administrative details related to the color book when it was a Smithsonian project. Smithsonian Institution Archives, MAH-27361.

'bit off more than I can chew.'"[30] This problem was not unlike the nomenclatural quandaries faced by bird workers attempting to reconcile new names with old ones. Earlier names simply could not be ignored, because they had been offered up to the world as valid forms, and the old needed to be reconciled with the new.

The rift between scientist and artist was implied at every turn in discussions with others about the work, beyond Langley's earlier sense of what would make for a useful publication. Many could not figure out if it was fish or fowl—was it a work on physics or a work on art or something of both? John Ridgway tried to get the Bureau of Standards in the Department of Commerce and Labor to take an interest in his brother's book, perhaps as a possible source of wide sales to a large national audience. A member of the bureau expressed some of this confusion about the nature of the work when he responded to John's inquiry, remarking that "I cannot judge from it the purpose or intent of your brother's book to say definitely whether or not it would be in conflict with the work of this bureau. Artists and Physicists have very opposing views as to color description and nomenclature. To which class does your brothers book belong—or is it apart from either?"[31]

In mid-March 1909, Ridgway took another monthlong furlough from the Smithsonian to work on the color book. At that moment he was still focused on the spectral colors. His technical research on color mixing, dye stability and other elements had progressed substantially. What he lacked, though, was funding. After being rebuffed by numerous publishers who questioned the commercial viability of the work, he had decided to publish it privately. But how to raise the money? It finally came from a somewhat surprising quarter: his good friend in Costa Rica, José Zeledón. The men had often talked about the book during Ridgway's two extended visits to the country in 1904–1905 and 1908. Writing to Ridgway during one of his furloughs, Zeledón urged him that there was no room for vacillation: "You *must* undertake the publication yourself." Zeledón then offered to pay for the bulk of the project, even though his own finances were not terribly stable. "I am not in a very flourishing financial position at present," he confessed, "but I can manage to furnish the $4500 required by using my credit." He thus borrowed the money to loan to Ridgway so the work could be completed. All told, Ridgway borrowed about $6,000 from Zeledón for the book.[32]

Ridgway understood that it took a village to produce such a work. He asked for advice from many others, as he typically did with his bird work, although most of his color collaborators had no particular interest in birds. In the foreword to the finished book, Ridgway thanked a dozen people for their help. He relied heavily on his brother John, who as an accomplished artist and scientific illustrator had fine sensibilities about

Image from Robert Ridgway's letter of April 4, 1904, to Edward William Nelson. Ridgway clipped this illustration from another source and noted, "R.R. after a day's work with the color-wheel." Smithsonian Institution Archives, RU 7364, box 8, fol. 6.

color. He especially praised John E. Thayer, a Massachusetts artist and bird collector with philanthropic tendencies. Thayer sponsored various natural history expeditions, donated collections, and wrote about birds. Along with Zeledón, Ridgway's highest praise is reserved for Thayer, even though his specific contribution is not noted. Because Thayer had inherited great wealth, he likely provided a subvention for some portion of the cost. Thayer and Ridgway may not have known each other well, as scant correspondence exists between them, but something in the work caught Thayer's fancy and elicited his support.

Free to push forward as he desired, Ridgway collected thousands of samples of named colors and studied a number of notable prior works. The book's production involved numerous technical difficulties. Ridgway wanted to produce a work with color plates that would retain their original hues, as he recognized the utility of having colors remain reliable between copies of the book, regardless of use, exposure to sun, or age. He rejected a variety of unstable pigments for his book, including most of the family of aniline and coal-tar dyes, as well as cochineal lakes (bug-derived pigments formed from an organic coloring matter fixed on a base that is usually inorganic), because they all tended either to fade or darken over time. Ridgway also spent considerable time obtaining sample dies that would not fade. German dye manufacturer Gruebler sent him one gram each of ninety different coal-tar pigments, donating the entire amount, for his testing purposes. Coal-tar or aniline colors were especially pure, and although some faded readily, others did not, so testing individual colors was critical. Dyes of vegetable origin, such as gamboges, violet carmine, and indigo, were all highly susceptible to fading. Despite all these efforts, Ridgway was forced in some cases to use more sensitive pigments and realized that the color samples were, in any case, not fade-proof. "CAUTION!!!" begins a note at the start of the plates. "Do not expose these plates to the light for a longer time than is necessary." My own examination of dozens of copies of this book in different parts of the world, however, shows them to be remarkably similar in their coloration from copy to copy—evidence, perhaps, that at least when the book was closed, the pigments did not fade.

By the end of May, Ridgway was ready to get the work into print. Several firms made serious offers or expressed substantial interest in the production work, including several German firms such as R. Friedlander and Son in Berlin and an unnamed English firm, but serious difficulties in getting the exacting preparations done for the color plates held up the process. Ridgway finally settled on the venerable Baltimore printing form of A. Hoen, an innovative company that was also the oldest lithographic house in America. The firm would have been already known to Ridgway; it did a wide variety of work, including projects produced for the federal government such as a series of cartography prints done for the US Geological Survey. Originally located in Virginia, it was a major supplier of lithographic products for the tobacco industry, among others.[33] It certainly didn't hurt that owner Albert B. Hoen was a laboratory chemist as well as a lithographer and that the firm also employed a talented colorist named Frank Portugal, who had a deep sympathy for Ridgway's work. Hoen had also done months of intensive experiments on Ridgway's behalf, at no cost, and Ridgway considered him a friend. Chromolithographer Julius Bien had also been suggested to Ridgway as a possible printer. Bien had done, among other works, the 1860 chromolithographic elephant folio edition of Audubon's *Birds of America*. But Ridgway did not feel that chromolithography was the solution to his problems of fading. "I know very well of Bien's reputation for some work in ordinary lines of chromo-lithography," Ridgway remarked, "but do not believe he could equal Hoen's work in this particular line."[34]

Ridgway requested a bid for a print run of five thousand copies—more than double the two thousand the Smithsonian had originally proposed—issued in a small octavo format, and each with 63 plates of 21 colors for a total of 1,323 colors. Each page had a row of pure white at the top and a row of pure black at the bottom to serve as index colors, and they weren't included in the total count of colors. (The finished work would have 10 fewer plates—53, with 21 colors, for a total of 1,115 colors, 2 of the index colors on the final plate being replaced with the actual colors "Black" and "White.") Stripped to its essence, it made use of 36 hues, each spaced approximately the same distance apart on the spectral chromatic scale. Once a satisfactory bid was in hand, Ridgway was ready to draw up a contract and have the work started immediately.[35] Hoen proved reliable and was the final printer of the work. The company was apparently in it for the long haul; some copies of the books have printed slips from Hoen and Company dated as late as the 1940s, noting the date and name of an examiner looking at the copy—presumably to check if it had faded.

Even in the midst of these plans, however, Ridgway corresponded with commercial publishers, keeping his options open as long as possible. Notably, he sent book magnate Henry Holt the manuscript and some of the original plates, as well as a copy of his 1886 book. He had corresponded with Holt in earlier years, but Holt had been lukewarm as to the book's prospects. But in the fall of 1909, not long after Ridgway had secured funding from Zeledón, Holt reemerged, now extremely eager to publish the book. He suggested publishing it as part of Holt's American Nature Series and made a point of visiting Ridgway at the Smithsonian that September as well as writing to him. Holt wrote that he would take a commission of somewhere between 10 percent and 50 percent on each copy. A charismatic raconteur, he corresponded with Ridgway for about a year, trying to talk him out of self-publishing. "I wonder what your friends know about publishing," Holt remarked. "Perhaps I know a little something about it, and I do know that generally when an author undertakes it at the advice of his friends, or of his own instance, he makes a terrible mux of it, mechanically and financially." Holt felt that Ridgway kept equivocating and observed, "I will go into it if you care to have me. But if you don't, for pity's sake don't force yourself to, for a reluctant client is as bad, I suspect, as a reluctant wife." At last, with funding from Zeledón in hand, Ridgway made up his mind to issue the work himself.[36]

Had Holt and Company—which would eventually become publishing giant Holt, Rinehart and Winston later in the century—shown interest in the project earlier, the book would have had a very different trajectory, because Ridgway would have been saved an enormous amount of labor and expense in getting the book colored, printed, and distributed. However, he would have also been required to again bend to someone else's will regarding the intent and audience for the book, and despite the hardships, he felt great pleasure in doing a publication exactly his own way—perhaps for the first and only time in his life.

No one works in a vacuum, and Ridgway was concerned about being beaten to the punch. He was especially worried about the appearance in 1908 of a French volume by Paul Klincksieck and Theodore Vallette entitled *Code des couleurs,* which California ornithologist Joseph Grinnell considered a better work than Ridgway's 1886 book. Grinnell put the squeeze on his friend in the summer of 1909 by commenting on how ratty his own copy of Ridgway's old book had become and what a great need existed for a new manual of colors, now that Ridgway's was long out of print. Such a work, noted Grinnell, would need to be published by a public institution that could bear the cost. "Here is an opportunity for some one properly situated to do systematic naturalists an invaluable service," he concluded, obviously referring to Ridgway and the Smithsonian. He then put his brother, Fordyce Grinnell Jr., up to writing a letter to the editor for the next issue. In it, Fordyce responded to the earlier piece and described the *Code des couleurs*. "I think it is a *better* book than Ridgway's," he wrote. "The colors are on heavy paper, and I think the book is more durable than Ridgway's." Fordyce also liked the clear numbering schema for individual colors: "There is nothing, I think, advantageous in writing *Van Dyke Brown* instead of *Orange 118*," he observed.[37]

Ridgway wasn't terribly amused by being served this oblique public notice to get his act together and get a new edition out. As he sometimes did when defending himself, he claimed that no details were necessary because he'd assumed the word was circulating that he was working on just this topic. But because the *Condor* had already said as much, he noted that it seemed proper to formally announce completion of the laborious work begun so long ago. Ridgway thus got in a little free advertising for his book. He described the expanded nature of the forthcoming volume, including one of its innovations: the ability to determine intermediate colors that fell between the actual colors shown in the plates. This would allow for some 5,300 colors to be identified.[38]

But more than just the Grinnell family rattled Ridgway's chain; others also mentioned the *Code des couleurs* around the same time. A University of Michigan–based morphologist named Jacob Reighard also thought that the *Code*, with its 720 color swatches, filled a notable gap in the literature. Lamenting the fact that Ridgway's 1886 book was long out of print, and echoing Fordyce Grinnell Jr., he suggested that users simply refer to the code accompanying each color in the French book. This would act like a telegraphic code, he noted, "by means of which men of different nations and professions may intercommunicate without risk of being misunderstood."[39]

Other potential European competitors also concerned Ridgway. "I will immediately send a note to the *Ibis* and the *Journ. für Ornithologie*, announcing the completion and early publication of the work so as to forestall, if possible, the English work to which Beebe refers," he wrote his brother in 1909.[40] He also sent out announcements in the next two years to American journals, including the *Auk, Condor, Ibis*, and the mushroom journal *Mycologia*. With the last journal, at least, Ridgway's publicity helped to head off a competitor. The American Mycological Society had begun to prepare its own publication for field use with mushrooms but scrapped these plans on hearing of Ridgway's new work. *Mycologia* also assisted Ridgway in heading off the French *Code des couleurs*, by describing it as too bulky and "entirely unsuited for field work."[41]

Despite his earlier disdain of colors that he claimed had no particular meaning, such as Elephant's Breath, Ridgway's selection of colors reflected some decidedly personal choices, and a few fanciful ones of his own, such as Dragons-blood Red. Others, such as Wall Green and Empire Yellow, had no obvious tie to anything. A number of shout-outs, however, are buried in the names. For instance, color theorist Ogden Rood, whose work influenced Ridgway's own and who consulted with the Smithsonian about the book, shows up in four names (Rood's Blue, Rood's Violet, Rood's Brown, and Rood's Lavender). Milton Bradley appears in Bradley's Blue and Bradley's Violet. Chapman's Blue is certainly a nod to Ridgway's friend Frank Chapman, whose own use of color in twentieth-century field guides was a notable innovation. Prout's Brown is an homage to Samuel Prout, a landscape colorist who wrote on color theory. Leitch's Blue saluted Richard P. Leitch, a painterly contemporary of Ridgway's. Some were probably family references, such as Deep Quaker Drab, Pale Quaker Drab, and the less-than-rousing Pallid Quaker Drab; his parents and his wife were Quakers, and one wonders if those choices were a criticism of the religion or simply an attempt at naming quieter colors. Vanderpoel's Blue, Green, and Violet all honor Emily Vanderpoel, who in 1902 published *Color Problems: A Practical Manual for the Lay Student of Color*, which would have been useful to Ridgway. He would have also shared her sensibilities about appealing to a broad audience. Previous works on color, Vanderpoel noted, had been too theoretical or too artistic to benefit "larger classes of people," and this would have aligned with Ridgway's own views of science and art.[42]

Many of these color names remain in wide use today, although attributions to Ridgway's work, other than sometimes noting the particular color was first used in 1912, have been lost. Skobeloff Green, for instance, a deep rich shade of spring green, is widely used today in interior design. Ridgway's choice of this particular word is a mystery, because although it is a known proper name, it has no known associations with color theorists, artists, or with anyone remotely related to his wide circle of colleagues and correspondents. A famous Russian general bears that name, but Ridgway never evidenced any interest in military history. Was it an acronym or anagram for something else, or an inside joke? Others intrigue, such as Pleroma Blue. The word "pleroma" seems to refer to a Gnostic idea related to the totality of divine powers. Its use was chiefly theological, and *The Oxford English Dictionary* describes it as "a state or condition of absolute fullness or plenitude."

In early 1913, the printed book was finally in his own hands, and shortly in those of his colleagues. It was priced at eight dollars, all of which went to Ridgway. Some of his correspondents considered the completion of the color book a matter of national pride. "Your task is a stupendous one . . . reflecting infinite merit not only on you personally but also on this country," Brewster noted. Writing in the *Condor,* Harry Oberholser, who worked in a birdy capacity for the federal government for decades, described the book as "the most important contribution to the subject of color standardization that has ever seen light." Although Oberholser had an almost sycophantic admiration of Ridgway and his descriptive taxonomic work, he was very well known, and his review would have helped to spur sales. Some larger commercial outlets certainly helped matters as well. The *Los Angeles Times,* in a 1913 Sunday review, called the color book "one of the most remarkable books of the whole world." It wouldn't be a best-seller, the article continued, because it wasn't really written for wide popular consumption, "but it is a volume that has been looked for by the scientists of the world for a long time."[43] The *Times*'s Sunday circulation was then about a hundred thousand, and would have reached a large readership on the West Coast.

Because he was self-publishing the book, Ridgway didn't have to worry about an intermediate editor. He was thus able to make tweaks right up to the printing of the actual pages. He finalized the color names late in the editorial process, doing so only after proofs of the plates had been printed and were going through the inevitable rounds of corrections and fixes. Ridgway described his work on the color books as "the most difficult work I ever undertook," but while engaged in the last throes of the book he was also trying to make steady progress on Part 6 of *Bulletin 50.* The color book work and his systematics work occupied fifteen hours a day for many days on end.[44] Part 6 of *Bulletin 50* was published in 1914, hard on the heels of the color book.

By the spring of 1913, *Color Standards* was racing off the shelves in the Ridgways' house and to buyers around the world. Though not widely reviewed in the popular press, it was helped along by notices in bird jour nals, such as the *Auk*'s write-up, which described the book as "about as nearly perfect as it can probably be made . . . That it will long remain the standard work of its kind goes without saying." Reviewers noted how well-suited Ridgway was to the task. "Probably no one in the country is better qualified than Mr. Ridgway as he combines the artist's knowledge and appreciation of color with a larger experience in matching colors in nature and a keen perception of minute differences in color tones," the *Auk* wrote.[45] Once again, all of those years studying small color differences in the course of differentiating species and subspecies had proven useful. *Science* magazine also published a detailed review in mid-1913 by William Jasper Spillman, a physicist and botanist. Spillman, already famous as having independently rediscovered Mendel's laws of genetics, understood the sacrifices Ridgway had endured and would have been more sympathetic than most to the challenges of merging physics with the natural sciences. "The task has been an enormous one, involving much pioneering in a little understood field of science," he noted, and his discussion of Ridgway's approach sounds much like that of the early geneticists working with thousands of hybridizations: "Many important problems had to be solved before the work reached the final stage, and a vast amount of work had to be done over and over again as advances were made." More important, though, Spillman the physicist was the only reviewer of the book who competently grasped and explained exactly how the schema worked and what method Ridgway had used. His detailed description of Ridgway's approach, running to four pages, provides the best surviving description of the painstaking nature of how the colors were selected.

In fact, he explained parts of the finer details of Ridgway's approach more clearly and accurately than did Ridgway himself. Ridgway had described the Maxwell color wheels used to create mixtures of colors but had not clearly explained his complicated methods, nor had he clearly described how to determine the colors that lay between the 1,115 colors shown in the book. Ridgway used both a chromatic scale, which was represented by three color swatches running along each row of each plate, and a tone scale, which was illustrated by the swatches running vertically. The system of identifying each color variation by means of symbols that indicated the location of the color in the scheme of classification was the most distinctive and original feature of the book, in Spillman's opinion. Its complexity was clear, even when simplified. "Each of the 36 segments of the fundamental series is designated by the odd numbers 1, 3, 5, etc., to 71, the even numbers being reserved for colors intermediate between them. These same numbers modified by the use of primes are used to designate the same colors modified by the admixture of neutral gray. Thus the color designated as 27′′′ is number 27 of the fundamental series weakened by the admixture of 77 per cent. of neutral gray (third series of broken colors.)" The book would probably not be the final solution to the problem of color standards, he concluded, but it was doubtful that anyone would soon top Ridgway's solution.[46]

The books arrived from Hoen in small lots and at irregular and sometimes long intervals. This occasionally made filling orders in a timely manner difficult. Evvie Ridgway was the business manager for the operation, taking in the money, wrapping the books, and getting them out the door. The firm of John A. Roebling and Sons, builders of the Brooklyn Bridge, ordered one copy and then ordered ten more. Ridgway sold many copies of the book to other publishers and book dealers, who in turn sold them to individuals or other firms. He sold at least fifty copies of the book in one shipment to William Wesley and Sons, a London book firm. Another unnamed New York firm bought twenty-five copies.[47]

The Ridgways' permanent move to Illinois in 1913 filled them with delight. For the first time in decades, they had no fixed address in Washington, although Robert sometimes returned to the capital on Smithsonian business. He and Evvie had lived in Washington in at least four locations, including for many years their house in Brookland, then moving three times in quick succession to a series of boardinghouses while selling their Brookland home. Not long after they arrived in Olney, however, southern Illinois experienced its worst drought in thirty-two years and the most incapacitating heat to date. Within two weeks, Ridgway was complaining about the occasional raw and disagreeable weather, which dipped down to freezing at least once during the fortnight after the hot weather ended. But he and Evvie were happy there, busy with rural tasks. They named their new place Larchmound, after two enormous larches growing at the front of the property. Judging from photographs, it was a spacious and elegant two-story home. Situated on eight acres, the large property, which had cost six hundred dollars when they purchased it in 1906, afforded both Ridgways the chance to commune more closely with their beloved nature.[48]

The Ridgways took the color books to Olney with them and continued to service sales from Illinois. The printed books consisted of the front matter and the printed pages but without the color swatches, all bound in a green cloth binding. The couple assembled the books as the orders arrived. Evvie cut the color swatches herself from the large printed sheets of the colors and painstakingly glued them down into the bound books. Inevitably, problems arose from adding the small swatches, which measured an inch by an inch and a half each. "I have been constantly bothered and worried by my color book, which has taken time I had hoped to devote to out-of-doors relaxation," Ridgway wrote Brewster. He often ended up mailing out repaired copies to replace damaged ones that had unwittingly been mailed out before errors were caught. "Owing to carelessness in cutting the colored papers and mounting them, every copy sold has required from an hour to half a day of my time in tedious retouching, replacing marred colors, etc. For this reason I have found it impossible to keep ahead of orders," he complained.[49]

Perhaps frustrations like these led to Ridgway's considering a commercial publisher to handle the assembly and distribution of the printed copies from the parts he and Evvie took to Olney. He investigated having the color book commercially published as late as a full year after its appearance. Frank Chapman, curator of birds at the American Museum of Natural History, made investigations on his behalf, and suggested the Macmillan Company. "Your book should be accepted not only as a standard by naturalists, but everyone having to do with color, whether in the arts or commerce; consequently, my first choice in this country would be the Macmillan Company, since they have a house in London and also in Australia, and of course connections everywhere." Although nothing came of it, Ridgway continued to try to increase the book's prominence and financial success through any means possible.[50]

Although the book did not sell out immediately, it is not known when the supply was gone (although in 1953 Herbert Friedmann, the head of the Bird Division, noted to a descendant of José Zeledón's that "the entire edition has long since been exhausted"). An order in 1921 from an employee of the Niagara Lithographic Company for three copies was promptly filled. As recipients of the work got their hands on copies in the years after publication, the work continued to be praised. "Since my connection with this company, my associates have manifested a great deal of interest in 'Color Standards,'" wrote R. H. Howland, the Niagara Lithographic employee, to Ridgway. "In fact my copy of your book is used continually, and is here considered the finest color book published." Coming from a commercial producer of lithography, this was high praise indeed, and would have delighted Ridgway. In any case, nearly a decade after its publication he apparently had no shortage of copies on hand. In a later letter, Howland—also an admirer of birds—noted that if Ridgway ever produced another edition of the color book, his own firm would be glad to bid on the job. Howland suggested lithography in place of Ridgway's tinted rectangles of paper and observed, incorrectly, that the lithography would not be susceptible to any fading. Ridgway must have found this of some interest, because the letter is heavily marked up in his hand.[51]

Despite the book's success, the weight of the loan from Zeledón hung over Ridgway's head; he greatly disliked carrying an obligation of any kind. In 1918, after a sudden revolution in Costa Rica in which one of Zeledón's brothers was imprisoned, the debt suddenly came due. Zeledón now needed the money repaid to him due to his family's political difficulties in Costa Rica.[52] The Zeledón family was disliked by dictator Federico Tinoco, who briefly usurped the presidency in 1917. This takeover was apparently the trigger for Zeledón's request for repayment. Tinoco was deposed in 1919 and political stability returned, but not before the family feared for their property and safety.

Ridgway paid off nearly half the debt in several months, trying and failing to sell his home in Olney to cover the remainder of the cost. "To tell the truth, for more than a year I have been ill, both in body and mind, mostly from discouragement over the financial outlook for the future," he wrote to his brother John in 1918. "That miserable color book has proven a veritable disaster, for it has involved me in difficulties that I may never be able to surmount." Ridgway was no financial wizard, which didn't help his situation. Alexander Wetmore, the assistant secretary of the Smithsonian, wrote to Ridgway's biographer Harry Harris in 1926 to note, "He has devoted himself so assiduously to his science that he has lost all business sense so that any investment he has tried to make has almost universally proved unfortunate." Ridgway would echo that sentiment in a letter to Charles W. Richmond the same year, observing that "mathematics in any form have always been more or less incomprehensible to me!"[53] To compound his concerns, the threat of a congressional vote mandating federal retirement by age sixty-five hung over his head, which, if passed, meant that—because he was sixty-eight that year—he would immediately be forced into retirement, and on a pension considerably lower than his salary.

Near the end of his life, Ridgway knew that the book had become very popular, and in 1927, he was pressed to issue a new edition. "This matter of a new edition of 'Color Standards' is a hard proposition for me," he wrote to his friend Casey Wood. More than ready to retire, he was peeved. "I had been hoping that I was *done forever* with color work, but Colonel Behrens seems so anxious to get out another edition and is so persistent that I am really seriously perplexed. I would not, under any circumstances, undertake to do any work in connection with a new edition . . . His offer of $500.00 is so ridiculously small that I hardly knew what to tell him."[54] Lieutenant Colonel T. T. Behrens, a member of the British Royal Engineers who had been introduced to Ridgway by Wood, was apparently wildly gung-ho to reprint the work. Nothing came to pass of the effort, but Behrens's enthusiasm is ample evidence of the book's popularity years later.

Others coopted Ridgway's system for their own uses. The Division of Geology and Geography of the US. National Research Council published a color chart for the field description of sedimentary rocks based directly on his system. Philatelists also made heavy use of Ridgway's colors in attempts to create a common language for the shades and tints used on collectible postage stamps. Although other works existed, such as B. W. Warhurst's 1899 A *Colour Dictionary . . . Specially Prepared for Stamp Collectors*, with its two hundred colors, assiduous collectors needed finer distinctions. Telling slightly different tints apart on stamps could be crucial to their identification and thus to their value; a common stamp could be worth only pennies if it was one particular shade of blue but valued at thousands of dollars if it was a different shade. That was because different inks could represent errors in printing or reflect rare variations due to very small print runs and other factors. The reliability of the colors in Ridgway's book was thus especially important to stamp connoisseurs. Richmond was interested in philately and corresponded with serious collectors who used Ridgway's book on color standards for solutions to a variety of color standardization problems. His correspondents included Stanley B. Ashbrook, one of the twentieth century's greatest experts on postage stamps. Unable to get his hands on a copy of Ridgway's book, Ashbrook sent Richmond a batch of thirty-one copies of an 1851 stamp to have him name the stamps' colors according to the color guide.[55] A box of cigars was more often than not the standard form of thanks for such help.

The color book also brought other inquiries to the Smithsonian. People who knew of Ridgway's book but did not have a copy often wrote to the institution to ask if they would perform color identification and matching on their behalf. Staff members usually obliged. Their primary technique was to cut a rectangle out of a piece of white paper and mask off the color strip in question. They could then compare the item at hand while avoiding any influence from a neighboring color or shade in the book.

The legacy of the book is better understood by scrutiny of just how often it was cited in subsequent years. The use of the personal computer to display and compare colors began to make printed color dictionaries less relevant after the mid-1980s. However, a study of Ridgway's color work done in 1985, relying on a search of an online database called the *Science Citation Index,* found that the 1912 book had been cited 82 times in the scientific literature between 1974 and 1977; 52 times between 1978 and 1980; and 53 times during 1981 through September 1983.These 187 citations were found in fifty-three different journals. Ridgway's color book was cited far more times than works by three contemporaries of his: Elliott Coues, described in detail in previous chapters; Steve Forbes (with whom he had coauthored *Birds of Illinois*); and Thomas Hunt Morgan, the Nobel Prize–winning geneticist who was easily the most famous of the four men. None of their works have been cited nearly as often as Ridgway's 1912 color dictionary, and in fact, the number of citations for any one of their publications is at least ten times less than that of *Color Standards.*[56] Its users spanned the physical and natural sciences, and although Ridgway had strongly resisted the more arcane physics aspects of the work that the Smithsonian tried to press on him, he had clearly absorbed and grasped enough to apply it with great competence to the final product.

Ridgway wanted his color work to standardize the use of color nomenclature. However, in the case of his color books, the issue of his authority was stood on its head. Although he was an unquestioned expert on bird matters, in the case of his 1912 color dictionary he was an amateur compared to the physicists who desired more technical details in the work when it was under Smithsonian influence. The encroachment of physics into natural history made even the most competent naturalists feel like amateurs. Furthermore, Ridgway's decision to use common-language names for colors meant that his terms included emotive elements, especially in cases where homage was paid to others through those names.

But this wasn't necessarily a bad thing. Aspects of the color work also let Ridgway relax and be less of a technician, despite his meticulous care in forming the color system he used for the book. Because he was able to create names for his colors without regard to any existing code of nomenclature, the names gave him a way to be lyrical at times—aesthetics in very small bites that didn't require a long soliloquy on the wonders of nature. Ridgway could have his natural beauty cake and eat it too, but in a very calculated way. It was (and still is) a common misperception that "popular" is somehow equal to "imprecise," and Ridgway's color books show just how incorrect that view is.

The color work also put Ridgway in a new position as an amateur among professionals. As a technical ornithologist used to arcane concepts, Ridgway was, for the only time in his career, on unsteady ground when faced with the physics aspects of the color work. A comparative look at astronomy, another discipline involving physics, is instructive here. In astronomy there were also at least two distinct camps. In that field, professionals were generally identified as having ties to a research institution, an advanced education in a related topic, and access to institutional funding. Amateurs, working with their own telescopes in their backyards and almost no outside money, not only lacked these elements but also strenuously resisted some of them in their work. Often, however, they lacked necessary knowledge. Some knew nothing about physics or had a faulty grasp of the principles of geometrical optics. Others had a particular interest (say, solar astronomy) but knew nothing about deep space, and thus lacked context. Just as ornithology began to see dramatic change in the professionalization of the field, so too did astronomy, which—like ornithology—rapidly grew more technical and saw the rise of organizations specific to the discipline. But just as amateur ornithologists resisted more technical innovations, so did amateur astronomers. Those with their own interests and little formal training could use the unaided eye or a minimum of instrumentation and thus rely on direct observation. In turn, this approach could produce reliable knowledge. Perhaps more than anything else, amateurs found the work satisfying. Theirs was not work for science as much as it was for the satisfaction of glimpsing the universe's mysteries. "Observation need not be filtered through the analytical medium of physics or mathematics," as one historian of astronomy has noted. And so was it for amateurs in the bird world.[57] It was what was done with that knowledge that was contested: how it was used, how it was coopted, what others thought about it, and what they said about it to a wider world.

Epilogue

To make a complete ornithology is impossible—we can
only arrive at a certain point, and our Successors
must do the rest.

JOHN LATHAM to T. W. Chevalier, May 4, 1923

ROBERT RIDGWAY succumbed to heart failure on March 25,
1929, at eight in the evening, at his home, Larchmound, in Olney,
Illinois. His sister Lida Palmatier, visiting from San Diego, was
with him when he died. His eulogy was given several days later by a local
Olney Presbyterian minister, John B. Farrell, who noted Ridgway's heroic
singleness of purpose above all things. "He literally gave his life to this
study, not primarily out of love for the science, but out of love for nature,"
Farrell noted. The second part of this statement was not very accurate;
Ridgway certainly loved nature, but science was much more on his mind
during his career, and he often felt an acute schism between the two. As
a minster, perhaps Farrell felt obliged to attribute a specific aesthetic to
Ridgway's life work. "There is so much of poetry, and of romance and
beauty about it all," he continued, which would have pleased Ridgway's
wife and perhaps simultaneously annoyed him. In any case, the ceremony was a simple and solemn affair. His colleague Stephen Forbes, with
whom he had written the two-volume history of the birds of Illinois, was
in attendance, as were his assistant Charles W. Richmond of the Smithsonian's Bird Division and many others from Illinois. A single cardinal in
an elm tree outside the open window of Ridgway's study fluted a loud,
clear, sweet song during the eulogy.[1]

Ridgway was buried in Bird Haven, the same location where his wife's
ashes had been scattered—a nearby eighteen-acre tract of land preserved
in his name by a large group of fellow ornithologists. He had developed
Bird Haven as a sanctuary for birds, as well as for experimentation with

the cultivation of trees and plants not native to the region. The securing of funding to purchase the land in perpetuity occupied a good deal of his last several years, as well as the energies of friends and family.

The press commented widely on Ridgway's life and career. One obituary noted that Ridgway's extreme shyness—his "objection to publicity"—amounted almost to eccentricity but that he nevertheless had a wonderful sense of humor. He was a great scientist and a staunch friend, and, the obituary continued, his ornithological work would live on "long after those who have enjoyed the charm of his personality will have become forgotten memories." The *Ibis* noted that he was without a doubt the most widely known American ornithologist in both his country and abroad.[2]

By the time of his death, Ridgway had become an anachronism: ornithology had progressed steadily forward in its study of bird behavior and environmental interactions. Even at the dawn of the twentieth century, his unwavering focus on anatomy, taxonomy, and classification in his publications meant that he was growing seriously out of step with broader trends in ornithology that related to the living bird. But his obsessions paid powerful dividends for science. Birds are now the most-studied and best-understood class of vertebrates, and Ridgway's contributions to this body of knowledge were essential to our modern understanding of birds. Ridgway also definitely considered himself an amateur in some aspects of science, even occasionally styling himself as such in print; one of his pieces in *Garden and Forest* was an 1897 four-page article entitled "An Amateur's Experiment." In it, he noted his efforts in planting trees to serve various purposes on his Brookland property. Although he knew a lot about the subject, he was no professional botanist, and he was glad to claim amateur status. And despite the intensity of his technical work, he never lost his love for birds in nature. "A bird lover he was born and a bird-lover he died," wrote ornithologist Frank Chapman in a piece about Ridgway for *Bird-Lore* shortly after his death.[3]

This book has tried to make sense of the competing notions of professionalism and amateurism through three elements—classification, language, and accountability—as a way to better understand the distinctions between the "two fires," as Joel Allen described the respective interests and passions of amateurs and professionals during the founding months of the American Ornithologists' Association. Trying to understand the evidence, as well as the mechanics, of evolution was extremely challenging work, as it repaired a methodic grasp of a process born of scattershot variations. Evolution is purposeless but efficient. It has no long-term goal. Tracing the diffuse influences of Darwin's work on these three topics is an elusive task, but their legitimacy and effect—as stated by the practitioners themselves—were undeniable.

The study of birds was in no way a monolithic undertaking late in the nineteenth century. The distinctions these three elements created were trends more than they were hard-edged categorizations, to be sure. Approaches to biology—both in museums and pedagogically through universities—were diverse. The casually interested bird observer stood at one pole and the hyperexperienced and privileged technical ornithologist at the other. In between them was a broad and contested territory, and where people were arranged on this continuum was the result of prejudices and attitudes about class and gender, resentments, financial well-being, insecurities, and situational abilities more than any rigid categorical evidence of "professional" or "amateur" status.

One criterion that marked the changes discussed in this book was that of location. Amateur ornithologists, such as those writing in the *Ornithologist and Oologist,* were most concerned about what something looked like: a nest, an egg, a particular bird found in a specific locale. But to the professional, the question of where a bird *fit* was of vital importance. How did it relate to other examples of the same species, and how could it be used comparatively? The specific uses of bird collections in the last quarter of the nineteenth century thus take on a special significance and bring us back to the birds themselves. Scientific practice—analysis, comparison, and writing—during Ridgway's time, though it might have appeared to be disconnected from birds and thus often disparaged by amateur naturalists, was in fact more tightly connected than ever to the actual specimens: not live ones, to be sure, but birds nevertheless.

As ornithology grew in complexity and further changed to emphasize the living bird after Ridgway's death, amateurs began to play an increasingly significant role. They also received a more sympathetic hearing by some prominent scientists. Ludlow Griscom, one of the deans of twentieth-century American ornithology, noted in 1929, two months before Ridgway's death, that unfortunate tensions had been set up by the "us versus them" aspects of the amateur-professional trope.[4] His own association with amateurs in both ornithology and botany, he observed, had been far more beneficial to him than it had probably been to them. Perhaps professionals got paid and amateurs did not, but it was apparent, he noted, that as far as ornithological excellence was concerned, there was no real distinction between the two. The quality of ornithological science available to a wide audience had risen to such a degree that the divide between the two groups had shrunk. It didn't follow that the men in paid museum positions were the best at their chosen profession. Griscom did feel that there was still a distinction between the two groups, though, and he suggested a new metric to distinguish them: time. Those working during all of their days were most likely to be professionals, regardless of income, training, or any institutional position, while those who did it avocationally were amateurs—ostensibly because they had other things on their minds besides birds, since they were doing other things with most of their daylight hours. Curiously, he noted, this issue of time was one that many amateurs overlooked. Many who claimed to have studied birds all their lives, when pressed would in fact reveal that their daily labors in ornithology took up less than half their time. It was perhaps like driving twenty or thirty miles an hour on the freeway to get somewhere instead of going the speed limit. Even if you drove for days, weeks, or months on end, by driving relatively slowly, your progress was necessarily limited. However, Griscom cautioned, *just* the fact of spending all your time studying birds didn't make you an ornithologist. Unlike his predecessors, he considered the term "scientific ornithologist" to be an absurd tautology. The quality of individual experience was also key.

In some ways Griscom prefigured by half a century Nathan Reingold's categories of cultivator, practitioner, and researcher, creating a stratum to define those who study the natural world and contribute to the store of knowledge about it. Griscom suggested three categories: first, the hobbyist, who had little time and no particular training. The next group, the scientist, was the largest. These people gave all of their time to their field of study, they gravitated to institutions that had the necessary gear and collections, and they had some special training. "They are primarily concerned with adding to the storehouse of definitely established facts," he noted. But there was a third group as well. This consisted of a handful of experts, the preeminent figures in the scientific world. They had extensive formal training, ample time to ply their trade in the field and in museums, excellent facilities, and, further, "an immeasurable superiority in intellect, personality, and character," which enabled them to digest the data furnished by the other two groups and to make synthetic generalizations and deductions of far-reaching importance. "Giants, they tower above the rest of us, beacons along the rough trail of human progress," wrote Griscom. The difference was distinctly qualitative as well as quantitative.

Ridgway's generation of professionals held a relatively unswerving attitude about amateurs. Once a dilettante, almost always a dilettante, they seemed to say. But Griscom felt differently, and his attitude opened up all kinds of possibilities to vault over the static nature the two categories imply. "At almost any moment the amateur can become a scientist, and every scientist started as an amateur. His amateur pupil or companion of to-day may become his superior tomorrow. I have seen it happen."

Solid evidence exists to support Griscom's claim that amateurs could gain in proficiency and contribute significantly at the highest scientific levels of discourse. Writing in 1979, one author noted that beginning in 1921, the AOU's Brewster Memorial Award, named for Ridgway's friend William Brewster—for "the most important recent work on the birds of the Western Hemisphere"—had been bestowed forty-five times, and one-fourth of the time the award had gone to people not earning a livelihood as biologists. It has still been possible recently, too, for amateurs to publish significant work in modern ornithology journals. A study has shown that 12 percent of the papers in the four leading American ornithological journals in 1975 were written by people not employed in biology. So amateur contributions have continued, despite the dramatic increases in the technical nature of ornithology.[5] These amateurs have also contributed financially to bird institutions and publications, supported research, and championed conservation efforts of many kinds.

By 1925, a tremendous amount had changed in the state of institutional ornithology. By the turn of the century, advanced training had emerged for a variety of scientific fields, notably the physical sciences. Ridgway, Allen, and Brewster all lacked a college education, yet they succeeded as full-time paid professionals working with birds. By 1910, however, the PhD had become the standard credential for study in a scientific field. And by the early 1930s, more than a hundred schools offered at least some classes in ornithology, and experimental biology began to emerge out of the country's college laboratories. Graduate education in biology and ornithology provided a new marker for professionalism, regularizing the ways in which people learned authoritatively about birds. By 1933, a pair of UC Berkeley zoologists—ornithologists Joseph Grinnell and Alden H. Miller—could boast a total of thirty-nine students writing doctoral dissertations on bird-related subjects. By that year, thirteen

major universities in the United States and Canada offered doctoral pro-
grams that allowed students to focus on ornithological topics.[6]

Coupled with these educational gains, the dramatic growth in size,
scope, and use of bird collections in the country also meant that the American Museum of Natural History in New York could boast an ornithology
department with five full-time professional ornithology staff members: a
curator-in-chief, a curator of oceanic birds, an associate curator of birds
of the eastern hemisphere, an associate curator of birds of the western
hemisphere, and a research associate.[7] Ornithology had reached a new
level of sophistication. Ridgway's style of intensely studying systematics
continued its decline, and the study of the living bird marked the next
phase of ornithology, coupled with advances in the use of genetics as an
analytical tool and the development of the modern evolutionary synthe-
sis. Leonhard Stejneger, a contemporary of Ridgway's, wrote in 1905 that
there would henceforth need to be two kinds of scientific ornithologists:
those who spent most of their time in the field and those who spent most
of their time in the closet. Neither should be held in prejudice, he wrote,
but both would be necessary, and field studies would prove critical. "In
the future no ornithologist who confines himself to the closet will ever
rise to the highest point in his science," Stejneger warned.[8] However, the
study of the living bird would continue to rest on the bedrock of the work
of "closet people" like Ridgway and his colleagues: the study collections,
publications, and checklists that described and quantified the bird world.

Perhaps the last word should come from Griscom, who observed a fact
that remains true: there will always be those who study birds for a living,
and many of them will invariably know more about birds than those with
avocational or other interests. However, the more that interested people
from all walks of life were available to study birds, the more data scientists would have to work with about some of the same topics that obsessed
Ridgway and the other members of the feathery tribe: geographic change,
speciation, subspecies, and distribution. Griscom remarked, "If scientists
can justly complain of their troubles, when science becomes too popular,
let them at least remember, that like everybody else they cannot get something for nothing. And what they have got is the assistance of the enthusiastic, reliable amateur."[9]

Robert Ridgway's Published Works

I have made every effort to make this a comprehensive list of Ridgway's published textual works, ranging from paragraph-long articles to multivolume books. These include solo writings, coauthored texts, and books to which he contributed a chapter or an introduction but was not the primary author. By far the largest quantity herein consists of articles, published in a very wide range of journals and magazines. Some minor items may well have eluded me.

I have specifically excluded all of his reports on the Department of Birds from within the Smithsonian's annual reports. These appeared every year from 1880 onward and are readily located within each of the institution's yearly publications. I have also left out citations for the artwork he created, either alone or jointly with his brother or others. While some of these are attributed, many (perhaps most) are unattributed, and any inclusion of these would be highly incomplete.

The bulk of these citations are drawn from two sources: Harry Harris's 1928 *Condor* biography of Ridgway and Alexander Wetmore's 1931 *National Academy of Sciences* biographical sketch of Ridgway. I have supplemented the list through my research, locating some items both earlier biographers missed. In a few cases I've corrected details in Harris's and Wetmore's original entries. All items within each year are in chronological order, and then in alphabetical order by title for items published the same month; if no specific month is known, those items appear last within that year's publications.

1869

"The Belted Kingfisher Again." *American Naturalist* 3, no. 1 (March): 53–54.

"Notices of Certain Obscurely Known Species of American Birds. (Based on Specimens in the Museum of the Smithsonian Institution)." *Proceedings of the Academy of Natural Sciences* (June): 125–135.

[With Spencer F. Baird.] "*Pelecanus trachyrhynchus* Yearly Sheds the Bony Process of Its Maxilla." *Ibis* 11, no. 3 (July): 350.

"A True Story of a Pet Bird." *American Naturalist* 3, no. 6 (August): 309–312.

1870
"A New Classification of the North American Falconidae, with Descriptions of Three New Species." *Proceedings of the Academy of Natural Sciences* (December): 138–150.

1872
"Relationship of the American White-fronted Owl." *American Naturalist* 6, no. 5 (May): 283–285.

"New Birds in Southern Illinois." *American Naturalist* 6, no. 7 (July): 430–431.

"On the Occurrence of a Near Relative of *Aegiothus flavirostris*, at Waltham, Mass." *American Naturalist* 6, no. 7 (July): 433–434.

"On the Occurrence of *Setophaga picta* in Arizona." *American Naturalist* 6, no. 7 (July): 436.

"Notes on the Vegetation of the Lower Wabash Valley. I. The Forests of the Bottom-Lands." *American Naturalist* 6, no. 11 (November): 658–665.

"Notes on the Vegetation of the Lower Wabash Valley. II. Peculiar Features of the Bottom-Lands." *American Naturalist* 6, no. 12 (December): 724–732.

"On the Relation between Color and Geographical Distribution in Birds, as Exhibited in Melanism and Hyperchromism." [Part 1 of 2.] *American Journal of Science*, 3rd ser., 4, no. 24 (December): 454–460.

1873
"Notes on the Vegetation of the Lower Wabash Valley. III. The Woods and Prairies of the Upland Portions." *American Naturalist* 7, no. 3 (March): 154–157.

[Description of *Centronyx ochrocephalus*.] *American Naturalist* 7, no. 4 (April): 237.

[Note on the *Pyranga roseogularis* of Cabot, by P. L. Sclater, with a description and plate by Ridgway.] *Ibis* 15, no. 2 (April): 126, pl. 3.

"The Prairie Birds of Southern Illinois." *American Naturalist* 7, no. 4 (April): 197–203.

"On the Relation between Color and Geographical Distribution in Birds, as Exhibited in Melanism and Hyperchromism." [Part 2 of 2.] *American Journal of Science*, 3rd ser., 5, no. 25 (September): 39–43.

"The Relation between the Color and the Geographical Distribution of Birds." *American Naturalist* 7, no. 9 (September): 548–555.

[With Spencer F. Baird.] "On Some New Forms of American Birds." *American Naturalist* 7, no. 10 (October): 602–619.

"Birds of Colorado Check List." *Bulletin of the Essex Institute* 5, no. 11 (November): 174–195.

"Notes on the Bird Fauna of the Salt Lake Valley and the Adjacent Portion of the Wahsatch Mountains." *Bulletin of the Essex Institute* 5, no. 11 (November): 168–173.

"The Grouse and Quails of North America, Discussed in Relation to Their Variation with Habitat." *Forest and Stream* 1, no. 19 (December 18): 289–290.

"Revision of the Falconine Genera *Micrastur, Geranospiza* and *Rupornis*, and the Strigine Genus *Glaucidium*." *Proceedings of the Boston Society of Natural History* 16 (December): 73–106.

[With Spencer F. Baird.] "Catalogue of the Ornithological Collection in the Museum of the [Boston] Society [of Natural History]. Part II. Catalogue of Falconidae." *Proceedings of the Boston Society of Natural History* 16 (December): 43–72.

[With Spencer F. Baird.] "On Some New Forms of American Birds." *Bulletin of the Essex Institute* 5, no. 12 (December): 197–201.

1874

[Letter to the editor of *Forest and Stream* concerning the question, being discussed by correspondents, as to whether snakes hiss.] *Forest and Stream* 1, no. 21 (January 1): 327.

"Nomenclature of American Game Birds." [Part 1 of 2.] *American Sportsman* 3, no. 14 (January 3): 210–211.

"Nomenclature of American Game Birds." [Part 2 of 2.] *American Sportsman* 3, no. 15 (January 10): 226–227.

"Catalogue of the Birds Ascertained to Occur in Illinois." *Annals of the Lyceum of Natural History of New York* 10 (January): 364–394.

"The Lower Wabash Valley Considered in Its Relation to the Faunal Districts of the Eastern Region of North America; with a Synopsis of Its Avian Fauna." *Proceedings of the Boston Society of Natural History* 16 (February): 304–332.

"Notes upon American Water Birds." *American Naturalist* 8, no. 2 (February): 108–111.

"Why and How Does the Ruffed Grouse Drum." *American Sportsman* 3, no. 21 (February 21): 322.

[With Spencer F. Baird and Thomas M. Brewer.] *History of North American Birds: Land Birds.* Vols. 1 and 2 (February), vol. 3 (March). Boston: Little,

Brown. [A special edition of fifty copies published the same year contained thirty-six hand-colored plates, colored by Ridgway, who also did the printed illustrations.]

[With Thomas M. Brewer.] "A New North American Bird." *American Naturalist* 8, no. 3 (March): 188–189.

"On Local Variations in the Notes and Nesting Habits of Birds." *American Naturalist* 8, no. 4 (April): 197–201.

"A Remarkable Peculiarity of *Centrocercus urophasianus*." *American Naturalist* 8, no. 4 (April): 240.

"Two Rare Owls from Arizona." *American Naturalist* 8, no. 4 (April): 239–240.

"The Dodo." *Forest and Stream* 2, no. 16 (May 28): 244.

"Discovery of a Burrowing Owl in Florida." *American Sportsman* 4, no. 14 (July 4): 216.

"Description of a New Bird from Colorado." *American Sportsman* 4, no. 16 (July 18): 241.

[Opinion as to whether the abundance of grasshoppers in Kansas and other locations has any connection with the decrease in the number of game birds in the same districts.] *American Sportsman* 4, no. 16 (July 18): 249.

"Story of a Wild Goose." *American Sportsman* 4, no. 17 (July 25): 258–259.

"Birds New to the Fauna of North America." *American Naturalist* 8, no. 7 (July): 434–435.

"Notice of a Species of Tern New to the Atlantic Coast of North America." *American Naturalist* 8, no. 7 (July): 433.

"An Unusually Large Wild Goose." *American Sportsman* 4, no. 18 (August 1): 274.

"Breeding Ground of White Pelicans at Pyramid Lake, Nevada." *American Sportsman* 4, no. 19 (August 8): 289–290, 297, figs. 1–3.

"Game Birds and Grasshoppers: A Reply to Vix." *American Sportsman* 4, no. 23 (September 5): 356.

[Editorial concerning a strange bird described by a correspondent in a previous number.] *Forest and Stream* 3, no. 6 (September 17): 85.

"List of Birds Observed at Various Localities Contiguous to the Central Pacific Railroad, from Sacramento City, California, to Salt Lake Valley, Utah." [Part 1 of 2.] *Bulletin of the Essex Institute* 6, no. 10 (October): 169–174.

"The Snow Goose." *American Naturalist* 8, no. 10 (October): 636–637.

"A Contribution to the 'Sparrow War.'" *American Sportsman* 5, no. 11 (December 12): 161.

[Editorial titled "The English Sparrow" includes matter contributed by Ridgway.] *Forest and Stream* 3, no. 20 (December 24): 309.

1875

"List of Birds Observed at Various Localities Contiguous to the Central Pacific Railroad, from Sacramento City, California, to Salt Lake Valley, Utah." [Part 2 of 2.] *Bulletin of the Essex Institute* 7, no. 1 (January): 10–40.

"Note on *Sterna longipennis* Nordmann." *American Naturalist* 9, no. 1 (January): 54–55.

"A Heronry in the Wabash Bottoms." *American Sportsman* 5, no. 20 (February 13): 312–313.

"Big Trees." [Part 1 of 4.] *American Sportsman* 5, no. 21 (February 20): 321–322.

"Big Trees." [Part 2 of 4.] *American Sportsman* 5, no. 22 (February 27): 337.

"Big Trees." [Part 3 of 4.] *American Sportsman* 5, no. 23 (March 6): 353–354.

"Big Trees." [Part 4 of 4.] *American Sportsman* 5, no. 24 (March 13): 369–370.

"Snow Birds and Little Owls." *American Sportsman* 5, no. 25 (March 20): 393.

"On *Nisus cooperi* (Bonaparte), and *N. gundlachi* (Lawrence)." *Proceedings of the Academy of Natural Sciences* (March): 78–88.

"On the Buteonine Subgenus *Craxirex* Gould." *Proceedings of the Academy of Natural Sciences* (March): 89–119.

"More about the Florida Burrowing Owl." *Rod and Gun* 6, no. 1 (April): 7.

"A Monograph of the Genus *Leucosticte*, Swainson; or Gray-crowned Purple Finches." *Bulletin of the United States Geological and Geographical Survey of the Territories*, 2nd ser., no. 2 (May): 51–82.

"Notice of a Very Rare Hawk." *Rod and Gun* 6, no. 5 (May): 65.

"Outlines of a Natural Arrangement of the Falconidae." *Bulletin of the United States Geological and Geographical Survey of the Territories*, 2nd ser., no. 4 (June): 1–18.

"Nesting of the Worm-Eating Warbler." *Field and Forest* 1, no. 2 (July): 10–12.

"The Sparrow Hawk or American Kestril." *Rod and Gun* 6, no. 14 (July): 109.

"Description of a New Wren from Eastern Florida." *American Naturalist* 9, no. 8 (August): 469–470.

"First Impressions of the Bird Fauna of California, and General Remarks on Western Ornithology." *Scientific Monthly* 1, no. 1 (October): 2–13.

"Our Native Trees." *Field and Forest* 1, no. 7 (December): 49–53.

"Studies of the American Falconidae: Monograph of Genus *Micrastur*." *Proceedings of the Academy of Natural Sciences* (December): 470–502.

1876

"The Genus *Glaucidium*." *Ibis* 18, no. 1 (January): 11–17.

"Second Thoughts on the Genus *Micrastur*." *Ibis* 18, no. 1 (January): 1–5.

"Studies of the American Falconidae: Monograph of the Polybori." *Bulletin of the United States Geological and Geographical Survey of the Territories* 1, no. 6 (February): 451–473.

[Letter to Walter Van Fleet giving Ridgway's views on the effect of the introduction of the House Sparrow into the United States.] *Watsontown (PA)
Record* (March 10): 1.

"Notes on the Genus *Helminthophaga.*" *Ibis* 18, no. 2 (April): 166–191.

"Ornithology of Guadalupe Island, Based on Notes and Collections Made by
Dr. Edward Palmer." *Bulletin of the United States Geological and Geographical Survey of the Territories* 2, no. 2 (April): 183–195.

"Regarding *Buteo vulgaris* in North America." *Bulletin of the Nuttall Ornithological Club* 1, no. 2 (April): 32–39.

"Studies of the American Falconidae." *Bulletin of the United States Geological and Geographical Survey of the Territories* 2, no. 2 (April): 91–182.

"Maximum Length of the Black Snake." *Forest and Stream* 6, no. 20 (June 22):
318.

"Giant Pear Trees." *Forest and Stream* 6, no. 21 (June 29): 337.

"The Black Snake Again." *Forest and Stream* 7, no. 2 (August 17): 20.

"'The Bank Swallow' Again." *American Naturalist* 10, no. 8 (August): 493–494.

"Notes on the Catalpa: *Catalpa bignonioides.*" *Field and Forest* 2, no. 2 (August): 27–29.

"'Sexual, Individual and Geographical Variation' in the Genus *Leucosticte.*"
Field and Forest 2, no. 3 (September): 37–43.

"On Geographical Variation in *Dendroica palmarum.*" *Bulletin of the Nuttall
Ornithological Club* 1, no. 4 (November): 81–87.

"The Little Cypress Swamp of Indiana." *Field and Forest* 2, no. 6 (December):
93–96.

1877

"On Geographical Variation in *Turdus migratorius.*" *Bulletin of the Nuttall
Ornithological Club* 2, no. 1 (January): 8–9.

"Mrs. Maxwell's Colorado Museum." [Part 1 of 2.] *Field and Forest* 2, no. 2
(May): 195–198.

"Mrs. Maxwell's Colorado Museum." [Part 2 of 2.] *Field and Forest* 2, no. 3
(June): 208–214.

"The Birds of Guadalupe Island, Discussed with Reference to the Present
Genesis of Species." *Bulletin of the Nuttall Ornithological Club* 2, no. 3
(July): 58–66.

"Mrs. Maxwell's Colorado Museum: Additional Notes." *Field and Forest* 3,
no. 1 (July): 11.

Report of the Geological Exploration of the Fortieth Parallel. Part III: "Ornithology," 303–669. Washington, DC: Government Printing Office.

1878

"Description of a New Wren from the Tres Marias Islands." *Bulletin of the Nuttall Ornithological Club* 3, no. 1 (January): 10–11.

"Eastward Range of *Chondestes grammaca*." *Bulletin of the Nuttall Ornithological Club* 3, no. 1 (January): 43–44.

"Three Additions to the Avifauna of North America." *Bulletin of the Nuttall Ornithological Club* 3, no. 1 (January): 37–38.

"Studies of the American Herodiones. Part I—Synopsis of the American Genera of Ardeidae and Ciconiidae; including Descriptions of Three New Genera, and a Monograph of the American Species of the Genus *Ardae* Linn." *Bulletin of the United States Geological and Geographical Survey of the Territories* 4, no. 1 (February): 219–251.

"On a New Humming Bird (*Atthis ellioti*) from Guatemala." *Proceedings of the U.S. National Museum* 1 (March): 8–10.

"Notes on Some of the Birds of Calaveras County, California, and Adjoining Localities." *Bulletin of the Nuttall Ornithological Club* 3, no. 2 (April): 64–68.

"Song Birds of the West." *Harper's New Monthly Magazine* 56, no. 336 (May): 857–880.

"A Review of the American Species of the Genus *Scops*, Savigny." *Proceedings of the U.S. National Museum* 1 (September): 85–117.

"Notes on Birds Observed at Mount Carmel, Southern Illinois, in the Spring of 1878." *Bulletin of the Nuttall Ornithological Club* 3, no. 4 (October): 162–166.

"Notes on the Ornithology of Southern Texas, Being a List of Birds Observed in the Vicinity of Fort Brown, Texas, from February, 1876 to June, 1878." By James C. Merrilll, assistant surgeon, US Army. [Edited by Ridgway.] *Proceedings of the U.S. National Museum* 1 (October): 118–173.

"Description of Two New Species of Birds from Costa Rica, and Notes on Other Rare Species from That Country." *Proceedings of the U.S. National Museum* 1 (December): 252–255.

"Descriptions of Several New Species and Geographical Races of Birds Contained in the Collection of the United States National Museum." *Proceedings of the U.S. National Museum* 1 (December): 247–252.

1879

"Descriptions of New Species and Races of American Birds, including Synopsis of the Genus *Tyrannus*, Cuvier." *Proceedings of the U.S. National Museum* 2 (April): 466–486.

[Letter to George Bird Grinnell referring to bird later described as *Seiurus naevius notabilis*.] *Forest and Stream* 12, no. 16 (May 22): 307.

"On the Use of Trinomials in Zoological Nomenclature." *Bulletin of the Nuttall Ornithological Club* 4, no. 3 (July): 129–134.

"Henslow's Bunting (*Coturniculus henslowi*) near Washington." *Bulletin of the Nuttall Ornithological Club* 4, no. 4 (October): 238.

[Letter to G. H. Ragsdale concerning *Peucaea illinoensis*.] *Temperance (Gainesville, TX) Vidette* (October 11).

"Note on *Helminthophaga gunnii*, Gibbs." *Bulletin of the Nuttall Ornithological Club* 4, no. 4 (October): 233–234.

"On a New Species of *Peucaea* from Southern Illinois and Central Texas." *Bulletin of the Nuttall Ornithological Club* 4, no. 4 (October): 218–222.

"Ueber den Gebrauch der Trinomina in der zoologischen Nomenclatur. (Übersetzt von Hermann Schalow)." *Journal für Ornithologie* 27, no. 148 (October), 410–417.

"Mrs. Maxwell's Colorado Museum. Catalogue of the Birds," 226–237. Contribution to Mary Dartt, *On the Plains, and among the Peaks; or, How Mrs. Maxwell Made Her Natural History Collection*. Philadelphia: Claxton, Remsen and Haffelfinger.

1880

"Description of an Unusual Plumage of *Buteo harlani*." *Bulletin of the Nuttall Ornithological Club* 5, no. 1 (January): 58–59.

"Late Breeding of the Blue Grosbeak." *Bulletin of the Nuttall Ornithological Club* 5, no. 1 (January): 53.

"Note on *Peucaea illinoensis*." *Bulletin of the Nuttall Ornithological Club* 5, no. 1 (January): 52.

"On Current Objectionable Names of North American Birds." *Bulletin of the Nuttall Ornithological Club* 5, no. 1 (January): 36–38.

"On Six Species of Birds New to the Fauna of Illinois, with Notes on Other Rare Illinois Birds." *Bulletin of the Nuttall Ornithological Club* 5, no. 1 (January): 30–32.

"Revisions of Nomenclature of Certain North American Birds." *Proceedings of the U.S. National Museum* 3 (March): 1–16.

"Description of the Adult Plumage of *Hierofalco gyrfalco obsoletus*." *Bulletin of the Nuttall Ornithological Club* 5, no. 2 (April): 92–95.

"The Northern Waxwing (*Ampelis garrulus*) in Southern Illinois." *Bulletin of the Nuttall Ornithological Club* 5, no. 2 (April): 118.

[Note concerning the capture of a specimen of the Greenfinch (*Ligurinus chloris*) at Lowville, Lewis County, New York. Refers to a note by Romeyn B. Hough entitled "The Greenfinch (*Ligurinus chloris*) in Northern New York."] *Bulletin of the Nuttall Ornithological Club* 5, no. 2 (April): 119.

"Notes on the American Vultures (Sarcorhamphidae), with Special Reference to Their Generic Nomenclature." *Bulletin of the Nuttall Ornithological Club* 5, no. 2 (April): 77–84.

"On the Moult of the Bill, or Parts of Its Covering, in Certain Alcidae." *Bulletin of the Nuttall Ornithological Club* 5, no. 2 (April): 126–127.

"On the Supposed Identity of *Ardea occidentalis*, Aud., and *A. würdemanni*, Baird." *Bulletin of the Nuttall Ornithological Club* 5, no. 2 (April): 123–124.

"The Little Brown Crane (*Grus fraterculus*, Cassin)." *Bulletin of the Nuttall Ornithological Club* 5, no. 3 (July): 187–188.

"On *Macrorhamphus griseus* (Gmel.) and *M. Scolopaceus* (Say)." *Bulletin of the Nuttall Ornithological Club* 5, no. 3 (July): 157–160.

"On a New Alaskan Sandpiper." *Bulletin of the Nuttall Ornithological Club* 5, no. 3 (July): 160–163.

"On *Rallus longirostrus*, Bodd., and Its Geographical Races." *Bulletin of the Nuttall Ornithological Club* 5, no. 3 (July): 138–140.

"*Scops flammeola* in Colorado." *Bulletin of the Nuttall Ornithological Club* 5, no. 3 (July): 185.

"A Catalogue of the Birds of North America." *Proceedings of the U.S. National Museum* 3 (August–September): 163–246.

"Catalogue of Trochilidae in the Collection of the United States National Museum." *Proceedings of the U.S. National Museum* 3 (October): 308–320.

"Description of the Eggs of the Caspian Tern (*Sterna caspia*)." *Bulletin of the Nuttall Ornithological Club* 5, no. 4 (October): 221–223.

"Note on *Helminthophaga cincinnatiensis*, Langdon." *Bulletin of the Nuttall Ornithological Club* 5 no. 4 (October): 237–238.

1881

"Swainson's Warbler (*Helonaea swainsoni*) in Texas." *Bulletin of the Nuttall Ornithological Club* 6, no. 1 (January): 54–55.

"The Caspian Tern in California." *Bulletin of the Nuttall Ornithological Club* 6, no. 2 (April): 124.

"A Hawk New to the United States." *Forest and Stream* 16, no. 11 (April 14): 206.

"On a Duck New to the North American Fauna." *Proceedings of the U.S. National Museum* 4 (April): 22–24.

"On *Amazilia yucatensis* (Cabot), and *A. cerviniventris*, Gould." *Proceedings of the U.S. National Museum* 4 (April): 25–26.

"Southern Range of the Raven on the Atlantic Coast of the United States." *Bulletin of the Nuttall Ornithological Club* 6, no. 2 (April): 118.

"An Unaccountable Migration of the Red-Headed Woodpecker." *Bulletin of the Nuttall Ornithological Club* 6, no. 2 (April): 120–122.

"A Revised Catalogue of the Birds Ascertained to Occur in Illinois." *Illinois State Laboratory of Natural History Bulletin* 1, no. 4 (May): 161–208.

"A Review of the Genus *Centurus,* Swainson." *Proceedings of the U.S. National Museum* 4 (July): 93–119.

"List of Species of Middle and South American Birds not Contained in the United States National Museum." *Proceedings of the U.S. National Museum* 4 (August–November): 165–203.

"On a Tropical Hawk to Be Added to the North American Fauna." *Bulletin of the Nuttall Ornithological Club* 6, no. 4 (October): 207–214.

"List of Special Desiderata among North American Birds." *Proceedings of the U.S. National Museum* 4 (November–December): 207–223.

"Nomenclature of North American Birds Chiefly Contained in the United States National Museum." Bulletin 21, *Bulletin of the U.S. National Museum,* 1–94.

1882

"Additions to the Catalogue of North American Birds." *Bulletin of the Nuttall Ornithological Club* 7, no. 1 (January): 61.

"The Great Black-Backed Gull (*Larus marinus*) from a New Locality." *Bulletin of the Nuttall Ornithological Club* 7, no. 1 (January): 60.

"Notes on Some of the Birds Observed near Wheatland, Knox County, Indiana, in the Spring of 1881." *Bulletin of the Nuttall Ornithological Club* 7, no. 1 (January): 15–23.

"On an Apparently New Heron from Florida." *Bulletin of the Nuttall Ornithological Club* 7, no. 1 (January): 1–6.

"On the Generic Name *Helminthophaga.*" *Bulletin of the Nuttall Ornithological Club* 7, no. 1 (January): 53–54.

[Correction of an erroneous identification of *Milvulus tyrannus* for *M. forficatus.* Item under the title "Fork-tailed Flycatcher. Correction." Consists of notes by four authors, one of whom is Ridgway.] *Ornithologist and Oologist* 6, no. 12 (February): 93.

"Catalogue of Old World Birds in the United States National Museum." *Proceedings of the U.S. National Museum* 4 (March): 317–333.

"Notes on Some Costa Rican Birds." *Proceedings of the U.S. National Museum* 4 (March): 333–337.

"Description of a New Fly-Catcher and a Supposed New Petrel from the Sandwich Islands." *Proceedings of the U.S. National Museum* 4 (April): 337–338.

"Description of a New Owl from Porto Rico." *Proceedings of the U.S. National Museum* 4 (April): 366–371.

[Description of the adult female of *Falco peregrinus pealei*. (Quoted matter from Ridgway, in a paper by J. H. Gurney, on "Notes on a 'Catalogue of the Acciptres in the British Museum' by R. Bowdler Sharpe (1874).")] *Ibis* 24, no. 2 (April): 297–298.

"Descriptions of Two New Thrushes from the United States." *Proceedings of the U.S. National Museum* 4 (April): 374–379.

"On Two Recent Additions to the North American Bird Fauna." *Proceedings of the U.S. National Museum* 4 (May): 414–415.

"Description of Several New Races of American Birds." *Proceedings of the U.S. National Museum* 5 (June): 9–15.

"Notes on the Native Trees of the Lower Wabash and White River Valleys, in Illinois and Indiana." *Proceedings of the U.S. National Museum* 5 (June): 43–46.

"On the Genera *Harporhynchus*, Cabanis, and *Methriopterus*, Reichenbach, with a Description of a New Genus of Miminae." *Proceedings of the U.S. National Museum* 5 (June): 43–46.

"Critical Remarks on the Tree-Creepers (*Certhia*) of Europe and North America." *Proceedings of the U.S. National Museum* 5 (July): 111–116.

"Descriptions of Some New North American Birds." *Proceedings of the U.S. National Museum* 5 (September): 343–346.

[With C. C. Nutting.] "On a Collection of Birds from the Hacienda 'La Palma,' Gulf of Nicoya, Costa Rica. By C. C. Nutting. With Critical Notes by R. Ridgway." *Proceedings of the U.S. National Museum* 5 (September): 382–409.

"Birds New to or Rare in the District of Columbia." *Bulletin of the Nuttall Ornithological Club* 7, no. 4 (October): 253.

"Distribution of the Fish Crow (*Corvus ossifragus*)." *Bulletin of the Nuttall Ornithological Club* 7 no. 4 (October): 250.

"List of Additions to the Catalogue of North American Birds." *Bulletin of the Nuttall Ornithological Club* 7, no. 4 (October): 257–258.

1883

"Geographical Variation in Size among Certain Anatidae and Gruidae." *Bulletin of the Nuttall Ornithological Club* 8, no. 1 (January): 62.

"On Le Conte's Bunting (*Coturniculus lecontei*) and Other Birds Observed in South-Eastern Illinois." *Bulletin of the Nuttall Ornithological Club* 8, no. 1 (January): 58.

"The Scissor-Tail (*Milvulus forficatus*) at Norfolk, Va." *Bulletin of the Nuttall Ornithological Club* 8, no. 1 (January): 59.

"On Some Remarkable Points of Relationship between the American King-Fishers." *Bulletin of the Nuttall Ornithological Club* 8, no. 1 (January): 59–60.

"Corrections." *Ornithologist and Oologist* 8, no. 2 (February): 13.

"Catalogue of a Collection of Birds Made in the Interior of Costa Rica by Mr. C. C. Nutting." *Proceedings of the U.S. National Museum* 5 (March): 493–502.

[With L. Belding.] "Catalogue of a Collection of Birds Made at Various Points along the Western Coast of Lower California, North of Cape St. Eugenio." *Proceedings of the U.S. National Museum* 5 (April): 527–532.

"Catalogue of a Collection of Birds Made near the Southern Extremity of the Peninsula of Lower California. By L. Belding." *Proceedings of the U.S. National Museum* 5 (April): 532–550.

"Description of a New Warbler from the Island of Santa Lucia, West Indies." *Proceedings of the U.S. National Museum* 5 (April): 525–526.

"Description of a Supposed New Plover from Chili." *Proceedings of the U.S. National Museum* 5 (April): 526–527.

"On the Genus *Tantalus* Linn., and Its Allies." *Proceedings of the U.S. National Museum* 5 (April): 550–561.

"Catalogue of the Aquatic and Fish-Eating Birds Exhibited by the United States National Museum [at the International Fisheries Exhibition, London]." No. 27, *Bulletin of the U.S. National Museum* (May): 139–184.

"Description of a New Petrel from Alaska." *Bulletin of the U.S. National Museum* 5 (May): 656–658.

[Letter to Dr. George Vasey, botanist of the US Department of Agriculture, inquiring whether there is any foundation for the somewhat prevalent popular belief that wheat will, under certain circumstances, turn into cheat, and Dr. Vasey's reply to the same.] *Olney (IL) Republican* (July 4).

"Descriptions of Some Birds, Supposed to Be Undescribed, from the Commander Islands and Petropaulovski, Collected by Dr. Leonhard Stejneger, U.S. Signal Service." *Proceedings of the U.S. National Museum* 6, no. 345 (August): 90–96.

"Notes on the Black Racer." *Forest and Stream* 21, no. 4 (August 23): 63.

"The Sparrow Controversy. Some Interesting Facts Bearing upon the Question." *(Washington, DC) Evening Star* (September 8): 2.

"The Sparrow Again." *Evening Star* [Washington, DC] (September 15): 2.

"'And Finally, Bretheren.' The English Sparrow Question Again." *(Washington, DC) Evening Star* (October 20): 2.

"*Anthus cervinus* (Pallas) in Lower California." *Proceedings of the U.S. National Museum* 6 (October): 156–157.

"Description of Some New Birds from Lower California, Collected by L. Belding." *Proceedings of the U.S. National Museum* 6 (October): 154–156.

[Letter to the editors concerning the US National Museum exhibit of aquatic and fish-eating birds at the London International Fisheries Exhibition]. *Ibis*, 5th ser., 1, no. 4 (October): 578–580.

"Note on *Merula confinis* (Baird)." *Proceedings of the U.S. National Museum* 6 (October): 158–159.

"Notes on Some Rare Species of Neotropical Birds." *Ibis*, 5th ser., 1, no. 4 (October): 399–401.

"On the Probable Identity of *Motacilla ocularis* Swinhoe and *M. amurensis* Seebohm, with Remarks on an Allied Supposed Species, *M. blakistoni* Seebohm." *Proceedings of the U.S. National Museum* 6 (October): 144–147.

"Flora of Sam's Point." *Bulletin of the Torrey Botanical Club* 10 (October–November): 121–122.

"Additions and Corrections to the List of Native Trees of the Lower Wabash." *Botanical Gazette* 8, no. 12 (December): 345–352.

1884

"List of Birds Found at Guaymas, Sonora, in December, 1882, and April, 1883. By L. Belding." [Edited by Ridgway.] *Proceedings of the U.S. National Museum* 6 (January): 343–344.

"Note on *Zenaidura yucatensis* Lawr." *Auk* 1, no. 1 (January): 96.

"Notes on Some Japanese Birds Related to North American Species." *Proceedings of the U.S. National Museum* 6 (January): 368–371.

"Notes on Three Guatemalan Birds." *Ibis* 26, no. 1 (January): 43–45.

"On a New *Carpodectes* from South-Western Costa Rica." *Ibis* 26, no. 1 (January): 27–28.

"Second Catalogue of a Collection of Birds Made near the Southern Extremity of Lower California. By L. Belding." [Edited by Ridgway.] *Proceedings of the U.S. National Museum* 6 (January): 344–352.

"*Ortyx virginianus* Not in Arizona." *Forest and Stream* 22, no. 7 (March 13): 124.

"Description of a New American Kingfisher." *Proceedings of the Biological Society of Washington* 2 (April): 95–96.

"Descriptions of Some New North American Birds." *Proceedings of the Biological Society of Washington* 2 (April): 89–95.

"Note on *Anas hyperboreus*, Pall., and *Anser albatus*, Cass." *Proceedings of the Biological Society of Washington* 2 (April): 107–108.

"Note on *Psaltriparus grindae*, Belding." *Proceedings of the Biological Society of Washington* 2 (April): 96

"Note on the Generic Name *Calodromas*." *Proceedings of the Biological Society of Washington* 2 (April): 97.

"Note Regarding the Earliest Name for *Carpodacus haemorrhous*." *Proceedings of the Biological Society of Washington* 2 (April): 110–111.

"On a Collection of Birds from Nicaragua. By Charles C. Nutting." [Edited by Ridgway.] *Proceedings of the U.S. National Museum* 6 (April): 372–410.

[Remarks concerning *Phalacrocorax violaceus* and *v. resplendens*.] *Auk* 1, no. 2 (April): 165.

[Remarks concerning two Central American species of birds commonly referred to the genus *Compsothlypis* Cabanis.] *Auk* 1, no. 2 (April): 169.

"Remarks on the Type Specimens of *Musciapa fulvifrons*, Giraud, and *Mitrephorus pallescens*, Coues." *Proceedings of the Biological Society of Washington* 2 (April): 108–110.

"A Review of the American Crossbills (*Loxia*) of the *L. curvirostra* type." *Proceedings of the Biological Society of Washington* 2 (April): 101–107.

"On Some Costa Rican Birds, with Descriptions of Several Supposed New Species." *Proceedings of the U.S. National Museum* 6 (April): 410–415.

"Southern Limit of Quail and Grouse." *Forest and Stream* 22, no. 13 (April 24): 243.

"Description of a New Snow Bunting from Alaska." *Proceedings of the U.S. National Museum* 7 (July): 68–70.

"*Melanetta fusca* (Linn.) in Alaska." *Proceedings of the U.S. National Museum* 7 (July): 68.

"Note on *Astur atricapillus striatulus*." *Auk* 1, no. 3 (July): 252–253.

"Note on *Selasphorus torridus* Salvin." *Proceedings of the U.S. National Museum* 7 (July): 14.

"On the Possible Specific Identity of *Buteo cooperi* Cass. with *B. harlani* (Aud.)." *Auk* 1, no. 3 (July): 253–254.

"Probable Breeding of the Red Crossbill (*Loxia curvirostra americana*) in Central Maryland." *Auk* 1, no. 3 (July): 292.

"The Probable Breeding Place of *Passerculus princeps*." *Auk* 1, no. 3 (July): 292–293.

"Remarks upon the Close Relationship between the White and Scarlet Ibises (*Eudocimus albus* and *E. ruber*)." *Auk* 1, no. 3 (July): 239–240.

"Description of a New Species of Field-Sparrow from New Mexico." *Proceedings of the U.S. National Museum* 7 (September): 259.

"On a Collection of Birds Made by Messrs. J. E. Benedict and W. Nye, of the United States Fish Commission Steamer 'Albatross.'" *Proceedings of the U.S. National Museum* 7 (September): 172–180.

"Another Kirtland's Warbler from Michigan." *Auk* 1, no. 4 (October): 389.

"Description of a New Species of Coot from the West Indies." *Proceedings of the U.S. National Museum* 7 (October): 358.

[Note concerning Bird Exhibit of the US National Museum at the New Orleans Exposition.] *Auk* 1, no. 4 (October): 403.

[Note relative to extent of Bird Collection in the US National Museum.] *Auk* 1, no. 4 (October): 403–404.

"The Bird-Collection of the U.S. National Museum." *Science* 4, no. 95 (November): 496–497.

1885

"Description of a New Race of the Red-Shouldered Hawk from Florida." *Proceedings of the U.S. National Museum* 7 (February): 514.

[Extract from letter to the editor concerning the name, etc., of *Spizella wortheni.*] *Ornithologist and Oologist* 5, no. 2 (February): 24.

"On Two Hitherto Unnamed Sparrows from the Coast of California." *Proceedings of the U.S. National Museum* 7 (February): 516–518.

"The European Sparrow." *American Field* 23, no. 13 (March): 295.

[Letter to Dr. J. B. Holder, in answer to request for criticism of a paper describing a bird called Sky-lark by the writer.] *(NY) Examiner* (March 19).

"Gurney's 'List of the Diurnal Birds of Prey.'" *Auk* 2, no. 2 (April): 203–205.

"Note on *Sarcorhamphus aequatorialis* Sharpe." *Auk* 2, no. 2 (April): 169–171.

"On *Buteo harlani* (Aud.) and *B. cooperi* Cass." *Auk* 2, no. 2 (April): 165–166.

"Remarks on the California Vulture (*Pseudogryphus californianus*)." *Auk* 2, no. 2 (April): 167–169.

"Where Did It Come From?" *Forest and Stream* 24, no. 11 (April 9): 204.

"Description of a New Species of *Contopus* from Tropical America." *Proceedings of the U.S. National Museum* 8 (May): 21.

"Description of a New Warbler from Yucatan." *Proceedings of the U.S. National Museum* 8 (May): 23.

"Description of Two New Birds from Costa Rica." *Proceedings of the U.S. National Museum* 8 (May): 23–24.

"Descriptions of Three Supposed New Honey Creepers from the Lesser Antilles, with a Synopsis of the Species of the Genera *Certhiola*." *Proceedings of the U.S. National Museum* 8 (May): 25–30.

"*Icterus cucullatus*, Swainson, and Its Geographical Variations." *Proceedings of the U.S. National Museum* 8 (May): 18–19.

"Note on the *Anser leucopareius* of Brandt." *Proceedings of the U.S. National Museum* 8 (May): 21–22.

"On *Cathartes burrovianus*, Cassin, and *C. urubitinga*, Pelzeln." *Proceedings of the U.S. National Museum* 8 (May): 34–36.

"On *Oestrelata fisheri* and *Oe. defilippiana.*" *Proceedings of the U.S. National Museum* 8 (May): 17–18.

"On *Onychotes gruberi.*" *Proceedings of the U.S. National Museum* 8 (May): 36–38.

"Remarks on the Type Specimens of *Buteo oxypterus,* Cassin." *Proceedings of the U.S. National Museum* 8 (May): 75–77.

"Description of a New Hawk from Cozumel." *Proceedings of the U.S. National Museum* 8 (June): 94–95.

"Description of a New Species of Boat-Billed Heron from Central America." *Proceedings of the U.S. National Museum* 8 (June): 93–94.

"On *Peucaea mexicana* (Lawr.), a Sparrow New to the United States." *Proceedings of the U.S. National Museum* 8 (June): 98–99.

"A Review of the American 'Golden Warblers.'" *Proceedings of the U.S. National Museum* 8 (September): 348–350.

"Some Emended Names of North American Birds." *Proceedings of the U.S. National Museum* 8 (September): 354–356.

"Description of a New Cardinal Grosbeak from Arizona." *Auk* 2, no. 4 (October): 343–345.

"Description of an Apparently New Species of *Dromococcyx* from British Guiana." *Proceedings of the U.S. National Museum* 8 (October): 559.

"*Helminthophila leucobronchialis.*" *Auk* 2, no. 4 (October): 359–363.

"A New Petrel for North America." *Auk* 2, no. 4 (October): 386–387.

"On *Junco cinereus* (Swains.) and Its Geographical Races." *Auk* 2, no. 4 (October): 363–364.

"Catalogue of a Collection of Birds Made on the Island of Cozumel, Yucatan, by the Naturalists of the U.S. Fish Commission Steamer Albatross, Capt. Z. L. Tanner, Commander." *Proceedings of the U.S. National Museum* 8 (October–November): 560–583.

[With Spencer Baird and Thomas M. Brewer.] *Water Birds of North America.* Memoirs of the Museum of Comparative Zoology at Harvard College, Vols. XII and XIII. Issued in continuation of the publications of the Geological Survey of California. Little, Brown.

1886

"Arizona Quail." *Forest and Stream* 25, no. 25 (January 14): 484.

[Under the pseudonym "Patoka."] "Birds and Bonnets." *Forest and Stream* 26, no. 1 (January 28): 5.

"On the Proper Name for the Prairie Hen." *Auk* 3, no. 1 (January): 132–133.

"The Scissor-Tailed Flycatcher (*Milvulus forficatus*) at Key West." *Auk* 3, no. 1 (January): 134.

"The Vernacular Name of *Plectrophenax hyperboreus*." *Auk* 3, no. 1 (January): 135.

"Is the Dodo an Extinct Bird?" *Science* 7, no. 160 (February): 190.

[Letter to Dr. B. H. Warren giving by request the writer's views concerning the food habits of hawks and owls with particular reference to the question as to whether these birds should or should not be protected by state laws.] *(West Chester, PA) Daily Local News*, 14, no. 91 (March 5): 1.

"Discovery of the Breeding Place of McKay's Snowflake (*Plectrophenax hyperboreus*)." *Auk* 3, no. 2 (April): 276–277.

[Note announcing the departure of the Fish Commission Steamer *Albatross* on a scientific cruise among the Bahamas and other islands of the West Indies.] *Auk* 3, no. 2 (April): 286–287.

"On Two Abnormally Colored Specimens of the Bluebird (*Sialia sialis*)." *Auk* 3, no. 2 (April): 282–283.

"*Tringa damacensis* in Alaska; a Sandpiper New to the North American Fauna." *Auk* 3, no. 2 (April): 275.

"'Water Birds of North America'—'A Few Corrections' Rectified." *Auk* 3, no. 2 (April): 266–268.

"Description of a New Genus of Oceanitidae." *Auk* 3, no. 3 (July): 334.

"Description of a New Genus of Tyrannidae from Santo Domingo." *Auk* 3, no. 3 (July): 382–383.

"Description of a New Species of Elf Owl from Socorro Island, Western Mexico." *Auk* 3, no. 3 (July): 333–334.

"Description of a New Species of Oyster-Catcher from the Galapagos Islands." *Auk* 3, no. 3 (July): 331.

"Description of Four New Species of Birds from the Bahama Islands." *Auk* 3, no. 3 (July): 334–337.

"Descriptions of Two New Species of Birds Supposed to Be from the Interior of Venezuela." *Auk* 3, no. 3 (July): 333.

"On the Glaucous Gull of Bering's Sea and Contiguous Waters." *Auk* 3, no. 3 (July): 330–331.

"Preliminary Descriptions of Some New Species of Birds from Southern Mexico, in the Collection of the Mexican Geographical and Exploring Commission." *Auk* 3, no. 3 (July): 331–333.

[Remarks concerning certain "Corrections" of alleged errors in "Water Birds of North America."] Auk 3, no. 3 (July): 403–404.

"Descriptions of Some New Species of Birds Supposed to Be from the Interior of Venezuela." *Proceedings of the U.S. National Museum* 9 (September): 92–94.

"On *Aestrelata sandwichensis* Ridgw." *Proceedings of the U.S. National Museum* 9 (September): 95–96.

[With Fernando Ferrari-Perez.] "Catalogue of Animals Collected by the Geographical and Exploring Commission of the Republic of Mexico. By Fernando Ferrari-Perez, Chief of the Natural History Section. II.—Birds. By F. Ferrari Perez. With Descriptions of Five New Species, and Critical Remarks on others of Great or Less Rarity or Interest. By Robert Ridgway." *Proceedings of the U.S. National Museum* 9 (September–October): 130–182.

"Description of a Melanistic Specimen of *Buteo latissimus* (Wils.)." *Proceedings of the U.S. National Museum* 9 (October): 248–249.

"Description of a New Species of the Genus *Empidonax* from Guatemala." *Ibis* 28, no. 4 (October): 459–460.

"On *Empidochanes fuscatus* (Max.) and *Empidonax brunneus* Ridgw." *Ibis* 28, no. 4 (October): 460–461.

"On the Species of the Genus *Empidonax*." *Ibis* 28, no. 4 (October): 461–468.

[With Lucien McShann Turner.] *Contributions to the Natural History of Alaska. Results of Investigations . . . in the Yukon District and the Aleutian Islands; Conducted under the Auspices of the Signal Service, United States Army, Extending from May, 1874, to August, 1881.* Washington, DC: Government Printing Office.

"Description of Some New Species of Birds from Cozumel Island, Yucatan." *Proceedings of the Biological Society of Washington* 3, 21–24.

Nomenclature of Colors for Naturalists and Compendium of Useful Information for Ornithologists. Boston: Little, Brown.

1887

"A Singularly Marked Specimen of *Sphyrapicus thyroideus*." *Auk* 4, no. 1 (January): 75–76.

"Description of a New Species of *Myiarchus,* Presumably from the Orinoco District of South America." *Proceedings of the U.S. National Museum* 4 (February): 520.

"Description of a New Subspecies of *Cyclorhis* from Yucatan." *Proceedings of the U.S. National Museum* 4 (February): 519.

"Description of a Recently-New Oyster-Catcher (*Haematopus galapagensis*) from the Galapagos Islands." *Proceedings of the U.S. National Museum* 4 (February): 325–326.

"Description of an Apparently New Species of *Picolaptes,* from the Lower Amazon." *Proceedings of the U.S. National Museum* 4 (February): 523.

"On a Probable Hybrid between *Dryobates nuttalli* (Gamb.) and *D. pubescens gairdnerii* (Aud.)." *Proceedings of the U.S. National Museum* 4 (February): 521–522.

"Description of a New Plumed Partridge from Sonora." *Forest and Stream* 28, no. 6 (March 3): 106.

"The Coppery-Tailed Trogon (*Trogon ambiguus*) Breeding in Southern Arizona." *Auk* 4, no. 2 (April): 161–162.

"Description of a New Form of *Spindalis* from the Bahamas." *Proceedings of the U.S. National Museum* 5 (April): 3.

"Description of a New Species of Cotinga from the Pacific Coast of Costa Rica." *Proceedings of the U.S. National Museum* 5 (April): 1–2.

"Description of the Adult Female of *Carpodectes antoniae* Zeledon; with Critical Remarks, Notes on Habits, etc., by Jose C. Zeledon." *Proceedings of the U.S. National Museum* 5 (April): 20.

"Feathered Songsters. Great Western Bird Center. A List of the Birds Found Breeding within the Corporate Limits of Mt. Carmel [Illinois]." *Mount Carmel (IL) Register* (April 25): 1.

"The Imperial Woodpecker (*Campephilus imperialis*) in Northern Sonora." *Auk* 4, no. 2 (April): 161.

"List of Birds Found Breeding within the Corporate Limits of Mt. Carmel, Illinois." *Bulletin of the Ridgway Ornithological Club* no. 2 (April): 26–35.

"Clarke's Nutcracker from the Kowak River, Alaska." *Auk* 4, no. 3 (July): 256.

"Clarke's Nutcracker (*Picicorvus columbianus*) in the Bristol Bay Region, Alaska." *Auk* 4, no. 3 (July): 255.

"Description of a New Genus of Dendrocolaptine Bird from the Lower Amazon." *Proceedings of the U.S. National Museum* 5 (July): 151.

"Description of a New Plumed Partridge from Sonora." *Proceedings of the U.S. National Museum* 5 (July): 148–150.

"Description of a New Species of *Phacellodomus* from Venezuela." *Proceedings of the U.S. National Museum* 5 (July): 152.

"Description of a New Species of *Porzana* from Costa Rica." *Proceedings of the U.S. National Museum* 5 (July): 111.

[Letter to the editors concerning the supposed breeding plumage of *Podiceps occidentalis* Lawrence.] *Ibis* 29, no. 3 (July): 361–362.

"Note on *Spizella monticola ochracea* Brewst." *Auk* 4, no. 3 (July): 258–259.

"Notes on *Ardea wuerdemanni* Baird." *Proceedings of the U.S. National Museum* 5 (July): 112–115.

"*Trogon ambiguus* Breeding in Arizona." *Proceedings of the U.S. National Museum* 5 (July): 147.

"Yellow-Headed Blackbird (*Xanthocephalus xanthocephalus*) in Maine." *Auk* 4, no. 3 (July): 256.

"Description of Two New Species of Kaup's Genus *Megascops*." *Proceedings of the U.S. National Museum* 5 (August): 267–268.

"Description of Two New Races of *Pyrrhuloxia sinuata* Bonap." *Auk* 4, no. 4 (October): 347.

"On the Correct Subspecific Title of Baird's Wren (no. 719b, A.O.U. Checklist)." *Auk* 4, no. 4 (October): 349–350.

"A Correction." *Ornithologist and Oologist* 12, no. 11 (November): 192.

"Description of a New Muscisaxicola, from Lake Titicaca, Peru." *Proceedings of the U.S. National Museum* 5 (November): 430.

"On *Phrygilus gayi* (Eyd. and Gerv.) and Allied Species." *Proceedings of the U.S. National Museum* 5 (November): 431–435.

Manual of North American Birds, Illustrated by 464 Outline Drawings of the Generic Characters. Philadelphia: J. B. Lippincott.

1888

"Grouse and Mallard Plumage." *Forest and Stream* 29, no. 24 (January 5): 463.

"Spencer Fullerton Baird." *Auk* 5, no. 1 (January): 1–14.

"Notes on Some Type-Specimens of American Troglodytidae in the Lafresnaye Collection." *Proceedings of the Boston Society of Natural History* 23 (March): 383–388.

"Description of a New Tityra from Western Mexico." *Auk* 5, no. 3 (July): 263.

"Catalogue of a Collection of Birds made by Mr. Chas. H. Townsend, on Islands in the Caribbean Sea and in Honduras." *Proceedings of the U.S. National Museum* 5 (August): 572–597.

"Descriptions of Some New Species and Genera of Birds from the Lower Amazon." *Proceedings of the U.S. National Museum* 5 (August): 516–528.

"Descriptions of Some New Species and Subspecies of Birds from Middle America." *Proceedings of the U.S. National Museum* 5 (August): 505–510.

"Note on the Generic Name *Uropsila*, Scl. and Salv." *Proceedings of the U.S. National Museum* 5 (August): 511.

"Remarks on *Catharus berlepschi* Lawr." *Proceedings of the U.S. National Museum* 5 (August): 504.

"Review of the Genus *Dendrocincla* Gray." *Proceedings of the U.S. National Museum* 5 (August): 488–497.

"A Review of the Genus *Psittacula* of Brisson." *Proceedings of the U.S. National Museum* 5 (August): 529–548.

[Charles Wickliffe Beckham: Obituary.] *Auk* 5, no. 4 (October): 445–446.

"Description of a New *Psaltriparus* from Southern Arizona." *Proceedings of the U.S. National Museum* 5 (October): 697.

"Supplementary Remarks on the Genus *Psittacula* Brisson." *Auk* 5, no. 4 (October): 460–462.

"Description of a New Pigeon from Guayaquil, Ecuador." *Proceedings of the U.S. National Museum* 6 (November): 112.

"Description of a New Western Subspecies of *Accipiter velox* (Wils.) and Subspecific Diagnosis of *A. cooperi mexicanus* (Swains.)." *Proceedings of the U.S. National Museum* 6 (November): 92.

"Note on *Aestrelata sandwichensis* Ridgw." *Proceedings of the U.S. National Museum* 6 (November): 104.

"Review of the Genus *Dendrocincla* Gray." *Proceedings of the U.S. National Museum* 5 (November): 488–497.

1889

"Description of the Adult Male of *Acanthidops bairdi*." *Proceedings of the U.S. National Museum* 11 (March 12): 196.

"Spring Notes on Migratory Birds." *Forest and Stream* 32, no. 21 (June 13): 420.

"Notes on Costa Rican Birds, with Descriptions of Seven New Species and Subspecies and One New Genus." *Proceedings of the U.S. National Museum* 11 (September): 537–546.

[With Stephen Forbes.] *Birds of Illinois: Part I. Descriptive Catalog / Part II. Economic Ornithology.* [Ridgway wrote the first volume; Forbes, the second, in 1895.]

1890

"*Buteo brachyurus* and *B. fuliginosus*." *Auk* 7, no. 1 (January): 90.

"A Chart of Standard Colors." *Garden and Forest* 3, no. 98 (January): 22–23.

"Intergradation between *Zonotrichia leucophrys* and *Z. intermedia*, and between the Latter and *Z. gambeli*." *Auk* 7, no. 1 (January): 96.

[With J. A. Allen, William Brewster, Elliott Coues, and John H. Sage.] "Second Supplement to the American Ornithologists' Union Check-List of North American Birds." *Auk* 7, no. 1 (January): 60–66.

"A Review of the Genus *Sclerurus* of Swainson." *Proceedings of the U.S. National Museum* 12, no. 762 (February): 21–31.

"A Review of the Genus *Xiphocolaptes* of Lesson." *Proceedings of the U.S. National Museum* 7, no. 761 (February): 1–20.

"Scientific Results of Explorations by the U.S. Fish Commission Steamer Albatross. Birds Collected on the Galapagos Islands in 1888." *Proceedings of the U.S. National Museum* 12, no. 767 (February): 101–128.

"Scientific Results of Explorations by the U.S. Fish Commission Steamer Albatross / No. 2: Birds Collected on the Island of Santa Lucia, West Indies, the Abrolhos Islands, Brazil, and at the Straits of Magellan, in 1887–'88." *Proceedings of the U.S. National Museum* 12, no. 768 (February): 129–139.

"A Northern Station for *Quercus lyrata*." *Garden and Forest* 3, no. 107 (March): 129.

"Harlan's Hawk a Race of the Red-tail, and Not a Distinct Species." *Auk* 7, no. 2 (April): 205.

"Salvin and Godman's Biologia Centrali-Americana—Aves." *Auk* 7, no. 2 (April): 189–195.

"Further Notes on the Genus *Xiphocolaptes* of Lesson." *Proceedings of the U.S. National Museum* 13, no. 796 (July): 47–48.

"*Junco hyemalis shufeldti* in Maryland." *Auk* 7, no. 3 (July): 289.

"A Yellow-Crowned *Regulus calendula.*" *Auk* 7, no. 3 (July): 292.

"Allen on Birds Collected in Bolivia." *Auk* 7, no. 4 (October): 381–382.

"Allen on Birds from Quito." *Auk* 7, no. 4 (October): 380–381.

"Allen on Individual and Seasonal Variation in the Genus *Elainea.*" *Auk* 7, no. 4 (October): 385–386.

"Allen on the Genus *Cyclorhis.*" *Auk* 7, no. 4 (October): 382–384.

"Allen on the Maximilian Types of South American Birds." *Auk* 7, no. 4 (October): 386–387.

"Allen's Descriptions of New South American Birds." *Auk* 7, no. 4 (October): 384–385.

"Observations on the Farrallon Rail (*Porzana jamaicensis coturniculus* Baird)." *Proceedings of the U.S. National Museum* 13, no. 828 (November): 309–311.

1891

"*Falco dominicensis* Gmel. versus *Falco sparveriodes* Vig." *Auk* 8, no. 1 (January): 113–114.

"A New Name Necessary for *Selasphorus floresii* Gould." *Auk* 8, no. 1 (January): 114.

"Note on the Alleged Occurrence of *Trochilus heloisa* (Less. and De Latt.) within the North American Limits." *Auk* 8, no. 1 (January): 115.

[With D. G. Elliot, John H. Sage, J. A. Allen, William Brewster, Elliott Coues, and H. W. Henshaw.] "Third Supplement to the American Ornithologists' Union Check-List of North American Birds." *Auk* 8, no. 1 (January): 83–90.

"*Cistothorus marianae, Buteo lineatus alleni,* and *Syrnium nebulosum alleni* in South Carolina." *Auk* 8, no. 2 (April): 240.

"Fulvous Tree Duck in Missouri." *Forest and Stream* 36, no. 22 (June 18): 435.

"Description of a New Sharp-Tailed Sparrow from California." *Proceedings of the U.S. National Museum* 14, no. 872 (October): 483–484.

"Description of a New Species of Whippoorwill from Costa Rica." *Proceedings of the U.S. National Museum* 14, no. 867 (October): 465–466.

"Description of Two Supposed New Forms of *Thamnophilus.*" *Proceedings of the U.S. National Museum* 14, no. 871 (October): 481.

"List of Birds Collected on the Bahamas Islands by the Naturalists of the Fish Commission Steamer Albatross." *Auk* 8, no. 4 (October): 333–339.

"Note on *Pachyrhamphus albinucha*, Burmeister." *Proceedings of the U.S. National Museum* 14, no. 870 (October): 479–480.

"Notes on Some Birds from the Interior of Honduras." *Proceedings of the U.S. National Museum* 14, no. 868 (October): 467–471.

"Notes on Some Costa Rican Birds." *Proceedings of the U.S. National Museum* 14, no. 869 (October): 473–478.

"Notes on the Genus *Sittasomus* of Swainson." *Proceedings of the U.S. National Museum* 14, no. 877 (October): 507–510.

"Directions for Collecting Birds." *Bulletin of the U.S. National Museum* Part A, no. 39: 1–27.

1892

[With D. G. Elliot, J. A. Allen, William Brewster, Frank M. Chapman, Charles B. Cory, Elliott Coues, H. W. Henshaw, and C. Hart Merriam.] "Fourth Supplement to the American Ornithologists' Union Check-List of North American Birds." *Auk* 9, no. 1 (January): 105–108.

"Transplanting the Trailing Arbutus." *Garden and Forest* 5, no. 218 (April): 202.

"Descriptions of Two New Forms of *Basileuterus rufifrons*, from Mexico." *Proceedings of the U.S. National Museum* 15, no. 895 (July): 119.

"The Humming Birds." *Report of the National Museum for 1890* (July): 253–383.

"Spring Arrivals at Washington, D.C." *Auk* 9, no. 3 (July): 307–308.

"*Zonotrichia albicollis* in California." *Auk* 9, no. 3 (July): 302.

"The Systematic Position of Humming-Birds: Reply to Dr. Shufeldt's 'Discussion.'" *Popular Science News* 26, no. 11 (November): 164–165.

"Nocturnal Songsters and Other Bird-Notes." *Science* 20, no. 515 (December): 343–344.

"Shufeldt on the Anatomy of the Humming-Birds and Swifts." *American Naturalist* 26, no. 313 (December): 1040–1041.

1893

[With D. G. Elliot, J. A. Allen, William Brewster, Frank M. Chapman, Charles B. Cory, Elliott Coues, H. W. Henshaw, and C. Hart Merriam.] "Fifth Supplement to the American Ornithologists' Union Check-List of North American Birds." *Auk* 10, no. 1 (January): 59–63.

"Destruction of Crows during the Recent Cold Spell." *Science* 21, no. 523 (February): 77.

"On the Local Segregation of Trees." *Garden and Forest* 6, no. 266 (March): 148–149.

"The American Plane Tree." *Meehan's Monthly* 3, no. 5 (May): 69–70.

"Age of Guano Deposits." *Science* 21, no. 543 (June): 360.

"Description of Two Supposed New Species of Swifts." *Proceedings of the U.S. National Museum* 16, no. 923 (June): 43–44.

"Description of a Supposed New Species of *Odontophorus* from Southern Mexico." *Proceedings of the U.S. National Museum* 16, no. 945 (July): 469–470.

"On a Small Collection of Birds from Costa Rica." *Proceedings of the U.S. National Museum* 16, no. 956 (October): 609–614.

[Remarks concerning the type specimen of *Malacoptila fuliginosa* Ridgway.] *Proceedings of the U.S. National Museum* 16, no. 947 (October): 513.

"Remarks on the Avian Genus *Myiarchus*, with Special Reference to *M. yucatensis* Lawrence." *Proceedings of the U.S. National Museum* 16, no. 955 (October): 605–608.

[With W. L. Abbott.] "Descriptions of Some New Birds Collected on the Islands of Aldabra and Assumption, Northwest of Madagascar." *Proceedings of the U.S. National Museum* 16, no. 953 (October): 597–600.

"Description of a New Storm Petrel from the Coast of Western Mexico." *Proceedings of the U.S. National Museum* 16, no. 962 (November): 687–688.

"A Revision of the Genus *Formicarius* Boddaert." *Proceedings of the U.S. National Museum* 16, no. 961 (November): 667–686.

"Scientific Results of Explorations by the U.S. Fish Commission Steamer Albatross. . . . No. XXVII—Catalog of a Collection of Birds made in Alaska by Mr. C. H. Townsend during the Cruise of the U.S. Fish Commission Steamer Albatross, in the Summer and Autumn of 1888." *Proceedings of the U.S. National Museum* 16, no. 960 (November): 663–665.

1894

"Allen's List of Birds Collected in Northeastern Sonora and Northwestern Chihuahua." *Auk* 11, no. 1 (January): 66–67.

"Allen's Notice of Some Venezuelan Birds, Collected by Mrs. H. H. Smith." *Auk* 11, no. 1 (January): 66.

"Chapman's Notes on Birds Observed near Trinidad, Cuba." *Auk* 11, no. 1 (January): 67.

"Note on *Rougetius aldabranus*." *Auk* 11, no. 1 (January): 74.

[With Elliott Coues, J. A. Allen, William Brewster, and C. Hart Merriam.] "Sixth Supplement to the American Ornithologists' Union Check-List of North American Birds." *Auk* 11, no. 1 (January): 46–51.

"Description of a New *Geothlypis* from Brownville, Texas." *Proceedings of the U.S. National Museum* 16, no. 964 (February): 691–692.

"Chapman on the Birds of the Island of Trinidad." *Auk* 11, no. 2 (April): 173.

[Description of *Pipilo orizabae* Cox.] *Auk* 11, no. 2 (April): 161.

"On Geographical Variation in *Sialia mexicana* Swainson." *Auk* 11, no. 2 (April): 145–160.

"*Picicorvus* an Untenable Genus." *Auk* 11, no. 2 (April): 179.

"Geographical, versus Sexual, Variation in *Oreortyx pictus*." *Auk* 11, no. 3 (July): 193–197.

"Nester und Eier der Kolibris." [Translated by Oscar Haase.] *Zeitschrift für Oologie* 4, no. 6 (September): 23–24.

"*Colinus virginianus cubanensis* Not a Florida Bird." *Auk* 11, no. 4 (October): 324.

"We, also, Take Exceptions." *Nidologist* 2, no. 2 (October): 29.

"Descriptions of Some New Birds from Aldabra, Assumption and Gloriosa Islands, Collected by Dr. W. L. Abbott." *Proceedings of the U.S. National Museum* 17, no. 1008 (November): 371–373.

"Descriptions of Twenty-Two New Species of Birds from the Galapagos Islands." *Proceedings of the U.S. National Museum* 17, no. 1007 (November): 357–370.

1895

"Additional Notes on the Trees of the Lower Wabash Valley." Pls. X–XV (reproduced photographs). *Proceedings of the U.S. National Museum* 17, no. 1010 (January): 409–421.

[With Elliott Coues, J. A. Allen, William Brewster, and C. Hart Merriam.] "Seventh Supplement to the American Ornithologists' Union Check-List of North American Birds." *Auk* 12, no. 2 (April): 163–169.

[Letter to the editor concerning the near annihilation of bluebirds in the District of Columbia by the blizzard of February 7–9, 1895.] *Christian Register* 74, no. 19 (May 9): 301.

"Klugheit der Kolibris" [Translated by Oscar Haase.] *Zeitschrift für Oologie* 5, no. 5 (August): 19.

"On the Correct Subspecific Names of the Texan and Mexican Screech Owls." *Auk* 7, no. 4 (October): 389–390.

"On Fisher's Petrel (*Aestrelata fisheri*)." Pl. IV. *Auk* 12, no. 4 (October): 319–322.

"*Junco phaeonotus* Wagler, not *J. cinereus* (Swainson)." *Auk* 12, no. 4 (October): 391.

"Ornithology of Illinois." *Proceedings of the U.S. National Museum* 7, 1–314.

1896

Manual of North American Birds. Illustrated by 464 Outline Drawings of the Generic Characters. 2nd ed. rev., with new preface and appendix; published March 7.

"Description of a New Species of Ground Warbler from Eastern Mexico." *Proceedings of the U.S. National Museum* 18, no. 1045 (April): 119–120.

"Preliminary Descriptions of Some New Birds from the Galapagos Archipelago." *Proceedings of the U.S. National Museum* 18, no. 1067 (April): 293–294.

"Description of a New Subspecies of the Genus *Peucedramus,* Coues." *Proceedings of the U.S. National Museum* 18, no. 1074 (May): 441.

[Letter to the editor and publisher of the *Nidologist* concerning his contemplated return to California.] *Nidologist* 3, no. 9 (May): 99.

"On Birds Collected by Doctor W. L. Abbott in the Seychelles, Amirantes, Gloriosa, Assumption, Aldabra, and Adjacent Islands. With Notes on Habits, etc., by the Collector." *Proceedings of the U.S. National Museum* 18, no. 1079 (June): 509–546.

"Characters of a New Subspecies of Passerine Birds." *Proceedings of the U.S. National Museum* 18, no. 1076 (June): 449–450.

"Have We Two Native Species of Trumpet Flower?" *Garden and Forest* 9, no. 455 (November): 453–454.

[Comparative characters of *Aestrelata fisheri* Ridgway and *Ae. gularis* (Peale).] In *Catalogue of the Birds in the British Museum* 25, 415–416.

[Description of *Aestralata longirostris* Stejneger.] In *Catalogue of the Birds in the British Museum* 25, 418.

[Description of *Oceanodroma macrodactyla* (Bryant).] In *Catalogue of the Birds in the British Museum* 25, 351.

[Description of *Oceanodroma socorreoensis* Townsend.] In *Catalogue of the Birds in the British Museum* 25, 352.

[Description of *Oceanodroma tristami* Stejneger, MS.] In *Catalogue of the Birds in the British Museum* 25, 354–355.

[Letter to Dr. G. Browne Goode concerning Ridgway's opinion of the services to ornithology of eminent English ornithologist Dr. Philip Lutley Sclater.] "Published Writings of Philip Lutley Sclater, 1844–1896." *Bulletin of the U.S. National Museum* 49, 18–19.

[List of private collections of birds, containing more than 1000 specimens, which have been presented at various times to the US National Museum.] *Report of the U.S. National Museum for 1893–94,* 48–49.

[Results of comparison of a specimen of *Aestrelata affinis* (Buller) with the types of *Ae. gularis* (Peale) and *Ae. fisheri* Ridgway.] In *Catalogue of the Birds in the British Museum* 25, 415.

1897
"Correct Nomenclature of the Texas Cardinal." *Auk* 14, no. 1 (January): 95.

"*Dendroica caerula* vs. *Dendroica rara.*" *Auk* 14, no. 1 (January): 97.

"*Melopelia leucoptera* in Osceola County, Florida." *Auk* 14, no. 1 (January): 88–89.

"Note on *Junco annectens* Baird and *J. ridgwayi* Mearns." *Auk* 14, no. 1 (January): 94.

[With William Brewster, J. A. Allen, Elliott Coues, and C. Hart Merriam.] "Eighth Supplement to the American Ornithologists' Union Check-List of North American Birds." *Auk* 14, no. 1 (January): 117–135.

"Birds of the Galapagos Archipelago." *Proceedings of the U.S. National Museum* 19, no. 1116 (March): 459–670.

"Where Junco Roosts." *Wilson Bulletin* 9, no. 3 (May): 25–27.

"Chapman's 'Bird Life.'" *Auk* 14, no. 3 (July): 336–338.

"Description of the Nest and Eggs of Bachman's Warbler (*Helminthophila bachmanii*)." *Auk* 14, no. 3 (July): 309–310.

"An Earlier Name for *Ammodramus leconteii*." *Auk* 14, no. 3 (July): 320.

"On the Status of *Lanius robustus* Baird as a North American Bird." *Auk* 14, no. 3 (July): 323.

[Remarks concerning *Megascops pinosus* Nelson and Palmer.] Aves, *Biologia Centrali-Americana* 3 (November): 17.

[Article on color of fruits of *Cissus ampelofsis* (=*Ampelofsis cordata*) and *C. stans* (=*A. arborea*).] *Garden and Forest* 5, no. 513 (December): 498.

"An Amateur's Experiment." *Garden and Forest* 5, no. 513 (December): 504–507. [Article about Ridgway's trees and plantings at Brookland.]

1898

"Birds of the Galapagos Islands." *American Naturalist* 32, no. 377 (May): 386–389.

[Note on *Polypodium polypodoides* as observed in Southern Florida.] *Plant World* 1, no. 9 (June): 137.

"Descriptions of Supposed New Genera, Species, and Subspecies of American Birds. I. Fringillidae." *Auk* 15, no. 3 (July): 223–230.

"Description of a New Species of Humming-Bird from Arizona." *Auk* 15, no. 4 (October): 325–326.

"*Hemithraupis:*—A Correction." *Auk* 15, no. 4 (October): 330–331.

"New Species, etc., of American Birds.—II. Fringillidae (continued)." *Auk* 15, no. 4 (October): 319–324.

"The Home of the Ivory-Bill." *Osprey* 3, no. 3 (November): 35–36.

1899

"New Species, etc., of American Birds.—III. Fringillidae (continued)." *Auk* 16, no. 1 (January): 35–37.

"On the Generic Name *Aimophila* versus *Peucaea*." *Auk* 16, no. 1 (January): 80–81.

"On the Genus *Astragalinus* Cabanis." *Auk* 16, no. 1 (January): 79–80.

"A Fraud—Look Out for Him!" *Osprey* 3, no. 6 (February): 94.

"New Species, etc., of American Birds.—IV. Fringillidae (concluded); Corvidae (Part)." *Auk* 16, no. 3 (July): 254–256.

[With Witmer Stone, Joel Allen, C. Hart Merriam, et al.] "Hints to Young Bird Students." *Bird-Lore* 1, no. 4 (August), 125–127.

1900

"New Species, etc., of American Birds—V. Corvidae (concluded)." *Auk* 17, no. 1 (January): 27–29.

"New Species, etc., of American Birds—VI. Fringillidae (Supplement)." *Auk* 17, no. 1 (January): 29–30.

"Concerning the Use of Scientific Names." *Condor* 2, no. 2 (April): 41.

"Song Birds of Europe and America." *Bird-Lore* 2, no. 3 (June): 69–75.

[Description of *Buteo borealis socorroensis*.] Aves, *Biologia Centrali-Americana* 3 (November): 64.

Manual of North American Birds. Illustrated by 464 Outline Drawings of the Generic Characters. 4th ed. [Second reprinting of the second edition, with no changes or additions.]

1901

"New Birds of the Families Tanagridae and Icteridae." *Proceedings of the Washington Academy of Science* 3 (April): 149–155.

The Birds of North and Middle America. A Descriptive Catalogue of the Higher Groups, Genera, Species, and Subspecies of Birds Known to Occur in North America, from the Arctic Islands to the Isthmus of Panama, the West Indies and Other Islands of the Caribbean Sea, and the Galapagos Archipelago. Part I: Family Fringillidae—The Finches. No. 50, Part I. [i–iv] v–xxx, 1–715, pls. I–XIX (outlines of generic characters). Washington, DC: US National Museum. (October): 1–745. [Cited as *Bulletin* 50.]

1902

"Descriptions of Three New Birds of the Families Mniotiltidae and Corvidae." *Auk* 19, no. 1 (January): 69–70.

"The Elf Owl in California." *Condor* 4, no. 1 (January): 18–19.

[With Willliam Brewster.] "Birds of the Cape Region of Lower California." *Bulletin of the Museum of Comparative Zoology* 41, no. 1 (September), 1–241.

The Birds of North and Middle America. A Descriptive Catalogue of the Higher

Groups, Genera, Species, and Subspecies of Birds Known to Occur in North America, from the Arctic Islands to the Isthmus of Panama, the West Indies and Other Islands of the Caribbean Sea, and the Galapagos Archipelago. Part II: Family Tanagridae—The Tanagers. Family Icteridae—The Troupials. Family Coerebidae—The Honey Creepers. Family Mniotiltidae—The Wood Warblers. No. 50, Part II. [i–vi], vii–xx, 1–834, pls. I–XXII (outlines generic characters). [Bulletin 50.] (October): 1–854.

1903

"Pycraft's Classification of the Falconiformes." *Science,* n.s., 17, no. 430 (March): 509–511.

"*Lophophanes* vs. *Baeolophus.*" *Auk* 20, no. 3 (July): 308.

"Relationships of the Madagascar Genus *Hypositta* Newton." *Proceedings of the Biological Society of Washington* 16 (September): 125.

"Diagnosis of Nine New Forms of American Birds." *Proceedings of the Biological Society of Washington* 16 (November): 167–170.

1904

"*Nannorchilus,* a New Name for *Hemiura,* Preoccupied." *Proceedings of the Biological Society of Washington* 17 (April): 102.

"Descriptions of Seven New Species and Subspecies of Birds from Tropical America." *Smithsonian Miscellaneous Collections* 47 (August): 112–113.

The Birds of North and Middle America. A Descriptive Catalogue of the Higher Groups, Genera, Species, and Subspecies of Birds Known to Occur in North America, from the Arctic Islands to the Isthmus of Panama, the West Indies and Other Islands of the Caribbean Sea, and the Galapagos Archipelago. Part III: Family Motacillidae—The Wagtails and Pipits. Family Hirundinidae—The Swallows. Family Ampelidae—The Waxwings. Family Ptilogonatidae—The Silky Flycatchers. Family Dulidae—The Palm Chats. Family Vireonidae—The Vireos. Family Laniidae—The Shrikes. Family Corvidae—The Crows and Jays. Family Paridae—The Titmice. Family Sittidae—The Nuthatches. Family Certhiidae—The Creepers. Family Troglodytidae—The Wrens. Family Cinclidae—The Dippers. Family Chamaeidae—The Wren-Tits. Family Sylviidae—The Warblers. No. 50, Part III. i–xx, 1–801, pls. I–XIX. [*Bulletin* 50.] (December): 1–840.

1905

[Notes on nesting boxes.] *Bird-Lore* 7, no. 1 (January): 18.

[Bibliography of the publications by Robert Ridgway.] *Indiana University Bulletin* 2, no. 6 (March): 125–142.

"Description of Some New Genera of Tyrannidae, Pipridae, and Cotingidae."

Proceedings of the Biological Society of Washington 18 (September): 207–210.

"Description of an Adult Female Euphonia Supposed to Be *Euphonia gnatho* (Cabanis)." *Proceedings of the Biological Society of Washington* 18 (October): 225–226.

"New Genera of Tyrannidae and Turdidae and New Forms of Tanagridae and Turdidae." *Proceedings of the Biological Society of Washington* 18 (October): 211–214.

"A Winter with the Birds in Costa Rica." *Condor* 7, no. 6 (November): 151–160. [With full-page frontispiece and seven reproduced photographs.]

1906

"Some Observations concerning the American Families of Oligomyodian Passeres." *Proceedings of the Biological Society of Washington* 20 (January): 7–16.

"*Atratus* versus *Megalonyx*." *Condor* 8, no. 2 (March): 53.

"*Atratus* versus *Megalonyx*." *Condor* 8, no. 4 (July): 100.

"Descriptions of Some New Forms of Oligomyodian Birds." *Proceedings of the Biological Society of Washington* 19 (September): 115–120.

1907

"*Cinclus mexicanus* Not a Costa Rican Bird." *Auk* 24, no. 1 (January): 105.

The Birds of North and Middle America. A Descriptive Catalogue of the Higher Groups, Genera, Species, and Subspecies of Birds Known to Occur in North America, from the Arctic Lands to the Isthmus of Panama, the West Indies and Other Islands of the Caribbean Sea, and the Galapagos Archipelago. Part IV. Family Turdidae—Thrushes. Family Zeledoniidae—Wren-Thrushes. Family Mimidae—Mocking-birds. Family Sturnidae—Starlings. Family Ploceidae—Weaver Birds. Family Alaudidae—Larks. Family Oxyruncidae—Sharp-bills. Family Tyrannidae—Tyrant Flycatchers. Family Pipridae—Manakins. Family Cotingidae—Chatterers. No. 50, Part IV. i–xxii, 1–973, pls. I–XXXIV. [*Bulletin* 50.] (July): 1–1029.

[With C. H. Ames.] "A Novel Theory of Extinction." Chapter 4 in W. B. Mershon, *The Passenger Pigeon*, 173–178. New York: Outing.

1908

"Red-Spotted Bluethroat in Alaska." *Auk* 25, no. 2 (April): 226.

"Type Locality of *Vireo pusillus*." *Auk* 25, no. 2 (April): 224–225.

"Diagnoses of Some New Forms of Neotropical Birds." *Proceedings of the Biological Society of Washington* 21 (October): 191–195.

1909

"New Genera, Species, and Subspecies of Formicariidae, Furnariidae, and Dendrocolaptidae." *Proceedings of the Biological Society of Washington* 22 (April): 69–74.

"Hybridism and Generic Characters in the Trochilidae." *Auk* 26, no. 4 (October): 440–442.

[Letter to the editor announcing successor to *Nomenclature of Colors.*] "New Edition of Ridgway's 'Nomenclature of Colors.'" *Auk* 26, no. 4 (October): 450.

[Letter to the editors regarding the publication of *Color Standards and Color Nomenclature.*] *Ibis* 51, no. 4 (October): 714–715.

[Transfer of the Division of Birds to the New U.S. National Museum.] *Auk* 26, no. 4 (October): 454–455.

[Letter to the editor announcing particulars in connection with the publication of *Color Standards and Color Nomenclature.*] *Condor* 11, no. 6 (November): 210.

[Frank H. Knowlton. Edited by Ridgway.] *Birds of the World: A Popular Account.* New York: Henry Holt.

1910

"Concerning Three Alleged 'Erroneous Georgia Records.'" *Auk* 27, no. 1 (January): 88.

"Diagnoses of New Forms of Micropodidae and Trochilidae." *Proceedings of the Biological Society of Washington* 24 (April): 53–56.

1911

"Diagnoses of Some New Forms of Picidae." *Proceedings of the Biological Society of Washington* 1911 24 (February): 31–36.

The Birds of North and Middle America. A Descriptive Catalogue of the Higher Groups, Genera, Species, and Subspecies of Birds Known to Occur in North America, from the Arctic Lands to the Isthmus of Panama, the West Indies and Other Islands of the Caribbean Sea, and the Galapagos Archipelago. Part V. Family Pteroptochidae — The Tapaculos. Family Formicariidae — The Antbirds. Family Furnariidae — The Ovenbirds. Family Dendrocolaptidae — The Woodhewers. Family Trochilidae — The Humming Birds. Family Micropopideae — The Swifts. Family Trogonidae — The Trogons. No. 50, Part V. i–xxiii, 1–859, pls. I–XXXIII. [*Bulletin* 50.] (November): 1–892.

1912

"Descriptions of Some New Species and Subspecies of Birds from Tropical America." *Proceedings of the Biological Society of Washington* 25 (May): 87–92.

"Diagnoses of Some New Genera of American Birds." *Proceedings of the Biological Society of Washington* 25 (May): 97–102.

Color Standards and Color Nomenclature, with Fifty-Three Colored Plates and Eleven Hundred and Fifteen Named Colors. Privately published.

1914

"Bird Life in Southern Illinois. I. Bird Haven." *Bird-Lore* 16, no. 6 (November): 409–420. [Seven illustrations.]

The Birds of North and Middle America. A Descriptive Catalogue of the Higher Groups, Genera, Species, and Subspecies of Birds Known to Occur in North America, from the Arctic Lands to the Isthmus of Panama, the West Indies and Other Islands of the Caribbean Sea, and the Galapagos Archipelago. Part VI. Family Picidae—The Woodpeckers. Family Capitonidae—The Barbets. Family Ramphastidae—The Toucans. Family Bucconidae—The Puff Birds. Family Galbulidae—The Jacamars. Family Alcedinidae—The Kingfishers. Family Todidae—The Todies. Family Momotidae—The Motmots. Family Caprimulgidae—The Goatsuckers. Family Nyctibiidae—The Potoos. Family Tytonidae—The Barn Owls. Family Bubonidae—The Eared Owls. No. 50, Part VI. i–xx, 1–882, pll. I–XXXVI. [*Bulletin 50.*] (April): 1–902.

1915

"Bird Life in Southern Illinois. II. Larchmound: A Naturalist's Diary." *Bird-Lore* 17, no. 1 (January): 1–7.

"Bird Life in Southern Illinois. III. Larchmound: A Naturalist's Diary." *Bird-Lore* 17, no. 2 (March): 91–103.

"Bird Life in Southern Illinois. IV. Changes Which Have Taken Place in Half a Century." *Bird-Lore* 17, no. 3 (May): 191–198.

"Descriptions of Some New Forms of American Cuckoos, Parrots, and Pigeons." *Proceedings of the Biological Society of Washington* 28 (May): 105–108.

"A New Pigeon from Chiriqui, Panama." *Proceedings of the Biological Society of Washington* 28 (June): 139.

"A New Pigeon from Jamaica." *Proceedings of the Biological Society of Washington* 28 (November): 177.

1916

The Birds of North and Middle America. A Descriptive Catalogue of the Higher Groups, Genera, Species, and Subspecies of Birds Known to Occur in North America, from the Arctic Lands to the Isthmus of Panama the West Indies and Other Islands of the Caribbean Sea, and the Galapagos Archipelago. Part VII. Family Cuculidae. The Cuckoos. Family Psittacidae. The Parrots. Family Columbidae. The Pigeons. No. 50, Part VII. i–xiii, 1–543, pls. I–XXIV. [*Bulletin* 50.] (May): 1–556.

1919

The Birds of North and Middle America. A Descriptive Catalogue of the Higher Groups, Genera, Species, and Subspecies of Birds Known to Occur in North America, from the Arctic Lands to the Isthmus of Panama the West Indies and Other Islands of the Caribbean Sea, and the Galapagos Archipelago. Part VIII. Family Jacanidae—The Jacanas. Family Oedicnemidae—The Thick-knees. Family Haematopodidae—The Oyster-catchers. Family Arenariidae—the Turnstones. Family Aphrizidae—The Surf Birds. Family Charadriidae—The Plovers. Family Scolopacidae—The Snipes. Family Phalaropodidae—The Phalaropes. Family Recurvirostridae—The Avocets and Stilts. Family Rynchopidae—The Skimmers. Family Sternidae—The Terns. Family Laridae—The Gulls. Family Stercorariidae—The Skuas. Family Alcidae—The Auks. No. 50, Part VIII. i–xvi, 1–852, pls. I–XXXIV. [*Bulletin* 50.] (June): 1–868.

1920

"Diagnoses of Some New Genera of Birds." Publication 2588, *Smithsonian Miscellaneous Collections* 72, no. 4 (December): 1–4.

1922

"Introduction." In *Check-List of the Birds of Illinois,* by Benjamin T. Gault. Chicago: Illinois Audubon Society.

1923

"'Generic Subdivision'—'The Genus Debased.'" *Auk* 40, no. 2 (April): 371–375.

"A Plea for Caution in Use of Trinomials." *Auk* 40, no. 2 (April): 375–376.

"What Is *Buteo rufescentior* Salvin and Godman?" *Auk* 40, no. 2 (April): 325.

"In Memoriam: José Castulo Zeledón." *Auk* 40, no. 4 (October): 682–689.

1924
"Additional Notes on *Hyla phaeocrypta* (?)." *Copeia* 128 (March): 39.
"A Scene at Bird Haven." *Audubon Bulletin* (Chicago) (Spring–Summer): 46.

1925
"Is the Love of Trees and Flowers a Sign of Effeminacy?" *Olney (IL) Times* (March 28): 5.
"Dr. Ridgway Tells of His Early Days." *Olney (IL) Times* (June 4): 1–2.
"Spring Notes at 'Larchmound.'" *Audubon Bulletin* (Chicago) (Summer): 46.
"The Birds of 'Larchmound'—A Resumé." *Bird-Lore* 27, no. 5 (September): 305–309. [Three illustrations.]
"Diagnosis of a New Genus of Buteonine Hawks (*Coryornis*, gen. nov.)." *Auk* 42, no. 4 (October): 585.
"Introduction." In *Bird Companions*, by Angela K. Main. Boston: Gorham Press.
"Introduction." In *Our Bird Friends and Foes*, by William Atherton DuPuy. Philadelphia: J. C. Winston Co.

1926
"As to the Type of *Falco peregrinus pealei*." *Condor* 28, no. 5 (September): 240.

1927
"The Advancing House Wren." *Cardinal* 2, no. 2: 34.
"Birds in Their Relation to the Farmer and Fruit Grower." *Audubon Bulletin* (Chicago), no. 18 (Spring–Summer): 29–33.
"Bird Haven: What It is and What It Is not." *Olney (IL) Advocate* (October 27).

1929
"Bird Haven—Its Purpose and Present Status." *Bird-Lore* 31, no. 1 (February): 1–6. [Three illustrations.]
[Open letter to Frank M. Chapman.] *Bird-Lore* 31, no. 3 (June): 176–178.

Abbreviations

AL	Alexander Library, Edward Grey Institute of Field Ornithology, Oxford University [now part of the Radcliffe Science Library, Bodleian Libraries, Oxford]
AMNH	American Museum of Natural History, Ornithology Department, New York
B-W	Blacker-Wood Library (now the Blacker-Wood Collection in Zoology and Ornithology), Special Collections Libraries, McGill University, Montreal
BAN	University of California, Berkeley, Bancroft Library. All items are from BANC MSS 67/125Z, Joel Allen Papers
HEHL	Henry E. Huntington Library, San Marino, California
MCZ	Museum of Comparative Zoology Archives, Harvard University, Cambridge, Massachusetts. All items are from the incoming correspondence files of William Brewster
NHM	Natural History Museum, Archives, Department of Zoology Collections, London, England
NOC *Bulletin*	*Bulletin of the Nuttall Ornithological Club*
RR	Robert Ridgway
SIA	Smithsonian Institution Archives, Washington, DC
USU	Utah State University, Logan. All items are from one collection: Caine Manuscript Collection 8, Papers of Robert and John Ridgway (Caine MSS 8)

Notes

Preface

"Prejudiced": Robert Ridgway (hereafter "RR") to William Brewster, Mar. 10, 1906, MCZ. "Field, not closet": Nathaniel Stickley Goss to Joel Allen, Oct. 30, 1883, B-W. "Please do not consider it necessary": Montague Chamberlain to Allen, Nov. 18, 1883, BAN; ignoramuses: "D.R." to Henry Eliot Howard, Dec. 10, 1915, AL; wastepaper basket: Helen M. Rait Kerr to H. F. Witherby, n.d., AL; Farber, *Discovering Birds,* 129; Desmond, "Redefining the X Axis," 4; Allman, "Present Aspects of Biology," 35.

"We are everything and everybody": Elliott Coues to Allen, Nov. 4, 1883, B-W. As one historian of science has noted, "One revisionist approach that might work is a careful examination of the language—the particular vernacular—by which American men of science described themselves and their communities." Lucier, "Professional and Scientist," 703. "The average 'Yankee'": Clinton Hart Merriam to Allen, Mar. 17, 1884, B-W. "Professionals were men of science who engaged in commercial relations": Lucier, "Professional and Scientist," 704.

Chapter 1. The Making of a Bird Man

Epigraph: Fannie Gunn to Robert Ridgway (hereafter "RR"), Mar. 10, 1867, box 1, fol. 30, USU.

1. Kohler, *All Creatures,* 230.

2. The only substantial (118 pp.) biographical sketch of RR was written in 1928 by Harry Harris, a California writer who published an account of him as a special issue of *The Condor,* the journal of the Cooper Ornithological Club, in a piece entitled "Robert Ridgway: With a Bibliography of His Published Writings and Fifty Illustrations" (vol. 30, no. 1). Ridgway provided most of the information to Harris, who interviewed a number of his colleagues as well as his brother John for the piece. The useful bibliography of Ridgway's publications goes up to 1927. The other published account of his life is a brief (12 pp.) account written in 1932 by Smithsonian secretary Alexander Wetmore, "Biographical Memoir of Robert Ridgway," which adds some new elements to his bibliography (which see in Appendix).

3. US Bureau of Census, *Seventh Population Census*, 1850, and *Eighth Population Census*, 1860.

4. Harris, "Ridgway," 41.

5. Ibid., 10.

6. RR's longtime friend Henry W. Henshaw, one of the founders of the Nuttall Ornithological Club and later chief of the US Biological Survey, whistled so much that one day while working for the US Geological Survey at the Smithsonian, Robert E. C. Stearns, the assistant curator of mollusks, walked up to Henshaw, waved a hammer in his face, and shouted, "Young man, if you don't stop that whistling there will be a vacancy in the Geological Survey!" Ibid., 32, 33.

7. Henrietta Ridgway to RR, Nov. 24, 1872, Caine MSS 8, box 1, fol. 25, USU; RR to John Ridgway, Dec. 7, 1926, B-W.

8. *Mt. Carmel Democrat*, June 16, 1877; Joel Allen to RR, June 22, 1877, B-W.

9. Harris, "Ridgway," 16; Wetmore, "Biographical Memoir," 58. Ridgway first used that description of Holloway in a letter to the *Olney (IL) Times*, June 4, 1925, 5.

10. For example, see Baird, "Distribution and Migrations of North American Birds," an abstract of a talk by Baird to the National Academy of Sciences the previous year.

11. Ridgway, "Spencer Fullerton Baird" (1888): 6.

12. RR to Spencer Fullerton Baird, Aug. 7, 1864, box 1, fol. 1, USU.

13. RR to Baird, Oct. 17, 1864, RU 7167, SIA.

14. "Warm and good heart": RR to Henrietta Ridgway, June 14, 1869, box 1, fol. 22, USU; Turner, "Umbrella."

15. Circumstances: RR to Baird, Nov. 13, 1865, RU 7167, SIA; toy paints: RR to Baird, Apr. 7, 1865, Jan. 27, Feb. 13, 1866, ibid.; elegance and accuracy: Ridgway, "Spencer Fullerton Baird," 13. Ridgway was not the only young teen of great seriousness of purpose trying to start a life in biology. George Sudworth was just thirteen when he wrote to Elliott Coues in 1877, sending bird specimens and asking for information. Sudworth would go on to be the dean of American forestry—a distinguished authority on American trees with few peers in the field. See Sudworth to Coues, n.d., March 1877, B-W.

16. Goodrich: RR to Baird, Oct. 3, 1866, RU 7167, SIA; *Birds of North America*: Baird to RR, Oct. 26, 1866, box 1, fol. 13, USU.

17. "Child's play": RR to Baird, Nov. 13, 1865, RU 7167, SIA; "thanks": RR to Baird, n.d. but first half of 1866, ibid.

18. Freeman, *Works*, 83.

19. "Discovery of time" and "knives in their brains": Goetzmann, *New Lands*, 400, 401.

20. Robert Wilson, *Explorer King*, 181-182.

21. For a detailed examination of the uses of modern study collections, see Remsen, "Importance of Continued Collecting."

22. Goetzmann, "Paradigm Lost," 24. This notion is also echoed in Goldstein, "Yours for Science." Goetzmann's *Exploration and Empire* is the standard—and still unsurpassed—work on scientific exploration of the American West. Another seminal work is Bartlett, *Great Surveys*.

23. Harris, "Ridgway," 18, citing letter from Baird to RR, Mar. 30, 1867.

24. A. E. Verrill to Allen, Mar. 21, 27, 1867, AMNH.

25. David Ridgway to Baird, Apr. 15, 1867, as quoted in Davenport, "Robert Ridgway," 276.

26. Salary and RR's choice: Harris, "Ridgway," 19; Fannie Gunn to RR, Mar. 10, 1867, box 1, fol. 30, USU.

27. "Peculiar talent": Henrietta Ridgway to RR, Aug. 4, 1867, ibid., fol. 20; "well contented": RR to David Ridgway, Apr. 18, 1867, ibid., fol. 2.

28. RR to Henrietta Ridgway, May 5, 1867, ibid.

29. "Striking personality": Bailey, HM 39965, HEHL; lemon-yellow gloves: Goetzmann, *New Lands,* 401; "clothing without a wrinkle": Century Association, *Clarence King Memoirs,* 345; "kind and agreeable": RR to Henrietta Ridgway, May 5, 1867, box 1, fol. 2, USU; "musical, infectious laugh": Bailey, HM 39965, HEHL; sunset: Maslin, "Sometimes One Man Can Have Two Lives."

30. Fearlessness: Clarence King, notes in James D. Hague's manuscript, "C.K.'s notes for my biographical sketch of him for Appleton's Encycl[opedia]," n.d., A-1, box 1, Clarence King Papers, Hague Collection, HEHL; King as black man: Sandweiss, *Passing Strange.*

31. Sources give two different dates for the departure of the expedition. The Harris memoir (p. 21) notes May 10 as the day of the second group's departure, although Ridgway was reconstructing details more than sixty years later. Bailey, in a manuscript memoir written more than forty years later, notes May 11 as the date, as does Ridgway in a letter written home a day and a half before departing. RR to David Ridgway, May 9, 1867, box 1, fol. 2, USU. Wilkins also notes May 1 and 11 as the departure dates. Letter of introduction for Ridgway: Baird, May 8, 1867, ibid., fol. 9.

32. RR to David Ridgway, May 17, 1867, ibid., fol. 3.

33. RR to Baird, Oct. 17, 1864, RU 7167, SIA.

34. William W. Bailey to William M. Bailey, June 3, 1867, HM 39966, HEHL.

35. "Appalled": Bailey, HM 39965, ibid.; Asa Gray: Wilkins, *Clarence King,* 104. Although unexamined by me, Bailey's diary on the trip is available in the New York Botanical Garden's archives, http://library.nybg.org/finding_guide/archv/bailey_ppb.html#series2.

36. RR to David Ridgway, May 9, 1867, box 1, fol. 2, USU.

37. Rivinus and Youssef, *Baird,* 83.

38. "Belligerent roosters": William W. Bailey to Loring Bailey, June 16, 1867, HM 27839, HEHL; "enjoying myself finely": RR to Henrietta Ridgway, June 24, 1867, box 1, fol. 4, USU; Ridgway fit and trim: RR to Henrietta Ridgway, Aug. 11, 1867, ibid., fol. 5.

39. Wilkins, *Clarence King,* 111. "Frequent and gratuitous service": Ridgway, *Report of the Geological Exploration,* 307; Parker's humor: William W. Bailey to Loring Bailey, Aug. 14, 1867, HM 27841, and Nov. 19, 1867, HM 27847, HEHL.

40. "No head and no discipline": William W. Bailey to Loring Bailey, Mar. 1, 1868, HM 27854, HEHL; "time flies": William W. Bailey to Loring Bailey, Feb. 2, 1868, HM 27852, ibid. Also see Hague Papers, box 23 Addenda, HEHL, for James Hague's diaries, which detail members' churchgoing and other activities in 1867 and 1868 while on the trip. The metaphor of the railroad is also relevant here as part of the back story for one key purpose of the trip: railroad surveying. For sophisticated commentary on the notion of the railroad as an annihilator of time and space, see Schivelbusch, *Railway Journey.*

41. Ridgway, "Notes on the Bird Fauna of the Salt Lake Valley" (1873); Ridgway, "List of Birds Observed at Various Localities" (1875): 12–13.

42. William W. Bailey to Loring Bailey, Oct. 27, 1867, HM 27845, Sept. 29, 1867, HM 27843, HEHL. See also other Bailey letters from his time on the Central American portion of the trip in this grouping, HEHL.

43. These synonyms are detailed in printed form in Ridgway's *Report of the Geological Exploration*, with Washoe names noted under specific birds' descriptions, and itemized in a letter from RR to David Ridgway, Oct. 15, 1867, box 1, fol. 5, USU; Paiute peaceable: RR to Henrietta Ridgway, Aug. 11, 1867, ibid.

44. Ridgway, *Report of the Geological Exploration*, 333-334; Wilkins, *Clarence King*, 105, 116.

45. "My own Greenland": William W. Bailey to Loring Bailey, Jan. 19, 1868, HM 27851, HEHL; "slept with Grant and Meade": William W. Bailey to Loring Bailey, Aug. 14, 1867, HM 27841, ibid.

46. "Annihilating space": William W. Bailey to Loring Bailey, Mar. 1, 1868, HM 27854, HEHL; "Hell": Wilkins, *Clarence King*, 123; "peevish": William W. Bailey to William M. Bailey, Jan. 4, 1868, HM 39968, ibid.

47. War Department demurs: Becker, "Biographical Notice of Samuel Franklin Emmons," 645; "jewel of a darkie cook": William W. Bailey to Loring Bailey, Sept. 6, 1867, HM 27842, HEHL; "you would not know me": RR to Henrietta Ridgway, Aug. 11, 1867, box 1, fol. 5, USU; Ridgway's height and weight: RR to Baird, Sept. 11, 1870, RU 7002, box 32, fol. 10, SIA.

48. Family illnesses: Henrietta Ridgway to RR, Mar. 23, 1868, box 1, fol. 21, USU; "wild beastes": Henrietta Ridgway to RR, Aug. 4, 1867, ibid., fol. 10.

49. References to this segment of the trip can be found in Ridgway's volume of the *Geological Exploration of the Fortieth Parallel*, 366-390, as well as in several pieces of his correspondence: RR to David Ridgway, Jan. 20, 1869, box 1, fol. 7, USU; Frederick A. Clark to RR, May 2, 1869, RU 105, ser. 2, box 15, fol. 1, SIA; and Samuel F. Emmons to RR, May 6, 1869, ibid., box 17, fol. 2. Also see Ketner and Ketner, "Comparisons of Avian Populations." In general, the avian community in the Salt Lake area now appears to be more diverse than it was in 1869, although the reasons for this are not clear. Ridgway "fallen in love with Terra": King to James T. Gardiner, June 22, 1869, HM 27826, HEHL; tired of being on the road: RR to Baird, June 16, 1870, RU 7002, box 32, fol. 10, SIA.

50. Mother and sister write to Baird: Henrietta and Fannie Ridgway to Baird, Sept. 7, 1870, ibid.; "not only foolish": Henrietta Ridgway to RR, Feb. 13, 1869, box 1, fol. 22, USU.

51. Rivinus and Youssef, *Spencer Baird*, 106-108; original is letter from Baird to H. W. Elliott, Nov. 30, 1869, RU 7002, box 3, SIA.

52. The only exception I found about Ridgway's views in the course of reading thousands of his private letters to friends, family, and colleagues was a single religiously tinged comment in 1903 to one of his best friends, William Brewster, when he mentions considering a matter "most earnestly—even prayerfully." RR to Brewster, Apr. 2, 1903, MCZ. Baird's views: Rivinus and Youssef, *Spencer Baird*, 108.

53. See Wilkins, *Clarence King*, 220-223, who notes that aspects of King's theory were strikingly similar to later descriptions of the evolutionary theory of punctuated equilibrium brought forth in the middle of the twentieth century. For Charles Darwin's famous quotation, see Darwin to Joseph Hooker, Jan. 11, 1844, http://www.darwinproject.ac.uk/entry-729.

54. William H. Brewer to Clarence King, Dec. 24, 1878, John Hay Papers, Brown University Library, http://worf.services.brown.edu/exhibits/archive/ames/55.html.

55. King's request for condensed report: King to RR, Sept. 22, 1876, B-W; Nuttall Ornithological Club review: *Bulletin of the Nuttall Ornithological Club* 3, no. 2

(April 1878): 81–83; *Forest and Stream* 10, no. 8 (1878): 135; "quintessence of stupidity": Charles Hallock to RR, Apr. 1, 1878, RU 105, box 19, fol. 1, SIA.

56. RR to Brewster, Aug. 7, 1906, MCZ.

57. Ridgway, *Birds of North and Middle America* 1 (1901): viii.

58. Kohler, "Subspecies Classification and Biological Survey," 38.

59. "Land Birds," *Boston Daily Advertiser,* May 5, 1874.

Chapter 2. The Smithsonian Years

1. Bairdians: Oesler, *Smithsonian,* 46–47; European school: Ridgway, "Spencer Fullerton Baird" (1888): 6. Ridgway credits his friend Leonhard Stejneger as the namer and describer of this Bairdian school; see *Proceedings of the U.S. National Museum* 7 (1884): 76.

2. Flack, *Desideratum.*

3. Working in Illinois: Spencer Fullerton Baird to Robert Ridgway (hereafter "RR"), Aug. 21, 1873, RU 7002, Henry Papers control no. 024776, box 5, fol. 1, vol. 14, SIA; salary: Baird to William Jones Rhees, Aug. 22, 1873, ibid.; Ridgway's importance: Baird to Joseph Henry, July 14, 1874, ibid., box 6, fol. 1, vol. 16.

4. Hellman, "Profiles: The Enigmatic Bequest," *New Yorker,* Dec. 10, 1966, 66.

5. Ewing, *Lost World,* 315–342.

6. Mary Helen Churchill Baird to RR, Feb. 4, 1888, RU 105, SIA.

7. Visitors: see, e.g., the Smithsonian's annual report of 1896, which notes museum visitors as totaling 201,744 in 1895. Smithsonian Institution, *Annual Report . . . to 1895,* 40. Description of Ridgway's office from photograph of him at work, RU 95, box 19, fol. 25, SIA.

8. Harris, "Robert Ridgway," 32.

9. Baird to George Perkins Marsh, Mar. 28, 1879, George Perkins Marsh Online Research Center, http://bailey2.uvm.edu/specialcollections/bairdlog2.html.

10. Morris Gibbs to RR, July 17, 1883, SIA.

11. Skull: Ridgway's handwritten annual report for 1890, RU 158, ser. 1, box 8, fol. 1, SIA; number of specimens on exhibit: Smithsonian Institution, *Annual Report for the Year Ending June 30, 1889,* 38. The Smithsonian printed sixteen thousand copies of its annual report that year, which meant that besides members of Congress and other politicians, a wide swath of the public had access to these details. Merriam and Maynard: RR to Frederick William True, June 20, 1898, RU 242, box 5, fol. 6, SIA.

12. RR to Edwin Irvine Haines, Dec. 6, 1897, B-W.

13. Morris Gibbs to RR, Mar. 1, 1888, RU 105, box 15, fol. 5 (copy also exists in box 18, fol. 2), SIA.

14. Baptist minister: J[ames] Benjamin Clayton to RR, undated, ibid., fol. 2; Smithsonian correspondents: Goldstein, "Yours for Science," 573–599; painting flowers: Mabel Aaron to RR, January 23, 1889, RU 105, box 12, fol. 1, SIA. Reference questions that addressed the broader social and cultural life of the nation were common, and have persisted. For example, in 1937, the US Bureau of Navigation wrote to curator Herbert Friedmann to ask about possible names of birds for use on new navy minesweepers. Friedmann responded with a long list of "pugnacious birds," many not found outside of the United States. His letter closed, "And lastly I may mention the Humming Bird, which is also noted for its pugnacious habits, but there again the

name might be inappropriate as denoting too small a size for a vessel." Friedmann to Mr. Henkel, Sept. 1, 1937, ibid., box 19, fol. 3.

15. Helping schools: Randolph I. Geare to RR, Feb. 1, 1894, RU 105, box 18, fol. 1, SIA; Pennsylvania collectors: Josiah Hoopes to RR, July 17, 1888, ibid., box 19, fol. 3.

16. William W. Bailey to RR, Jan. 16, 1873, ibid., ser. 2, box 12, fol. 5.

17. See, e.g., José Castulo Zeledón to RR, Oct. 6, 1886, B-W, as well as Harris's discussion of Ridgway's relationship to Zeledón. A series of letters from Zeledón to Ridgway also exists at the Blacker-Wood Library in Montreal. RR's possible dog: Zeledón to RR, Sept. 8, 1883, B-W.

18. *Richmond (VA) Dispatch*, Sept. 20, 1885.

19. Nathan Clifford Brown to RR, May 5, 1884, RU 105, ser. 2, box 14, fol. 2, SIA.

20. See the several folders of correspondence from Henry K. Coale to RR in RU 105, box 15, SIA, which relate to the activities of the Ridgway Ornithological Club. Coale's hopes for Smithsonian job: Coale to RR, Mar. 27, 1884, ibid. fol. 2; Coale's position at Field Museum: Coale to Joel Allen, Jan. 30, June 15, 1894, AMNH.

21. Concerns about feathers for plumage: George C. Henning to RR, Aug. 29, 1884, RU 105, box 19, fol. 3, SIA; Charles F. Amery to RR, Nov. 27, Dec. 21, 1886, ibid., box 12, fol. 4; "Bird Friends and Foes," *Washington Post*, Jan. 27, 1907, 4.

22. Amery to RR, Mar. 1, 1887, RU 105, box 12, fol. 4, SIA.

23. Science definition from 1819: Rees, *Cyclopaedia*, s.v. "Science"; science definition from 1867: Brande and Cox, *Dictionary of Science*, 365; Packard et al., "Introductory," 124; science definition from 1896: Fouillee, "Hegemony of Science," 143.

24. "Wonderful bird eyes": Badè, *Life and Letters*, 332; Frank Chapman's description of Ridgway's hair, moustache, and eyes from Chapman, *Autobiography of a Bird-Lover*, 44; "dandy": Harris, "Robert Ridgway," 21; Miss Perkins fascinating: Henrietta Ridgway to RR, Apr. 12, 1874, box 1, fol. 26, USU; quiet section of city: RR to Allen, Nov. 28, 1883, BAN.

25. Letters not arriving; Henrietta Ridgway to RR, Feb. 5, 1871, box 1, fol. 24, USU; "dreadfully scientific": RR to Henrietta Ridgway, Apr. 28, 1871, ibid.; two months' vacation: Baird to William Jones Rhees, Aug. 22, 1873, RU 7002, Henry Papers control no. 024777, box 5, fol. 1, vol. 14, SIA; Christianity and temperance: Henrietta Ridgway to RR, Jan. 31, 1869, box 1, fol. 22, USU, and Apr. 12, 1874, ibid., fol. 26.

26. Smithsonian Institution, *Annual Report . . . for the Year 1874*, [vi].

27. Author's correspondence with Bradley Cook, university archivist, Indiana University, May 2009. Details on Ridgway's honorary master's degree are found in the IU Archives' Board of Trustees minutes for 1884, 81.

28. Japanese birds: RR to Baird, Mar. 21, 1884, RU 105, SIA; "great opportunities," Allen to RR, Jan. 18, 1884, B-W.

29. Baird to Joseph Henry, July 14, 1874, and Baird to Ridgway, July 30, Aug. 28, 1874, RU 7002, Henry Papers control no. 024800, box 6, fol. 1, Private Outgoing Correspondence, vol. 16: May 14, 1874–November 8, 1874.

30. RR to Baird, Sept. 11, 1870, RU 7002, box 32, fol. 10, SIA.

31. Selling three hundred birds: Allen to RR, Dec. 7, 1880, B-W, and RR to Allen, Dec. 13, 1880, AMNH; patriotism: Brewster to Sharpe, Feb. 15, 1890, B-W; Henshaw's collection best served in Europe: Henshaw to Sharpe, July 7, 1888, and Brewster to Sharpe, Feb. 15, 1890, B-W; Sclater's personal collection: Sclater to Albert Carl Günther, Mar. 19, 1884, DF200/26, NHM.

32. RR to Baird, June 28, 1873, RU 7002, box 32, fol. 10, SIA; RR to Count Hans Hermann Carl Ludwig von Berlepsch, Dec. 9, 1882, RU 105, ser. 2, box 13, fol. 5, SIA.

33. RR to Baird, Jan. 18, 1881, RU 7002, box 32, fol. 10, SIA.

34. RR to Anne Taylor, June 26, 1887, outgoing letter copybooks, RU 105, box 1, vol. 3, pp. 406–407, SIA.

35. Zeledón to RR, Oct. 15, 1885, B-W.

36. "Frugal table": Dyche, "Science for a Livelihood," 303; Ridgway threatens to leave: Baird to RR, Aug. 28, 1874, RU 7002, box 6, fol. 1, vol. 16, p. 251 of letterbook, SIA; agitating for raises: RR to Baird, Aug. 12, 1882, RU 105, SIA.

37. "Government pauper": RR to Brewster, Mar. 10, 1906, MCZ. For Ridgway's salary during his last three years at the Smithsonian, see document by Alexander Wetmore, June 12, 1926, RU 105, box 25, fol. 6, SIA; and RR to Wetmore, July 11, 1928, RU 7006, box 56, fol. 3, SIA.

38. Christopher Greaves to RR, April 26, 1900, RU 105, box 18, fol. 5, SIA. For a brief review of Greaves's book, see [Jones], "Review," 19–20. Family jobs at Smithsonian: RR and John Ridgway to George A. Bates, June 4, 1881, RU 105, SIA.

39. RR to G. Brown Goode, June 25, 1889, RU 201, fol. 10, SIA.

40. Richmond working on Ridgway's behalf: RR to Charles Wallace Richmond, Dec. 18, 1906, RU 105, box 25, fol. 6, SIA. Also see Richmond's obituary: Stone, "In Memoriam: Charles Wallace Richmond," 1–22. Richmond was appointed the head of the division in 1929 but asked to step down within the year into a subordinate position and was replaced by Herbert Friedmann.

41. RR to Baird, Oct. 9, 1884, RU 7002, box 32, fol. 10, SIA. For the article, see Ridgway, "Bird-Collection of the U.S. National Museum" (1884).

42. RR to Brewster, Feb. 4, 22, 1878, MCZ.

43. "Rusty and discontented": RR to Brewster, Jan. 2, 1880, ibid.; RR to Baird, Mar. 28, 1883, RU 105, SIA.

44. RR to Baird, July 9, 1883, RU 201, box 1, fol. 7, SIA.

45. RR to Richmond, Feb. 7, 1898, RU 105, box 25, fol. 6, SIA.

46. RR to Richmond, Feb. 23, 1898, ibid.

47. Chowder, "North to Alaska," 92.

48. RR to Brewster, Apr. 27, 1899, MCZ.

49. RR to Baird, Aug. 12, 1882, RU 105, SIA.

50. "Ithuriel," i.

51. Unlocked bird cases: RR to Goode, Feb. 11, 1889, RU 201, box 1, fol. 9, SIA; J. W. Twig to Goode, June 20, 1884, and RR to Goode, June 21, 1884, ibid.

52. RR to Richard Rathbun, Sept. 7, 1900, RU 31, box 19, fol. 16, SIA.

53. Hot weather: RR to Allen, June 19, 1884, B-W; scientist in danger of roasting to death: Field, Stamm, and Ewing, Castle, 140.

54. Cases damaged: RR to Goode, July 9, 1884, RU 201, box 1, fol. 9, SIA; "grimy walls": Coues to William Jones Rhees, Sept. 23, 1884, RH 2940, box 52, Rhees Collection, HEHL.

55. Bird Division at world's fairs: Henson, "Spencer Baird's Dream," 119; role of evolution: see, e.g., Rydell, All the World's a Fair, 40–41.

56. Acorn: RR to R. Edward Earll, Oct. 21, 1891, RU 70, box 8, fol. 6, SIA; Ridgway family visit to fair: RR to Earll, Aug. 19, 1893, ibid.

57. "Great ornithological centre": Allen to RR, Mar. 9, 1886, B-W; Herbert Friedmann to Frank M. Chapman, Dec. 14, 1932, RU 105, box 15, SIA.

58. Ridgway, "Spencer Fullerton Baird" (1888): 6.

59. Ritterbush, "Biology and the Smithsonian Institution," 28.

60. Bruce, Modern American Science, 217.

61. Smithsonian contributions to world's fairs: Rydell, *All the World's a Fair,* 43; Goode, "Museums of the Future," 249.

62. Goode, "Museums of the Future," 248.

Chapter 3. To Have or Have Not

Epigraph: Elliott Coues to Joel Allen, Jan. 10, 1878, B-W.

1. "On the Descent of Man," 401; Wright, *North American Review* 113, no. 232 (July 1871): 64.

2. Bates, *Scientific Societies,* esp. chap. 3, "The Triumph of Specialization, 1866–1918."

3. Bruce, *Modern American Science,* 327.

4. Menand, *Metaphysical Club,* 439.

5. William Brewster to Robert Ridgway (hereafter "RR"), July 10, 1872, B-W.

6. "Vegetating at Harvard": Thomas, "Nuttall," 164; also Leighton, *American Gardens,* 62; overwork of name: Batchelder, *Nuttall Ornithological Club,* 16.

7. *Bulletin of the Nuttall Ornithological Club* 1, no. 2 (July 1876): 30–31.

8. Gill, "Blue-Winged Warbler," doi:10.2173/bna.584.

9. "Hard work writing for the people": Alpheus S. Packard Jr. to Allen, Dec. 6, 1866, AMNH; "older and younger student of nature": Packard et al., "Introductory," 1; "wisdom and goodness of the Creator": ibid., 2. For a run of correspondence from Packard to Allen on the subject of the new journal and some of the difficulties faced by the publication, which is still in print today, see the folder "Packard-Allen, 1866–," AMNH.

10. Addition of sixty-five new members: Batchelder, *Account of the NOC,* 26; American Chemical Society membership: Reese, *Century of Chemistry;* Allen to RR, Feb. 10, 1877, B-W.

11. Allen to RR, May 2?, 1876, BAN; "The Nuttall Ornithological Club," *Bulletin of the Nuttall Ornithological Club* 1, no. 2 (July 1876): 31.

12. For biographical and bibliographical details, see Allen, *Autobiographical Notes;* Chapman, "In Memoriam: Joel Asaph Allen"; and Sterling and Ainley, manuscript, chap. I, sec. ii, "Birth of the AOU." For an overview situating Allen in the American conservation movement, see Barrow, *Nature's Ghosts,* 79–84.

13. Short articles: Allen to RR, Aug. 6, 1876, B-W; Audubon sisters: Ruthven Deane to Allen, Sept. 28, 1880, AMNH.

14. Coues to Allen, Jan. 10, 1878, B-W.

15. "More popular in character than subscription list": Allen to Ridgway, June 23, 1878, B-W; club has ninety-eight members: Batchelder, *Account of the NOC,* 26.

16. Allen to RR, Dec. 20, 1877, B-W.

17. *Bulletin of the Nuttall Ornithological Club* 3, no. 3 (July 1878): 104.

18. See the somewhat hagiographic details of key founders of German ornithology in the nineteenth century as detailed in Stresemann, *Ornithology;* their work is detailed in part 2 of that book. "Bulletin seems to thrive": Allen to RR, June 23, 1878, B-W.

19. On the sparrow wars, see Barrow, *Passion for Birds,* 47–50, as well as his endnotes detailing Boston newspaper sources. "Taboo subject": RR to Allen, Oct. 13, 1883, BAN; "wretched exotic": Brewster to RR, Sept. 5, 1883, B-W; misplaced sentiment: RR to William M. Heath, Oct. 8, 1884, RU 105, SIA; notoriously deficient": RR to

Allen, Jan. 21, 1878, AMNH. A 1991 argument observes that sparrows were also probably decried as undesirables for social reasons beyond their economic influence. Fine and Lazaros, "Dirty Birds." The authors document the ways that nonnative sparrows were framed by anti-"invasive species" ornithologists as a "menace to the American ecosystem," dirty and useless "immigrants" that competed unfairly with native birds and should be eliminated. Ibid., 375. They argue that the metaphor worked because of perceptions about the threat of new immigrants to the economy and social fabric of America of the era. Also see Coates, "Eastenders Go West," as well as the very useful work by Doughty, "English Sparrow."

20. Coues, "Ineligibility of the European House Sparrow."

21. Brewer, "European House-Sparrow," 587.

22. Coues to Allen, Jan. 10, 1878, B-W.

23. Danger of antagonizing Coues: Brewster to Allen, Mar. 14, 1882, BAN; Coues's evident self-importance: Ruthven Deane to Allen, June 19, 1881, AMNH. The story of Coues's life has been admirably recounted by Cutright and Brodhead, *Elliott Coues: Naturalist and Frontier Historian*.

24. Coues to Allen, Jan. 10, 1878. The name of the organization would later cause mild embarrassment in an unanticipated way: the use of the word "union," which by the 1950s had acquired connotations of labor unrest. "The American Ornithologists' Union . . . is a scientific organization devoted to the study of birds, and not, as the name might imply, a labor union type of organization," Smithsonian Curator of Birds Herbert Friedmann remarked in a letter to a freelance writer. Friedmann to Lyman Gaylord, Dec. 27, 1954, RU 105, box 18, fol. 1, SIA.

25. "Appeal to the Women of the Country in Behalf of the Birds."

26. RR to Brewster, Jan. 2, 1877, MCZ.

27. Club members call on Ridgway: Nathan Clifford Brown to RR, June 25, 1878, RU 105, box 14, fol. 2, SIA; "very great favor": Allen to RR, June 15, 1881, B-W.

28. Allen to RR, Mar. 10, 1880, B-W.

29. Allen to Montague Chamberlain, Nov. 23, 1883, BAN.

30. Coues to Allen, June 8, 1883, B-W.

31. Brewster to Ridgway, Sept. 5, 1883, ibid.

32. Unanimity in nomenclature: RR to Allen, Aug. 11, 1883, BAN; "baby born, baptized and dressed": Coues to Allen, Aug. 21, 1883, B-W.

33. Amateur's crude views: RR to Allen, Aug. 18, 1883, BAN; "limited and select number": Allen to RR, Aug. 21, 1883, B-W; stability to nomenclature: Allen to RR, Oct. 5, 1883, ibid.

34. RR to Brewster, Sept. 10, 1883, MCZ.

35. Coues to Allen, June 4, 1883, B-W.

36. Elliot to Allen, Sept. 4, 1883, and Brewster to Allen, Sept. 26, 1883, BAN.

37. Brewster to Allen, Sept. 27, 1883, ibid.

38. Coues to Allen, Jan. 1, 1884, B-W.

39. Catholic: Coues to Allen, June 4, 1883, ibid.; British naturalists: Coues to Allen, quoting Newton, Nov. 4, 1883, ibid. Also see *Bulletin of the Nuttall Ornithological Club* 8, no. 4 (October 1883): 224, for a list of the foreign members elected.

40. Coues to Allen, Oct. 5, 1883, B-W; RR to Allen, Oct. 13, 1883, BAN.

41. Daniel Elliot Giraud to Allen, Oct. 9, 1883, B-W.

42. "Having their confidence": Allen to RR, Oct. 14, 1883, ibid.; "the interests of the two may sometimes clash": Chamberlain to Allen, Nov. 18, 1883, BAN. For more

on Chamberlain contextualizing his contributions to his own country's ornithology, see Ainley, "From Natural History to Avian Biology."

43. "Rosy-cheeked young man": Chapman, *Autobiography of a Bird-Lover,* 41; dreads keeping of the accounts: Clinton Hart Merriam to Allen, Oct. 17, 1883, B-W. Merriam would go on to be one of the country's most decisive evolutionists, arguing that field evidence supported gradual evolution and emphasizing the important role of geographic isolation in divergence. "In all respects unfitted": RR to Allen, Oct. 13, 1883, BAN.

44. "Nervous prostration": Allen to RR, Aug. 21, 1883, B-W; "excessive nervousness" and "I am all broken up": Brewster to Allen, Aug. 3, 1883 [two letters of same date], BAN; "nervous irritability": José Castulo Zeledón to RR, Sept. 8, 1883, B-W; "brain weakness": Zeledón to RR, Feb. 16, 1887, ibid.; "rather doubtful": RR to Allen, Aug. 27, 1883, MCZ; "I do not think I have any actual disease": Coues to Allen, May 9, 1884, B-W; "I feel as if I had long lived": Coues to Allen, Aug. 26, 1882, ibid.

45. See, e.g., Brinkel, Khan, and Kraemer, "Systematic Review of Arsenic Exposure"; Morton and Caron, "Encephalopathy"; and Yoshihisa, "Mental Health Burden."

46. Merriam to RR, Dec. 21, 1883, B-W.

47. "Put up job": Merriam to Allen, Oct. 23, 1883, B-W; "cheapen the publication": Merriam to Allen, Oct. 18, 1883, ibid.

48. Allen to Ridgway, Nov. 10, 1883, ibid.

49. Circulation of the *O&O*: Barrow, *Passion for Birds,* 17; Barrow, "The Specimen Dealer."

50. Joseph M. Wade to Ernest Ingersoll, Jan. 10, 1880, B-W.

51. John Hall Sage incident noted in his obituary by Witmer Stone, "In Memoriam—John Hall Sage," 9; Merriam to Allen, Oct. 17, 18, 1883, and Merriam to RR, Oct. 18, 1883, B-W. Also see Barrow, *Passion for Birds,* 54–57, for a summary of the Wade/AOU issue.

52. "Undertake to cinch him": Coues to Allen, Aug. 31, 1883, B-W; "the best of the sciences": Wade to Allen, undated but late fall 1883, BAN.

53. Coues to Allen, Nov. 4, 1883, B-W.

54. Wade endorses Coues: Coues to Allen, Nov. 13, 1883, ibid.; "most valuable article": Merriam to Allen, Jan. 2, 1884, ibid.

55. "A pleasant line": Coues to Allen, Jan. 1, 1884, ibid.; "contemptuously silent": Coues to Allen, Feb. 5, 1884, ibid.; "cordially wish it success": Allen, "Notes and News," 106; "sneer or eloquent silence": Coues to Allen, Jan. 18, 1884, B-W.

56. See the folder of correspondence from Merriam to Allen for letters written between January and April 1884 detailing Merriam's efforts with the Migration Committee, B-W.

57. Allen to Chamberlain, Nov. 23, 1883, BAN.

58. "Matter of great delicacy": Allen to RR, Oct. 14, 1883, B-W; "mortgaged and preempted": Coues to Allen, Oct. 20, 1883, B-W.

59. Merriam to Allen, Oct. 26, 1883, ibid.

60. "Outer doors": Batchelder, *Account,* 47n; "flops down promptly": Coues to Allen, Oct. 30, 1883, B-W.

61. Coues to Allen, Oct. 20, 1883.

62. "Servile imitation": Merriam to Allen, Nov. 5, 1883, B-W; RR to Allen, Nov. 13, 1883, BAN; Merriam to Allen, Nov. 7, 1883, B-W; Chamberlain to Allen, Nov. 18, 1883, BAN; and Chamberlain to RR, Nov. 20, 1883, RU 105, box 14, fol. 5, SIA.

63. "Talked over too easily": Coues to Allen, Nov. 8, 1883, B-W; "neat, not to say gaudy": Coues to Allen, Nov. 16, 1883, ibid.

64. Allen to Ridgway, Nov. 10, 1883, ibid.

65. "Good antithesis": Spencer Fullerton Baird to Allen, Nov. 9, 1883, BAN; indebted to Baird for name: Allen to RR, May 14, 1884, B-W.

66. "aukward organ": Allen to RR, May 14, 1884; "Ibis occidentalis": Coues to Allen, Jan. 18, 1884, B-W; Donald, "Notices of Recent Ornithological Publications," 203-204.

67. Brewster to RR, Dec. 23, 1883, B-W; Chamberlain to Allen, Dec. 25, 1883, BAN; Allen, "Notes and News," 105.

68. Allen, "Notes and News," 105.

69. Samuel Wells Willard to Allen, Jan. 22, 1884, B-W; Walter E. Bryant to Ridgway, Mar. 5, 1884, RU 105, ser. 2, box 14, fol. 2, SIA.

70. "An entire failure": RR to Allen, Jan. 15, 1884, B-W; "sickest and most forlorn": Coues to Allen, Jan. 18, 1884, ibid.; Allen describing Auk imagery: Allen to RR, Jan. 12, 1884, ibid.; details of printing: Merriam to Allen, Jan. 8, 28, 1884, and Allen to RR, Dec. 14, 1885, ibid.

71. Ridgway dissatisfied with magazine's appearance: RR to Allen, Jan. 21, 1884, BAN; "supreme disgust": RR to Allen, Apr. 22, 1884, ibid.; AOU business takes too much time: Merriam to Allen, undated but after Sept. 26, 1883, B-W.

72. Reingold, "Professionalization of Science," 33.

73. Rossiter, *Women Scientists*, 54-55.

74. Flint, Letter to Editor, 18.

75. Wade, "Gastro-oological," 57.

Chapter 4. Bird Study Collections

Epigraph: Richard Bowdler Sharpe to Charles Walter de Vis, Apr. 3, 1897, DF230/20 [carbon outletters], 48, NHM.

1. Species debate nonexistent: McOuat, "Cataloging Power," 2; "Darwinian": Robert Ridgway (hereafter "RR") to Joel Allen, Dec. 13, 1880, AMNH.

2. RR to William Brewster, Jan. 9, 1886, MCZ; Ridgway, "Descriptions of Twenty-Two New Species of Birds from the Galapagos Islands" (1894).

3. Richard Bowdler Sharpe to Mr. Haddock?, Jan. 23, 1900, DF230/20, 157-158, NHM.

4. "Kindly give me your opinion": Eugene Pintard Bicknell to RR, June 27, 1881, RU 105, box 13, fol. 5, SIA; Ridgway names bird: Bicknell to RR, Oct. 30, 1881, ibid. Also see http://www.ns.ec.gc.ca/wildlife/bicknells_thrush/e/species_is_born.html for useful background on Bicknell and his discovery, despite several minor factual errors about Ridgway.

5. Burroughs: *Writings of John Burroughs*, 9:51; Ridgway, "Descriptions of Two New Thrushes" (1882): 377-378.

6. "Scarcely intitled": Bicknell to RR, Jan. 1, 1882, RU 105, box 13, fol. 5, SIA; benefit of your opinion": Bicknell to Allen, May 29, 1884, B-W.

7. For a useful overview, see Haffer, "History of Species Concepts," and McKitrick and Zink, "Species Concepts." For modern invertebrate zoology approaches that attempt to align with the species concept, see Vellai et al., "Genome Economization."

8. The meaning of the term "type specimen" has been contested—see a late

twentieth-century call for standardization: Banks et al., "Type Specimens," 413. "Keeping order in nature's complex household": Kohler, "Subspecies Classification," 36.

9. RR to Brewster, Nov. 29, 1874, MCZ; see also Ridgway, "Discovery of a Burrowing Owl in Florida" (1874).

10. "A most unfortunate accident": Brewster to RR, Dec. 4, 1874, B-W; mouse chewing off legs of Ridgway's skins: Brewster to RR, September 22, 1882, ibid.

11. Smithsonian collections comprise just one case: RR to Brewster, Nov. 5, 1896, MCZ; examples of other missing or lost specimens: RR to Allen, Apr. 7, 1884, BAN; changes in Smithsonian loan policies: RR to Spencer Fullerton Baird, Feb. 10, 1886, RU 105, ser. 2, box 14, fol. 5, SIA; RR to Frederick William True, Nov. 6, 1889, RU 201, fol. 10, ibid.; RR to Brewster, Feb. 12, 1886, MCZ; Joseph Henry's and Baird's attitudes and practices: personal correspondence with Pamela Henson, director, Smithsonian Institutional History Division, February 2011.

12. Use of Smithsonian study specimens: RR to Brewster, Sept. 22, 1881, MCZ; Brewster to RR, Feb. 17, 1886, B-W.

13. Change in MCZ loan policies: *Annual Report of the Curator of the Museum of Comparative Zoology,* 8; Smithsonian loan policies: RR to Ernst Hartert, Mar. 30, 1896, TM 1/22/18, [penciled # 250], NHM; RR to Brewster, Nov. 15, 1893, MCZ; "How sorry I am": Sharpe to Charles Wallace Richmond, Apr. 10, 1899, RU 105, ser. 2, box 13, fol. 7, SIA.

14. "Last hours of some of the species drawing nigh": Henry Henshaw to Richmond, June 8, 1899, RU 105, box 19, fol. 3, SIA; "on the very verge of extinction": Henshaw to Richmond, Aug. 3, 1899, ibid. For a sophisticated, book-length discussion of the apparent contradictions in scientists reconciling their actions in killing birds with knowledge of their pending demise, see Barrow, *Nature's Ghosts.*

15. Henshaw to RR, June 7, 1898, B-W.

16. "Almost impossible to obtain them": RR to Hartert, Jan. 25, 1896, TM 1/22/18, [pencilled # 214], NHM. See Barrow, *Nature's Ghosts,* 129, for details on Ridgway's ownership of pet parakeets hatched from eggs captured during this 1896 expedition.

17. Goldstein, "Yours for Science," 574–575.

18. "My life is ornithology": Dr. Hornug to Richmond, Oct. 31, 1899, RU 105, box 19, fol. 5, SIA; "cannot bear to give it up": Charles K. Worshem to Allen, June 13, 1884, B-W.

19. J. H. Keen to O. V. Aplin, Sept. 21, 1893, AL.

20. "Hardest for collecting that I ever saw": Brewster to RR, June 2, 1874, B-W; Cherrie, *Dark Trails,* 112–125.

21. Sharpe to RR, Apr. 24, 1898, B-W.

22. RR to G. Brown Goode, July 14, 1888, RU 201, box 1, fol. 9, SIA.

23. Ludlow Griscom to Herbert Friedmann, Dec. 12, 1930, RU 105, box 18, fol. 7, SIA. It seems unlikely that this was the same Hasso von Wedel as the famous World War II German flying ace, but without more research it's difficult to say with certainty. For more details on Wedel, see Alexander Wetmore Papers, RU 7006, ser. 1, box 11, fol. 1, ibid.

24. Brewster to RR, Dec. 4, 1888, B-W.

25. "Policemen and other obstacles": Edward A. Colby to RR, Feb. 12, 1887, RU 105, ser. 2, box 15, fol. 4, SIA; "nothing to fire at": Brewster to RR, Dec. 31, 1879, B-W.

26. Griscom, "Notes," 54.

27. Walter E. Bryant to RR, Feb. 19, 1884, RU 105, box 14, fol. 2, SIA.

28. "Wasting your time out West": Brewster to RR, Feb. 24, 1874, B-W; California Condor "appears to be now practically extinct": RR to Brewster, Apr. 4, 1881, MCZ.

29. George L. Guy to RR, Sept. 1, 1889, RU 105, box 18, fol. 7, SIA.

30. Supplicant for publications: J. G. Hoskin to Ridgway, Apr. 6, 1890, RU 105, SIA; see also letter from E. C. Greenwood of Corpus Christi, Texas, offering local birds to Ridgway in exchange for publications: E. C. Greenwood to RR, Mar. 29, 1888, ibid., box 18, fol. 5; copy of *PRR* sent: docketed on letter from Y. E. Blaisdell to RR, Apr. 3, 1885, ibid., box 13, fol. 6.

31. AMNH's chronic limitations: Allen to RR, Sept. 28, 1893, B-W; "pleasure or the pain": Henshaw to RR, Oct. 22, 1884, B-W; frozen albatross: Reginald Barter to Arthur Butler, July 7, 1897, DF 230/2 (1897 letters, vertebrata), NHM.

32. "Seems to pervert the moral sense": Oswald Hawkins Latter to A. H. Cox, June 23, 1931, AL; "brutal and ignorant": Coues, *Field Ornithology,* 21; "law abiding ornithologist": Wilmot W. Brown to Hartert, Nov. 21, 1897, TM 1/25/20, NHM.

33. British Museum holdings: Mearns and Mearns, *Bird Collectors,* 84. Smithsonian collection size: personal correspondence with James Dean, collections manager, Division of Birds, Smithsonian Institution, Mar. 1, 2010. Detailed historical statistics are difficult to come by, but Dean notes probable ranges (plus or minus 10 percent) of 76,000 to 85,000 skins in 1875, and 175,000 to 195,000 total specimens in 1900. In both cases I've picked the middle range of the numbers he provided. Biddulph's collection details: Col. Biddulph to Sharpe, June 7, 1897, DF 230/2, NHM. "Great many species still wanting": Sharpe to de Vis, Apr. 3, 1897, DF 230/20 [carbon outletters], 48, ibid. For details on the British Museum's zoological acquisitions for 1897, see the bound volumes of correspondence to and from the British Museum, held by the London NHM, which include the examples cited in a volume entitled "1897 letters, vertebrata."

34. "Utterly impossible": Henshaw to RR, Oct. 7, 1873, B-W; "a considerable number got away wounded": Henshaw to RR, Dec. 19, 1880, ibid.; "slewed and skilled 50 feathered innocents": C. W. Beckham to RR, May 7, 1885, RU 105, ser. 2, box 12, fol. 8, SIA.

35. "A panacea for all ills": Brewster to Allen, Sept. 3, 1885, RU 7440, box 1, fol. 8, SIA; "bloody little birds": Mearns and Mearns, *Bird Collectors,* 29, quoting Coues.

36. Employing people: Lyman Belding to RR, May 9, 1880, RU 105, box 13, fol. 1, SIA: "I hope to do better at Murphys for I confidently count on the help of several boys." "Hundred-fold": Henshaw to RR, Dec. 6, 1884, B-W. In this same letter, Henshaw notes how, in ninety minutes, he also killed ten examples of *Geothlypis beldingi,* Belding's Yellowthroat—now critically endangered.

37. Hasbrouck, "Evolution and Dichromatism in the Genus *Megascops.*" See also Hrubant, "Analysis of the Color Phases of the Eastern Screech Owl," which builds directly on Hasbrouck's morphological data.

38. Henry K. Coale to RR, Jan. 25, 1884, RU 105, box 15, fol. 2, SIA.

39. "You will do very well": Coues, *Field Ornithology,* 31. After several months of intermittent effort and training, learning how to make study skins in the Los Angeles County Museum of Natural History's ornithology lab, I was able to do two birds a day. It's not easy. Waterton, *Wanderings,* 336.

40. Maynard's Dermal Preservative: RR to Baird, July 5, 1883, RU 201, fol. 7, SIA; *Ornithologist and Oologist,* January 1885, 15; Maynard, *Manual of Taxidermy,* 46. See

my argument in chapter 3 about the role of arsenic and widespread depression among naturalists. "Cheaper at double the price": RR to Baird, July 5, 1883; Winkler, "Obtaining, Preserving, and Preparing," 288.

41. "Disgusted with the skins": George F. Clingman to RR, July 15, 1878, RU 105, ser. 2, box 15, fol. 1, SIA; "first class in every respect": Alvin Seale to Hartert, June 28, 1897, TM 1/30/15 [penciled # 200], NHM; value of condors: Henshaw to RR, Oct. 22, 1884, B-W; collectors' approaches to selling: see, e.g., Seale to Hartert, June 28, Oct. 24, 1897, and Wilmot W. Brown to Hartert, Nov. 21, 1897, TM 1/30/15, NHM.

42. RR to Richmond, Feb. 7, 1905, RU 105, box 25, fol. 6, SIA. See also Ridgway, "Winter with the Birds in Costa Rica" (1905).

43. Keen's Mouse: Keen to Aplin, Sept. 21, 1893, AL; Keen's Bat: Merriam, "Bats of Queen Charlotte Islands."

44. Coues, Field Ornithology, 27.

45. RR to Sharpe, July 10, 1891, B-W.

46. Brewster to RR, Jan. 26, 1879, ibid.

47. "Subspecies research program": Barrow, Passion for Birds, 76. See also Kohler, "Subspecies Classification," 35. "Considerable majority": RR to Brewster, Jan. 22, 1879, MCZ.

48. Kohler, "Subspecies Classification," 38.

49. Reviews five hundred specimens: RR to Brewster, Feb. 2, 1899, MCZ; uses three hundred larks: Allen to RR, Oct. 31, 1889 (two letters), B-W; Oregon curator: Nathan Clifford Brown to RR, Mar. 6, 1882, RU 105, ser. 2, box 14, fol. 2, SIA.

50. Large series necessary to show limits of individuals: RR to Hartert, Oct. 21, 1894, TM 1/22/18 [pencilled # 251], NHM. British attitudes toward subspecies are detailed in chapter 6.

51. RR to Brewster, July 30, 1905, MCZ.

52. "A deep and far-reaching significance": Brewster to RR, Dec. 2, 1881, B-W; "there is so much variation": RR to Brewster, Feb. 8, 1882, MCZ.

53. Guzy and Ritchison, "Common Yellowthroat."

54. Coues, Field Ornithology, 46.

55. A variety of other labeling examples and details can be found in Winkler, "Obtaining, Preserving, and Preparing."

56. RR to Charles B. Cory, Aug. 23, 1882, RU 105, box 15, fol. 5, SIA.

57. See, e.g., Weiner, Beak of the Finch, for an explanation of how speciation can work in relatively short periods of time, using the Galápagos "finches" as an example.

58. The AOU Sub-Committee on Bird Measurements was formed at the direction of the organization's president in 1888 and consisted of Coues, Charles Cory, Merriam, Ridgway, and Leonhard Stejneger. The group worked for two years, submitting a final report in 1890. See RU 7150, box 39, SIA, for details.

Chapter 5. Nomenclatural Struggles, Checklists, and Codes

1. A useful synthetic overview of the history of avian classification, going into the twenty-first century, is found in Bruce, "Brief History of Classifying Birds," 1–43. The body of work on systematics and classification over the centuries is immense and speaks to botany, zoology, and geology, among other disciplines. Carol Kaesuk Yoon provides a lively discussion of the challenges of taxonomic work in Naming Nature.

2. Mayr, "Ridgway, Robert," 445.

3. "Very trying to my mind": José Castulo Zeledón to Robert Ridgway (hereafter "RR"), June 14, 1883, B-W; Ridgway claims priority for bird named for Zeledón's sister: Ridgway, "On a New *Carpodectes*" (1884): 27–28.

4. RR to Edward William Nelson, [n.d. but ca. 1906], RU 7364, box 8, fol. 6, SIA.

5. Naming a genus after Cory: RR to Charles Wallace Richmond, July 30, 1925, RU 105, box 25, fol. 6, SIA; new species: RR to William Brewster, Apr. 2, 1897, MCZ. In this letter, Ridgway notes new species he's discovered among study skins.

6. Joel Allen to RR, Apr. 15, 1884, B-W.

7. Ridgway, "Birds of Colorado Check List" (1873).

8. Bartram, "Descriptive Catalogue," 288–296.

9. For an exchange between Allen and Coues concerning Bartramian names, see Coues, "Fasti Ornithologiae Redevivi," including his remark, "consistency is a jewel," on p. 346, followed by Allen's response, "The Availability of Certain Bartramian Names," and finally by Coues's counterresponse, "Reply to Mr. J. A. Allen."

10. Allen to RR, Feb. 4, 1880, B-W.

11. Ridgway, "Catalogue of the Birds of North America" (1880); "a new list is so urgently called for": Allen to RR, June 11, 1880, B-W.

12. Ridgway, "Nomenclature of North American Birds" (1881).

13. "Sick and tired": Morris Gibbs to RR, June 7, 1881, RU 105, box 18, fol. 2, SIA; "I became reconciled": Gibbs to RR, July 17, 1883, ibid.

14. "The fate of list-makers": Elliott Coues to RR, Apr. 23, 1881, B-W; "previous to the appearance of this paper": Nathan Clifford Brown to RR, Apr. 15, 1881, RU 105, ser. 2, box 14, fol. 2, SIA.

15. Coues to Allen, Aug. 26, 1882, B-W.

16. Coues to Allen, Sept. 5, 1882, ibid.

17. Coues to Allen, Nov. 13, 1883, ibid.

18. Merriam, "Coues Lexicon."

19. "Pedantic hypercriticism": Coues, *Auk* 1, no. 1 (1884): 51.

20. Coues to Allen, Jan. 1, 1884, B-W.

21. Coues to Allen, Nov. 8, 1883, ibid.

22. Coues to Allen, Nov. 29, 1883, ibid.

23. Hazard, "Lay View of 'Ornithophilologicalities,'" 300–302.

24. "How queerly some things turn out": Coues to Allen, Jan. 1, 1884, B-W; requests for information: Henry Henshaw to Richmond, Oct. 7, 1900, RU 105, box 19, fol. 3, SIA.

25. Neither Canada nor Mexico had anything approaching a national systematic approach to nomenclature in the nineteenth century. In the 1880s Canadian ornithologists made a conscious decision to follow the AOU's approach (rather than that of the BOU or one of their own making), probably because prominent Canadian ornithologist Montague Chamberlain was an official of the AOU. For more details, see Ainley, "From Natural History to Avian Biology," 23 ff.

26. Reducing Bartramian names: Clinton Hart Merriam to Allen, Sept. 15, 1883, B-W; Allen's attendance record at AOU nomenclatural meetings: Allen to RR, Dec. 6, 1896, B-W; "Two persons decide to some extent": Coues to Allen, Apr. 24, 1884, B-W.

27. "I do not see where the matter is going to end": RR to Allen, Apr. 11, 1884, B-W; "in the interest of harmony and unity": Allen to RR, Oct. 5, 1883, ibid.

28. "The future is in your hands": Augustus Merriam to Allen, Feb. 22, 1884, B-W; gaining professional unanimity on scientific names: RR to Allen and RR to

Coues [identical letters], Aug. 11, 1883, BAN and RU 105, box 15, fol. 5, SIA; "crudest possible information": RR to Allen, Aug. 18, 1883, BAN; "wholly impersonal system": Brewster to RR, Sept. 5, 1883, B-W.

29. Confusion caused by competing checklists: Allen, "A.O.U. Check-List," 7; Allen talking Ridgway out of rewriting Introduction: Allen to RR, Jan. 14, 1886, B-W.

30. This arrangement described in a review of the AOU's work in "The American Ornithologists' Union," *Science*, Oct. 17, 1884, 374–376.

31. Auklet: AOU Committee on Nomenclature, "Committee on Classification and Nomenclature: Proceedings of Meetings, Fourth Session, 1885," Apr. 23, 1885, RU 7150, box 38, fol. 13, SIA; "honored our Pioneers little enough": Joseph M. Wade to Allen, [n.d. but ca. 1883], BAN.

32. Clinton Hart Merriam to Allen, Mar. 13, 1884, B-W.

33. Scudder, "Canons"; "most extensive branch of zoology": American Ornithologists' Union, *Code*, 8; Scudder, "Historical Sketch"; "not enough wheels within wheels": Samuel Scudder to Allen, Jan. 1, 1887, RU 7440, box 1, fol. 1, SIA.

34. Handwritten statement read at council meeting, 1906, unattributed, but probably by Allen, Nov. 15, 1906, RU 7440, box 4, fol. 6, SIA.

35. Bartramian names whittled down: RR to Allen, Apr. 11, 1884. B-W, and AOU *Check-list*, 8; "the list might be considerably extended": Allen to RR, June 15, 1880, B-W.

36. Allen to RR, Feb. 19, 1880, B-W.

37. "Let's hypothecate": RR to Coues [but with other AOU Nomenclature Committee members weighing in throughout the document], Oct. 26, 1885, RU 7150, box 38, fol. 12, SIA; "like the Spartan babes": Augustus Chapman Merriam to Allen, Feb. 22, 1884, B-W.

38. Carpenter, "American Ornithologists' Union Code of Nomenclature."

39. Allen to RR, Feb. 8, 1886, B-W.

40. Seebohm, "Remarks on Certain Points."

41. *Bulletin of the Nuttall Ornithological Club* 5, no. 3 (July 1880): 170–173.

42. French works: Allen to RR, Feb. 7, 1884, B-W. The French sources he cites are Congrès Géologique International, "Règles à suivre pour établir la nomenclature des espèces" (Paris: Société Zoologique de France, 1881), and "Règles applicables de la nomenclature des êtres organisés, proposées pour la Société zoologique de France," *Comptes Rendus de l'Academie des Sciences* (1882). "It will be well . . . if you can make yourself familiar": Allen to RR, Feb. 7, 1884, B-W. For the committee's detailed rationale for using the tenth edition of Linnaeus as the basis for their work, see the notes from the second meeting of the Nomenclature Committee, Dec. 12, 1883, RU 7150, box 38, SIA.

43. "Names of harsh and inelegant pronunciation": AOU *Code*, sec. C, "Recommendations for Zoölogical Nomenclature in the Future," rec. 3.12, 64. Ridgway criticizes Brewster's supposed longer-than-five-syllable name: RR to Brewster, Dec. 14, 1887, MCZ.

44. Handwritten statement read at council meeting, 1906, unattributed, but probably by Allen, Nov. 15, 1906, RU 7440, box 4, fol. 6, SIA.

45. Meyers, foreword, *Genera of Fishes and Classification of Fishes*, vii.

46. R.C.M., "'Amateur,'" 128.

47. See Gates and Shteir, *Natural Eloquence*, for details on women's reinscription of science, including a much-needed international context.

48. C. F. Bruch, "Ornithologische Beiträge," *Isis, eine encyclopädische Zeitschrift, vorzüglich für Naturgeschichte, vergleichende Anatomie und Physiologie* 21 (1828): col. 725, translated by Stresemann, *Ornithology from Aristotle to the Present*, 201.

49. Binomials "blunt and unhandy": Coues, "Progress of American Ornithology," 373; trinomials "an unwieldy instrument": Daniel Elliot Giraud to Allen, Oct. 9, 1883, B-W; "am glad you are agitating": Allen to RR, May 24, 1879, ibid.

50. *Journal für Ornithologie* passage as translated by Stresemann, *Ornithology from Aristotle to the Present*, 251. Kleinschmidt was a particularly fierce opponent of the subspecies concept. See Kleinschmidt, "Was ist die Subspecies?" and Williams, "Otto Kleinschmidt."

51. Browne, *Darwin*, 123.

52. "Truly British stupidity": Coues to Allen, Jan. 29, 1883, B-W; "they will surely fall in line": RR to Allen, Apr. 30, 1884, BAN.

53. During his trip, Coues felt that his English compatriots held him in the highest esteem. "*L'A.O.U. c'est nous*," he wrote to Allen after his arrival there, "and don't forget this, nor the fact that nobody in England now supposes it to be anything else!" Coues to Allen, June 1884, in Cutright and Brodhead, *Elliott Coues*, 282; "it will be best for our interests": Coues to Allen, Apr. 22, 1884, B-W; Frederick True's trip to British Museum: G. Brown Goode to Albert Carl Günther, Dec. 15, 1883, DF200/25, vol. 136, NHM; Ridgway's disapproval of Coues's visit: RR to Allen, Apr. 29, 1884, BAN. "I have, as you requested, shown your letter to Henshaw, and he at once expressed his unqualified disapproval," Ridgway also noted.

54. Brewster to Allen, Apr. 29, 1884, B-W.

55. Coues to Allen, May 10, 1884, ibid.

56. Sclater will settle British reluctance to embrace trinomials: Coues to Allen, June 11, 1884, B-W; "low opinion of all innovators": Stresemann, *Ornithology from Aristotle to the Present*, 262.

57. *Ibis* didn't discuss trinomials until 1912: Johnson, "*The Ibis*," 61; "dreamt I saw all the Hirundines": Christopher J. Alexander to Horace G. Alexander, Feb. 25, 1907, AL.

58. Rejection of Strickland Code: Stresemann, *Ornithology from Aristotle to the Present*, 263.

59. Allen to Sharpe, Feb. 7, 1891, B-W.

60. See, e.g., Meinertzhagen, "Preliminary Study"; "ornithological nomenclature is turned upside down": A. Halle MacPherson to O. V. Aplin, June 24, 1918, AL.

61. Coues, "Fasti Ornithologiae Redivivi," 339.

62. For a useful discussion of the valorization issue, see Bowker and Star, *Sorting Things Out*. An intriguing discussion of the problematic dominating nature of science appears in Harding, *Whose Science?* 35–36.

63. Allen, "Correspondence."

64. Brewster to RR, Jan. 15, 1897, B-W.

Chapter 6. Publications about Birds

Epigraph: Alpheus S. Packard Jr. to Joel Allen, Dec. 6, 1866, AMNH. How authors expressed themselves in both scientific and popular articles on ornithology bears scrutiny. The language writers used expressed their attitudes toward evidence. They could convey degrees of reliability through the use of such words as "maybe," "un-doubtedly," and "possibly"; signal when they were working with hearsay evidence by writing "it seems" or "apparently"; and express inductive ("must," "evidently") or deductive ("presumably" and "could") reasoning. An analysis of such expressions could signal the authority readers could attribute to the printed information. Such clio-metric analysis has not been done very much, other than by John Battaglio, *Rhetoric of Science,* and would provide fertile ground for research.

1. Eugene Bicknell to Allen, May 29, 1884, B-W.

2. Bicknell to Allen, May 31, 1884, ibid.

3. The original quotation by Pascal is: "Je n'ai fait celle-ci plus longue que parce que je n'ai pas eu le loisir de la faire plus courte" (*Lettres Provinciales,* letter 16), "The present letter is a very long one, simply because I had no time to make it shorter" (*The Provincial Letters* [New York: Modern Library, 1941]). The quotation has also been attributed (apparently erroneously) to Mark Twain.

4. Quantity of periodicals in England during the nineteenth century: http://www.sciper.org/introduction.html, "Science in the Nineteenth-Century Periodical"; quantity of new magazines in the United States: Mott, *History of American Magazines,* 4:11.

5. Robert Ridgway (hereafter "RR") to Frederick William True, May 26, 1900, RU 242, box 5, fol. 6, SIA. Also see Currie, "New Bird of Paradise."

6. RR to Brewster, January 19, 1877, MCZ.

7. Relationship between color and geographical distribution: Allen, "On the Mammals and Winter Birds."

8. Allen quoted in Darwin, *Descent of Man,* 212.

9. Ridgway, "Relation between Color and Geographical Distribution in Birds" (1872, 1873).

10. Allen to RR, Apr. 12, 1873, B-W; RR to Allen, Apr. 18, 1873, AMNH; Coues, "Color-Variation"; Ridgway, "Relation between Color and Geographical Distribution in Birds." Also see Cutright and Brodhead, *Elliott Coues,* 151–155, for a discussion of the altercation. Baird's comments on the topic of color variation by region within species were, in fact, very brief; see his "Distribution and Migrations," 78–90, 184–192, 337–47. He mentions this issue only on p. 191, noting that "specimens from the Pacific coast are apt to be darker in color than those from the interior, the latter frequently exhibiting a bleached or weather-beaten appearance, possibly the result of greater exposure to the elements and less protection by dense forests."

11. Coues to RR, Sept. 25, 1873, B-W.

12. Ibid.; "Notes," *American Naturalist.* For other details, see Barrow, *Passion for Birds,* 80–83; Cutright and Brodhead, *Elliott Coues,* 151–155; and other uncited and unpublished correspondence between Coues and Ridgway in the Blacker-Wood Library at McGill University.

13. Amos W. Butler to RR, May 12, 1881, RU 105, ser. 2, box 14, SIA; Ridgway, "Belted Kingfisher Again," 1869; Ridgway, "Bird Haven," 1929.

14. Allen to RR, June 20, 1876, B-W.

15. Lucius H. Cannon to RR, May 25, 1885, RU 105, box 14, fol. 4, SIA.

16. RR to Randolph Geare, Dec. 28, 1894, RU 201, box 1, fol. 13, ibid.

17. See, e.g., RR to Geare, Nov. 22, 1894, RU 105, box 18, fol. 1, ibid.

18. *Naturalists' Quarterly* recruits Ridgway's publications: George Bates to RR, Mar. 5, 10, 1880, ibid. "I have no connection whatever with the 'Naturalists' Agency' of which you complain," Bates wrote Ridgway on March 10; reader requests four free

publications noted in *Register:* James G. Crane to RR, Sept. 7, 1882, ibid., box 15, fol. 5.

19. Allen to Ridgway, Aug. 1, Dec. 8, 1885, B-W.

20. Ridgway to Editor, *Youth's Companion,* Apr. 9, 1887, ibid.; Ridgway to Editor, *Cosmopolitan,* Apr. 5, 1887, ibid.

21. "He has 'done' that region to death already": Brewster to Allen, Aug. 3, 1883, BAN; paper publishes three articles on sparrows: *(Washington, DC) Evening Star,* Sept. 8, 15, Oct. 20, 1883.

22. Arthur Humble Evans to O. V. Aplin, Oct. 15, 1904, AL.

23. Ridgway brothers share illustrative tasks: RR to Brewster, June 17, 1883, MCZ; RR and John L. Ridgway to George A. Bates, June 4, 1881, RU 105, SIA, and RR to John L. Ridgway, Aug. 6, 1927, B-W; "we have no stated price": Allen to RR, Mar. 10, 1894, ibid.

24. Details of *Birds of North America* production: RR to Spencer Fullerton Baird, Dec. 6, 1875, RU 7002, box 32, fol. 10, SIA, and RR to Brewster, May 28, 1889, MCZ. The most thorough bibliographical description of this book can be found in Zimmer, *Catalogue,* 34–35, which Zimmer describes as "a splendid work, authoritative and beautifully prepared." This letter implies that Ridgway has hand-colored fifty copies of seven of the plates: RR to Baird, July 30, 1875, RU 7002, box 32, fol. 10, SIA.

25. RR to Baird, July 1, 1875, RU 7002, box 32, fol. 10, SIA.

26. "I myself have a respect now": E. T. Adney to RR, Dec. 30, 1889, RU 105, ser. 2, box 12, fol. 1, SIA; hot weather and drawing, Ridgway's assessment of his abilities ("by far the handsomest"): both in RR to Baird, Aug. 18, 1875, RU 7002, box 32, fol. 10, ibid.; Ridgway exults over a specific image ("The Blue Jay is going to be the finest plate of the entire series"): RR to Baird, Sept. 21, 1875, ibid.; "not at all satisfactory to me": RR to Allen, Dec. 15, 1879, AMNH.

27. RR to Baird, Aug. 13, 1875, RU 7002, box 32, fol. 10, SIA.

28. Ridgway's son, Audie (then twelve years old), carrying pictures to Smithsonian from home: RR to Brewster, Mar. 15, 1890, MCZ; Ridgway and Brewster simultaneously exchanging skins: RR to Brewster, Dec. 1, 1890, ibid.

29. Brewer wrote most species descriptions for land birds: Rivinus and Youssef, *Spencer Baird,* 158; "The event of the year 1884": Allen, "Recent Literature."

30. Ridgway offers to cancel two-month trip west: Ridgway to Alexander Emanuel Agassiz, Mar. 9, 1883, MCZ; Ridgway's money woes related to the work: RR to Baird, Sept. 18, 26, 1879, RU 7002, box 32, fol. 11, SIA; Baird-Agassiz conflicts: Rivinus and Youssef, *Spencer Baird,* 98–105, and Cochrane, *National Academy of Sciences,* 62.

31. RR to "Gentlemen," Little, Brown and Co., Nov. 23, 1885, B-W.

32. The heron piece by Ridgway is "Studies of the American Herodiones" (1878); "the wood cut will be pretty expensive": Coues to RR, Nov. 15, 1877, B-W; size of print run of Ridgway's heron article: Coues to RR, Feb. 13, 1878, B-W.

33. Blum, *Picturing Nature,* 161–163, 194–195.

34. "Only a fair remuneration": RR to Stephen Alfred Forbes, June 22, 1883, B-W; fire: RR to Allen, Nov. 25, 1889, RU 7440, box 1, fol. 9, SIA; birdwatchers recruited: Davenport, "Robert Ridgway," 284.

35. Allen to RR, May 28, 1886, B-W.

36. *Olney (IL) Times,* Mar. 28, 1925, 5.

37. Adams, "Is Botany a Suitable Study for Young Men?"; Rudolph, "Women in Nineteenth Century American Botany."

38. "Reduced to the smallest compass": Allen, "Recent Literature," 334.

39. Ridgway tells Allen he will promote work to publisher: RR to Brewster, Apr. 6, 1886, MCZ.

40. Allen to RR, Aug. 11, 1887, B-W; Brewster to RR, Oct. 24, 1887, ibid.; and Montague Chamberlain to RR, Nov. 22, 1887, RU 105, box 14, fol. 5, SIA.

41. Goode to RR, Sept. 15, 1894, RU 201, box 1, fol. 5, SIA.

42. Examples include T. Egleston, *Catalogue of Minerals and Synonyms,* Bulletin 33 (Washington, DC: Smithsonian Institution, 1889); Charles D. Sherborn, *An Index to the Genera and Species of Foraminifera,* Smithsonian Miscellaneous Collections 37 (Washington, DC: Smithsonian Institution, 1886, 1889); and George Brown Goode, *Catalogue of the Fishes of the Bermudas,* Bulletin 5 (Washington, DC: Smithsonian Institution, 1876).

43. RR to Goode, Dec. 1, 1894, RU 201, box 1, fol. 5, SIA.

44. RR to Goode, Feb. 7, 1895, ibid.

45. Ridgway's working conditions detailed in letters to Goode, Oct. 9, 1895, RU 201, box 1, fol. 5, SIA, and to Brewster, Jan. 21, 1896, MCZ.

46. RR to Goode, Aug. 10, 1896, RU 201, box 1, fol. 5, SIA.

47. Outline characteristics described in letters to Goode, Sept. 7, 1894, Aug. 10, 1896, ibid.

48. Evelyn Ridgway to True, July 3, 1900, RU 7181, box 1, fol. 5, SIA.

49. On Dresser's work, Ridgway noted to Brewster, "I may add that it is my intention to publish all my future ornithological works in the same style (as to size, etc.) as Dresser's 'Birds of Europe' which to me seems a model of convenience and elegance, combined," Apr. 20, 1886, MCZ. Also see RR to Richard Rathbun, Sept. 7, 1900, RU 31, box 19, fol. 16, SIA, for details on the size and length of the work.

50. Audubon Ridgway obituary: Richmond, "Notes and News"; Audubon Ridgway as photographer: Harris, "Robert Ridgway," 36; Audubon Ridgway's parents' whereabouts upon his death, and other details: "Washingtonian Succumbs to Acute Pneumonia at Chicago."

51. RR to Brewster, Apr. 24, 1901, MCZ.

52. RR to Rathbun, May 6, 1901, RU 31, box 19, fol. 16, SIA.

53. "Science is not literature": Ridgway, *Bulletin 50,* xii; "we trust the author's life may be spared": [Jones], "Review."

54. T.D.A.C., "Review."

55. Ridgway, *Bulletin 50,* 1–2.

56. Allen, "Review," 1902.

57. J. Grinnell, "Review," 1902.

58. T.D.A.C., "Review."

59. Ridgway, *Bulletin 50,* viii.

60. [Jones], "Review."

61. Review: *Chicago Tribune,* Nov. 16, 1901, 2; Brewster to RR, Dec. 5, 1901, B-W.

62. See, e.g., Armitage, *Nature Study Movement,* 15–69, and Henson, "Through Books."

63. Wilfrid B. Alexander to Horace G. Alexander, Apr. 15, 1921, AL.

64. Opposed to inaccurate depictions of the natural world: Burroughs, "Summit of the Years," 616–617; "our nature writers are either all very dull, or else very romantic and unreliable": Burroughs to T. A. Coward, Apr. 15, 1908, AL.

65. Adams, *Humming Birds Described and Illustrated,* 3-5.

66. RR to Brewster, Mar. 6, 1911, MCZ.

67. Brewster to RR, Mar. 16, 1911, B-W.

68. RR to Brewster, Mar. 10, 1903, MCZ.

69. "Largest and most useful piece of systematic bird work": J. Grinnell, "Review," 1903; "stick to my task": RR to Brewster, Sept. 3, 1903, MCZ.

70. Ridgway explaining new species in Bulletin 50 needing publication elsewhere to claim priority: RR to Brewster, Sept. 3, 1903, MCZ; Ridgway noting he likes at least ten specimens: RR to Brewster, Aug. 8, 1905, ibid.

71. "Arduous labor involved in its preparation": RR to Brewster, Mar. 10, 1906, MCZ; "enormity of the task is well shown": "Curator Ridgway Publishes Part Four of Voluminous Work," *Washington Herald,* Aug. 3, 1907, 5; "the most thoro of all the systematic treatises . . . ever published": J. Grinnell, "Review," 1908, 53.

72. RR to True, July 25, 1910, RU 242, box 5, fol. 6, SIA.

73. RR to Brewster, Feb. 23, 1906, MCZ.

74. RR to Brewster, May 25, 1907, ibid.

75. RR to Brewster, Dec. 21, 1907, ibid.

76. Ibid.

77. Home sale: RR to Brewster, Aug. 13, 1908, MCZ; "I am so thankful": Julia Evelyn Ridgway to John L. Ridgway, Sept. 24, 1907, B-W.

78. "The greatest systematic work on birds ever published": Frank M. Chapman to RR, Apr. 17, 1914, box 1, fol. 14, USU.

79. "To glance at them . . . is pleasing": Brewster to RR, Feb. 5, 1918, B-W.

80. Influenza attack: RR to T. S. Palmer, July 6, 1920, B-W; ideas about others to take up Bulletin 50: RR to Charles Wallace Richmond, June 15, 1920, RU 105, box 25, fol. 6, SIA; asking for help on Bulletin 50: RR to Edward William Nelson, July 25, 1927, RU 7364, box 8, fol. 6, ibid.

81. Ridgway can't make meeting: Charles Greeley Abbot to RR, Apr. 5, 1921, RU 105, box 25, fol. 6, ibid.; likes to show medal to friends: RR to Joseph Harvey Riley, July 12, 1921, ibid.

82. Evelyn Ridgway's physician, treatment, and schedule: RR to Riley, Apr. 5, 1921, RU 105, box 25, fol. 6, SIA, and RR to John L. Ridgway, Nov. 6, 1925, B-W.

83. RR to John L. Ridgway, Dec. 17, 1923, B-W.

84. "Cross-word puzzle": RR to Richmond, July 11, 1925, RU 105, box 25, fol. 6, SIA; "to make a complete ornithology is impossible": John Latham to T. W. Chevalier, May 4, 1923, AL.

85. RR to Alexander Wetmore, Feb. 10, 1926, RU 7006, box 56, fol. 3, SIA; and Ridgway, "Plea for Caution" (1923).

86. Evvie's instructions for her memorial service: Evelyn Ridgway to RR, Nov. 5, 1926, box 1, fol. 19, USU; "many of our happiest and most care free days": RR to Nelson, June 23, 1927, RU 7364, box 8, fol. 6, SIA.

87. Future plans: RR to Nelson, June 23, 1927, RU 7364, box 8, fol. 6, SIA; eyesight: RR to Casey Albert Wood, Jan. 22, 1929, B-W.

88. "Recognized as standard through the world": Wetmore to RR, June 27, 1928, RU 7006, box 56, fol. 3, SIA; Ridgway's salary: RR to Wetmore, July 11, 1928, ibid. The modern equivalent of Ridgway's salary was calculated using the Consumer Price Index.

89. Herbert Friedmann responsible for the entire contents of three volumes published after Ridgway's death: Friedmann, *Birds of North and Middle America,* Pt. 11,

v, and Miller, "Review," 42; "somebody else's dead horse": Leon J. Cole to Friedmann, Jan. 13, 1947, RU 105, series 2, box 15, fol. 4, SIA.

90. Armitage, *Nature Study Movement*, 1–2; Wiebe, *Search for Order*, 111.

91. Battaglio, *Rhetoric of Science*, 11.

Chapter 7. Standardizing the Colors of Birds

Epigraph: *Los Angeles Times*, Apr. 13, 1913, III2.

1. Robert Ridgway (hereafter "RR") to Spencer Fullerton Baird, Oct. 17, 1864, RU 7167, SIA.

2. "Hand-colored plates are apt to vary": Joel Allen to RR, Dec. 14, 1888, B-W; technique to preserve bill color: RR to William Brewster, Jan. 31, 1878, MCZ; turpentine has changed the color of a skin: Eugene Bicknell to RR, Dec. 9, 1882, RU 105, series 2, box 13, fol. 5, SIA.

3. RR to Allen, May 14, 1884, BAN.

4. Brewster to RR, Nov. 12, 1882, B-W.

5. Issue of compensation: RR to Brewster, Nov. 21, 1882, MCZ; Brewster's offer to write letters of support: Brewster to RR, Nov. 22, 1882, B-W; Allen supports book: Allen to RR, Oct. 24 ("perhaps I might venture to 'feel the pulse' a little and put in a word"), Nov. 4, 10, 1884 ("Saw Mr. Bartlett (head of firm of Little, Brown & Co.) on Saturday, and left with him the Mss. and plates, which he said he would like to examine . . . I had a very pleasant interview with him, and he seemed satisfied with the assurances I gave him as to the merit of the work and the great need for such a publication"), all B-W.

6. Ridgway, *Nomenclature of Colors* (1886), [19].

7. "Much research [and] a nice display of judgement": Allen, "Review: Ridgway's Nomenclature," 153; "to amateurs the book will be especially welcome": Montague Chamberlain to RR, Jan. 25, 1887, RU 105, box 14, fol. 5, SIA.

8. William Eagle Clark to RR, Aug. 1, Oct. 5, 1887, RU 105, box 15, fol. 2, SIA.

9. Orcutt, "Wanted, a Chart of Standard Colors," and "A Chart of Standard Colors."

10. Ridgway's interactions with Langley regarding flight are quoted in Harris, "Ridgway," 63; most expensive: "The proposed project, perhaps the most expensive scientific publication the Institution has ever undertaken, is a large scheme and it has to be contemplated under its economic as well as its scientific side," Samuel Langley to Richard Rathbun, Sept. 14, 1898, RU 31, box 19, fol. 15, SIA.

11. Langley to Ogden N. Rood, Nov. 3, 1898, RU 31, box 19, fol. 15, SIA.

12. Langley to RR, Nov. 11, 1898, ibid.

13. RR to Langley, Dec. 16, 1898, ibid.

14. RR to Langley, Nov. 17, 1898, and Langley to RR, Dec. 14, 1898, ibid.

15. Langley cautiously concurs: "If [the Institution] is going to co-operate at all, the only course open seems to be to pay the entire cost." Frederick William True to Langley, Nov. 29, 1898, ibid.; cost of Bendire's work on birds as a point of comparison: A. Howard Clark to Rathbun, Dec. 1, 1898, ibid.

16. Bradley to Langley, Dec. 13, 1898, ibid.

17. A. B. Meyer to Langley, Sept. 19, 1899, ibid.

18. Langley to John La Farge, Nov. 25, 1899, ibid.

19. Robert Coleman Child to Charles F. Adams, Feb. 8, 1902, ibid., fol. 16.

20. Three-person committee suggested: Rathbun to Langley, Dec. 13, 1899, ibid.,

fol. 15; committee members suggested, and $100 approved: Langley to Rathbun, Dec. 4, 1899, ibid.; committee discussed: Langley to Rathbun, Jan. 10, 1902, ibid., fol. 16; purchasing books: Langley to Rathbun, Dec. 4, 1899, ibid., fol. 15.

21. Ridgway and Rathbun make a plan for supporting the color work: RR to Rathbun, Nov. 24, 1899, and Rathbun to Langley, Dec. 13, 1899, both RU 31, box 19, fol. 15, SIA; approving purchase of colors: Rathbun to Langley Oct. 18, 1899, ibid.; sending half-pans of colors to Ridgway: Fred A. Schmidt to Color Committee, Mar. 17, 28, 30, 1900, ibid., fol. 16; switches to watercolors: RR to Rathbun, Apr. 2, 1900, ibid.

22. RR to Rathbun, Oct. 29, 1900, RU 31, box 19, fol. 16, SIA.

23. Color Committee members to Langley, Jan. 7, 1901, ibid.

24. RR to Holmes, Dec. 28, 1900, ibid.

25. Color Committee members to Langley, Jan. 7, 1901, ibid.

26. Plan for color book "quite good": Ogden Rood to Langley, Jan. 18, 1901, ibid.; "no great essay written on color": Langley to RR, no day noted but approximately Feb. 8, 1901, ibid.

27. RR to Langley, Feb. 11, 1901, ibid.

28. The end of Ridgway's involvement with the work as a Smithsonian project: RR to Rathbun, May 6, 1901, ibid.; Langley to Rathbun, Jan. 10, 1902 (citing his letter of May 13, 1901, formally releasing Ridgway from the obligation), ibid.; Langley to Rathbun, Mar. 4, 1902, ibid.

29. RR to Rathbun, Jan. 17, 1903, ibid.

30. Furlough from Smithsonian: RR to Brewster, Nov. 29, 1906, MCZ; "I am surely 'up a stump' on the color work": RR to Edward William Nelson, June 13, 1907, RU 7364, box 8, fol. 6, SIA.

31. George W. Vinal to John Ridgway, Mar. 8, 1909, B-W.

32. José Castulo Zeledón to RR, April 4, 1909, B-W; and RR to John L. Ridgway, August 26, 1918, B-W, recounting the history of the loans from Zeledón. Also see Ridgway's obituary of Zeledón: "In Memoriam: José Castulo Zeledón" (1923).

33. Last, *Color Explosion*, 104.

34. Ridgway to Henry Holt, Aug. 3, 1909, B-W.

35. RR to John L. Ridgway, May 31, 1909, ibid.

36. RR to John L. Ridgway, Sept. 9, 1909; Holt to RR, June 4, Aug. 10, Sept. 6, 1909, ibid.

37. "Here is an opportunity": J. Grinnell, "Editor Notes and News." Grinnell would later claim that he knew nothing of Ridgway's planned book before penning this article. "I previously *knew nothing at all about [it]*, strange as it may seem. After all, the Pacific Coast is so remote from the Atlantic, that many pieces of news fail to reach us at all," Grinnell to RR, Sept. 12, 1909, B-W. I don't believe Grinnell's protestations, however, given the specificity of his comments. "I think it is a *better* book than Ridgway's": F. Grinnell Jr., "Correspondence."

38. Ridgway, [Letter to editor announcing particulars in connection with the publication of *Color Standards and Color Nomenclature*] (1909).

39. Reighard, "Code of Colors for Naturalists."

40. "I will immediately send a note": RR to John L. Ridgway, Sept. 9, 1909, B-W.

41. Ridgway's pending color book described in Ricker, "New Color Guide." "After a consultation with Dr. Ridgway, and later with the other members of the committee," noted Ricker, "it was decided to leave the field of color work in favor of Dr. Ridgway."

42. Vanderpoel, *Color Problems*, vii.

43. "Your task is a stupendous one": Brewster to RR, Jan. 15, 1912, B-W; "most

important contribution": Oberholser, "Review—Color Standards and Color Nomenclature," 132; "it is a volume that has been looked for": "Dictionary of Colors."

44. RR to Brewster, Apr. 1, 1912, MCZ.

45. Stone, "Review: Ridgway's 'Color Standards and Color Nomenclature.'"

46. Spillman, "Review." Another brief but accurate description of Ridgway's setup is "Dictionary of Colors."

47. RR to Brewster, May 21, 1913, MCZ.

48. Ridgways vacate apartment June 1 and will move to Olney: RR to Brewster, May 21, 1913, ibid.; Ridgway goes into great detail about his newly purchased Olney property, as well as general costs of real estate in Olney: RR to Brewster, July 5, 1906, ibid.

49. RR to Brewster, Oct. 28, 1913, ibid.

50. Frank M. Chapman to RR, Apr. 17, 1914, box 1, fol. 14, USU.

51. Entire edition long exhausted: Herbert Friedmann to Orlando Zeledón Castro, June 25, 1953, RU 105, box 14, SIA; "my copy of your book is used continually": R. H. Howland to Ridgway, Mar. 4, 1921, box 1, fol. 15, USU; Howland offers his firm's services if new edition ever published: R. H. Howland to RR, Mar. 12, 1921, ibid.

52. José Castulo Zeledón to RR, July 5, 1918, B-W.

53. "Difficulties I may never be able to surmount": RR to John L. Ridgway, Aug. 26, 1918, B-W; Ridgway has no business sense: Alexander Wetmore to Harry Harris, Dec. 14, 1926, RU 7006, box 28, fol. 2, SIA; math incomprehensible to Ridgway: RR to Charles Wallace Richmond, Feb. 25, 1926, RU 105, box 25, fol. 6, ibid.

54. RR to Casey Albert Wood, Sept. 28, 1927, B-W.

55. Stanley B. Ashbrook to Charles W. Richmond, Jan. 4, 1926, RU 105, box 12, fol. 3, SIA; Wallace to Ashbrook, January 19, 1926, ibid.

56. Davis, "Ridgway's 'Color Standards.'"

57. Lankford, "Amateurs versus Professionals," 20. See also Rothenberg, "Organization and Control."

Epilogue

Epigraph: John Latham to T. W. Chevalier, May 4, 1923, AL.

1. *Olney (IL) Mail*, Apr. 1, 1929, 1.

2. "Objection to publicity": Schantz, "Robert Ridgway"; Ridgway most widely known American ornithologist: *Ibis*, 12th ser., 5, no. 3 (1929): 526.

3. Chapman, "Robert Ridgway," 129.

4. Griscom, "Role of the Amateur."

5. Mayfield, "Commentary."

6. In particular, see Barrow, *Passion for Birds*, in the section "Graduate Training for Biology," 184–190, as well as a view from 1933 by Arthur A. Allen, "Ornithological Education in America," 215–229, in Sterling, *Contributions*.

7. Positions as noted on the museum's letterhead: Frank Chapman to Herbert Friedmann, Dec. 8, 1932, RU 105, box 15, fol. 1, SIA.

8. Closet ornithologist will never rise to highest point: Stejneger, "Letter to the Editor." However, note Robert Kohler's challenge to the widespread assumption that systematics suffered a widespread decline in the twentieth century (Kohler, *All Creatures*).

9. Griscom, "Role of the Amateur," 19.

Bibliography

Adams, Henry Gardiner. *Humming Birds Described and Illustrated*. London: Groombridge and Sons, 1872.

Adams, J. F. A. "Is Botany a Suitable Study for Young Men?" *Science* 9, no. 209 (1887): 116–117.

Ainley, Marianne. "From Natural History to Avian Biology: Canadian Ornithology, 1860–1950." PhD diss., McGill University, 1985.

Allen, Joel. "A.O.U. Check-List—Its History and Its Future." *Auk* 20, no. 1 (1903): 1–9.

———. *Autobiographical Notes and a Bibliography of Scientific Publications*. New York: American Museum of Natural History, 1916.

———. "The Availability of Certain Bartramian Names in Ornithology." *American Naturalist* 10, no. 1 (1876): 21–29.

———. "Correspondence." *Auk* 1, no. 3 (1884): 303–304.

———. "Notes and News." *Auk* 1, no. 1 (1884): 105–106.

———. "On the Mammals and Winter Birds of East Florida, with an Examination of Certain Assumed Specific Characters in Birds, and a Sketch of the Bird-Faunae of Eastern North America." *Bulletin of the Museum of Comparative Zoology* 2, no. 3 (1871): 161–451.

———. "Recent Literature." *Auk* 4, no. 4 (1887): 334.

———. "Review: Ridgway's 'Birds of North and Middle America' Part 1." *Auk* 19, no. 1 (1902): 97–102.

———. "Review: Ridgway's Nomenclature of Colors and Ornithologists' Compendium." *Auk* 4, no. 2 (1887): 153.

Allman, George James. "The Present Aspects of Biology and the Method of Biological Study." *American Naturalist* 8, no. 1 (1874): 34–43.

American Ornithologists' Union. *Code of Nomenclature and Check-List of North American Birds*. New York: American Ornithologists' Union, 1886.

"The American Ornithologists' Union." *Science* 4, no. 89 (1884): 374–376.

Annual Report of the Curator of the Museum of Comparative Zoology at Harvard College, to the President and Fellows of Harvard College, for 1889–90. Cambridge: University Press: John Wilson and Son, 1890.

"An Appeal to the Women of the Country in Behalf of the Birds." *Science* 7, no. 160 (1886): 204–205.

Armitage, Kevin C. *The Nature Study Movement: The Forgotten Popularizer of America's Conservation Ethic.* Lawrence: University Press of Kansas, 2009.

Badè, William Frederic. *Life and Letters of John Muir.* Vol. 2. Boston: Houghton Mifflin, 1924.

Baird, Spencer Fullerton. "The Distribution and Migrations of North American Birds." *American Journal of Science and Arts,* 2nd ser., 31 (1866): 78–90, 184–192, 337–347.

Banks, Richard C., et al. "Type Specimens and Basic Principles of Avian Taxonomy." *Auk* 110, no. 2 (1993): 413–414.

Barrow, Mark V., Jr. *Nature's Ghosts: Confronting Extinction from the Age of Jefferson to the Age of Ecology.* Chicago: University of Chicago Press, 2009.

———. *A Passion for Birds: American Ornithology after Audubon.* Princeton, NJ: Princeton University Press, 1998.

———. "The Specimen Dealer: Entrepreneurial Natural History in America's Gilded Age." *Journal of the History of Biology* 33 (2000): 493–534.

Bartlett, Richard. *Great Surveys of the American West.* Norman: University of Oklahoma Press, 1962.

Bartram, William. "Descriptive Catalogue of the Birds of the Eastern United States." *Travels through North and South Carolina, Georgia, East and West Florida.* Philadelphia: James and Johnson, 1791.

Batchelder, Charles Foster. *An Account of the Nuttall Ornithological Club, 1873 to 1919.* Cambridge, MA: Nuttall Ornithological Club, 1937.

Bates, Ralph S. *Scientific Societies in the United States.* Cambridge, MA: MIT Press, 1965.

Battaglio, John T. *The Rhetoric of Science in American Ornithological Discourse.* Stamford, CT: Ablex, 1998.

Becker, George F. "Biographical Notice of Samuel Franklin Emmons." *Transactions of the American Institute of Mining Engineers* 42 (1912): 645.

Blum, Ann Shelby. *Picturing Nature: American Nineteenth-Century Zoological Illustration.* Princeton, NJ: Princeton University Press, 1993.

Bowker, Geoffrey C., and Susan Leigh Star. *Sorting Things Out: Classification and Its Consequences.* Cambridge, MA: MIT Press, 1999.

Brande, W. T., and George W. Cox, eds. *A Dictionary of Science, Literature, and Art.* London: Longmans, Green, 1867.

Brewer, Thomas M. "The European House-Sparrow." *Atlantic Monthly* 21, no. 127 (1868): 583–588.

Brinkel, J., M. M. H. Khan, and A. Kraemer. "A Systematic Review of Arsenic Exposure and Its Social and Mental Health Effects with Special Reference to Bangladesh." *International Journal of Environmental Research and Public Health* 6, no. 5 (2009): 1609–1619.

Browne, Janet. *Charles Darwin: The Power of Place.* New York: Knopf, 2002.

Bruce, Murray. "A Brief History of Classifying Birds." In *The Handbook of the Birds of the World,* edited by Andrew Elliott and David Christie, vol. 8:1–43. Barcelona: Lynx, 2003.

Bruce, Robert V. *The Launching of Modern American Science, 1846–1876.* Ithaca, NY: Cornell University Press, 1988.

Burroughs, John. "The Summit of the Years." *Atlantic Monthly* 109 (1912): 616–617.

———. *The Writings of John Burroughs.* 15 vols. Boston: Houghton Mifflin, 1904–1909.

Carpenter, Frederic H. "American Ornithologists' Union Code of Nomenclature and Checklist of N. A. Birds." *Ornithologist and Oologist* 11, no. 5 (1886): 80.

Century Association, King Memorial Committee. *Clarence King Memoirs: The Helmet of Mambrino.* New York: G. P. Putnam's Sons, 1904.

Chapman, Frank M. *Autobiography of a Bird-Lover.* New York: D. Appleton-Century, 1933.

———. "In Memoriam: Joel Asaph Allen." *Auk* 39, no. 1 (1922): 1–14.

———. "Robert Ridgway, 1850–1929." *Bird-Lore* 31, no. 3 (1929): 173–178.

Cherrie, George K. *Dark Trails: Adventures of a Naturalist.* New York: G. P. Putnam's Sons, 1930.

Chowder, Ken. "North to Alaska." *Smithsonian Magazine* 34, no. 3 (2003): 92–101.

Cittadino, Eugene. "Ecology and the Professionalization of Botany in America, 1890–1905." *Studies in the History of Biology* 4 (1980): 171–198.

Coates, Peter. "Eastenders Go West: English Sparrows, Immigrants, and the Nature of Fear." *Journal of American Studies* 39, no. 3 (2005): 431–462.

Cochrane, Rexmond C. *The National Academy of Sciences: The First Hundred Years, 1863–1963.* Washington, DC: National Academy of Sciences, 1978.

Coues, Elliott. "Color-Variation in Birds Dependent upon Climatic Influences." *American Naturalist* 7, no. 7 (1873): 415–418.

———. "Fasti Ornithologiae Redevivi: No. I. Bartram's Travels." *Proceedings of the Academy of Natural Sciences of Philadelphia* 27, no. 2 (1875): 338–358.

———. *Field Ornithology.* Salem, MA: Naturalists' Agency, 1874.

———. "The Ineligibility of the European House Sparrow in America." *American Naturalist* 12, no. 8 (1878): 499.

———. "Ornithophilologicalities." *Auk* 1, no. 1 (1884): 49–58.

———. "Progress of American Ornithology." *American Naturalist* 5, no. 6 (1871): 364–373.

———. "Reply to Mr. J. A. Allen's 'Availability of Certain Bartramian Names in Ornithology.'" *American Naturalist* 10, no. 2 (1876): 98–102.

Currie, Rolla P. "A New Bird of Paradise." *Proceedings of the U.S. National Museum*, 22. Washington, DC: Government Printing Office, 1900.

Cutright, Paul R., and Michael J. Brodhead. *Elliott Coues: Naturalist and Frontier Historian.* 2001; orig. publ. Urbana: University of Illinois Press, 1981.

Darwin, Charles. *The Descent of Man, and Selection in Relation to Sex.* London: John Murray, 1874.

Davenport, F. Garvin. "Robert Ridgway: Illinois Naturalist." *Journal of the Illinois State Historical Society* 63, no. 3 (1970): 271–289.

Davis, Elisabeth B. "Ridgway's 'Color Standards': A Classic in Biology." *Bios* 56, no. 3 (1985): 143–152.

Desmond, Adrian. "Redefining the X Axis: 'Professionals,' 'Amateurs,' and the Making of Mid-Victorian Biology—A Progress Report." *Journal of the History of Biology* 34, no. 1 (2001): 3–50.

"Dictionary of Colors." *Los Angeles Times,* Apr. 13, 1913, III2.

Donald, Paul F. "Notices of Recent Ornithological Publications." *Ibis* 26, no. 2 (1884): 203–204.

Doughty, Robin. "The English Sparrow in the American Landscape: A Paradox in Nineteenth Century Wildlife Conservation." Oxford University, School of Geography, Research Papers 19. Oxford: Oxford Publishing, 1978.

Dyche, L. L. "Science for a Livelihood." *Science* 8, no. 191 (1886): 291–304.

Ewing, Heather. *The Lost World of James Smithson: Science, Revolution, and the Birth of the Smithsonian.* New York: Bloomsbury, 2007.

Farber, Paul. *Discovering Birds: The Emergence of Ornithology as a Scientific Discipline, 1760–1850.* Baltimore: John Hopkins University Press, 1982.

Field, Cynthia R., Richard E. Stamm, and Heather P. Ewing. *The Castle: An Illustrated History of the Smithsonian Building.* Washington, DC: Smithsonian Institution Press, 1993.

Fine, Gary Alan, and Lazaros Christoforides. "Dirty Birds, Filthy Immigrants, and the English Sparrow War: Metaphorical Linkage in Constructing Social Problems." *Symbolic Interaction* 14 (1991): 375–391.

Flack, J. Kirkpatrick. *Desideratum in Washington: The Intellectual Community in the Capital City, 1870–1900.* Cambridge, MA: Schenkman, 1975.

Flint, William G. Letter to editor. *Ornithologist and Oologist* 9, no. 2 (1884): 18.

Fouillee, Alfred. "The Hegemony of Science and Philosophy." *International Journal of Ethics* 6, no. 2 (1896): 137–164.

Freeman, Richard B. *The Works of Charles Darwin: An Annotated Bibliographical Handlist.* Hamden, CT: Archon Books, 1977.

Friedmann, Herbert. *The Birds of North and Middle America,* Parts 9–11. Washington, DC: Smithsonian Institution, 1941–1950. See Appendix for bibliographic details of each of the eight parts written by Ridgway, 1901–1919.

Gates, Barbara T., and Ann B. Shteir, eds. *Natural Eloquence: Women Reinscribe Science.* Madison, WI: University of Wisconsin Press, 1997.

Gill, Frank B., Ronald A. Canterbury, and John L. Confer. "Blue-Winged Warbler (*Vermivora pinus*)." *Birds of North America Online,* edited by A. Poole. Ithaca: Cornell Lab of Ornithology, 2001. doi:10.2173/bna.584.

Goetzmann, William H. *Exploration and Empire: The Explorer and the Scientist in the Winning of the American West.* New York: Alfred A. Knopf, 1966.

———. *New Lands, New Men: America and the Second Great Age of Discovery.* New York: Viking, 1986.

———. "Paradigm Lost." In *The Sciences in the American Context: New Perspectives,* edited by Nathan Reingold. Washington, DC: Smithsonian Institution, 1979.

Goldstein, Daniel. "'Yours for Science': The Smithsonian Institution's Correspondents and the Shape of Scientific Community in Nineteenth-Century America." *Isis* 85 (1994): 573–599.

Goode, G. Brown. "The Museums of the Future." *Annual Report of the United States National Museum: Year Ending June 30, 1897.* Washington, DC: Government Printing Office, 1901.

Grinnell F[ordyce], Jr. "Correspondence." *Condor* 11, no. 5 (1909): 177.

Grinnell, Joseph. "Editor Notes and News." *Condor* 11, no. 4 (1909): 141.

———. "Review." *Condor* 4, no. 1 (1902): 22–23.

———. "Review." *Condor* 5, no. 1 (1903): 22–23.

———. "Review." *Condor* 10, no. 1 (1908): 53.

Griscom, Ludlow. "Notes on the Collecting Trip of M. Abbott Frazar in Sonora and Chihuahua for William Brewster." *Auk* 50, no. 1 (1933): 54–58.

———. "The Role of the Amateur." *Bulletin of the Northeastern Bird-Banding Association* 5, no. 1 (1929): 16–20.

Guzy, Michael J., and Gary Ritchison. "Common Yellowthroat." In *The Birds of North America: Life Histories for the Twenty-First Century,* edited by A. Poole and Frank Gill, no. 448. Philadelphia: Academy of Natural Sciences of Philadelphia and American Ornithologists' Union, 1999.

Haffer, Jürgen. "The History of Species Concepts and Species Limits in Ornithology." *Bulletin of the British Ornithological Club Centenary Supplement 112A* (1992): 107–158.

Harding, Sandra. *Whose Science? Whose Knowledge? Thinking from Women's Lives.* Ithaca, NY: Cornell University Press, 1991.

Harris, Harry. "Robert Ridgway: With a Bibliography of His Published Writings and Fifty Illustrations." *Condor* 30, no. 1 (1928): 4–118.

Hasbrouck, E. M. "Evolution and Dichromatism in the Genus *Megascops.*" *American Naturalist* 27, no. 319 (1893): 638–639.

Hazard, R.G. "A Lay View of 'Ornithophilologicalities.'" *Auk* 1, no. 3 (1884): 300–302.

Hellman, Geoffrey T. "Profiles: The Enigmatic Bequest." *New Yorker,* Dec. 10, 1966.

Henson, Pamela M. "Spencer Baird's Dream: A U.S. National Museum." In *Cultures and Institutions of Natural History: Essays in the History and Philosophy of Science.* San Francisco: California Academy of Sciences, 2000.

———. "'Through Books to Nature': Anna Botsford Comstock and the Nature Study Movement." In Gates and Shteir, *Natural Eloquence,* 116–143.

Hrubant, H. Everett. "An Analysis of the Color Phases of the Eastern Screech Owl, *Otus asio,* by the Gene Frequency Method." *American Naturalist* 89, no. 847 (1955): 223–230.

"Ithuriel." *London: A Weekly Journal of the Time,* Aug. 30, 1890.

Johnson, Kristin. "*The Ibis,* Journal of the British Ornithologists' Union: A Pre-Synthesis Portrait." MA thesis, Oregon State University, 2000.

[Jones, Lynds.] "Review: Birds of North and Middle America." *Wilson Bulletin* 14, no. 1 (1902): 34–36.

Kelsey, Robin Earle. *Archive Style: Photographs and Illustrations for U.S. Surveys, 1850–1890.* Berkeley: University of California Press, 2007.

Ketner, Keith B., and Donna C. Ketner. "Comparisons of Avian Populations at Six Localities in Northern Nevada and Utah: 1867–1869 and 1994–2000." *Great Basin Birds* 5, no. 1 (2002): 6–12.

Kirsch, Scott. "The Allison Commission and the National Map: Towards a Republic of Knowledge in Late Nineteenth-Century America." *Journal of Historical Geography* 36, no. 1 (2010): 29–42.

Kleinschmidt, Otto. "Was ist die Subspecies?" *Ornithologische Monatsberichte* 5 (1897): 74–76.

Klincksieck, Paul, and Th. Valette. *Code des couleurs, à l'usage des naturalistes, artistes, commerçants et industriels, 720 échantillons de couleurs classes d'après la méthode Chevreul simplifée.* Paris: n.p., 1908.

Kohler, Robert E. *All Creatures: Naturalists, Collectors, and Biodiversity, 1850–1950.* Princeton, NJ: Princeton University Press, 2006.

———. "Subspecies Classification and Biological Survey, 1850s–1930s." In *Spaces of Science,* edited by Ursula Klein, 31–48. Berlin: Max-Planck-Institut für Wissenschaftsgeschichte, 2003.

Lankford, John. "Amateurs versus Professionals: The Controversy over Telescope Size in Late Victorian Science." *Isis* 72 (1981): 11–28.

Last, Jay T. *The Color Explosion: Nineteenth-Century American Lithography.* Santa Ana, CA: Hillcrest, 2005.

Leighton, Ann. *American Gardens of the Nineteenth Century.* Amherst: University of Massachusetts Press, 1987.

Lucier, Paul. "The Professional and the Scientist in Nineteenth-Century America." *Isis* 100 (2009): 699–732.

Maslin, Janet. "Sometimes One Man Can Have Two Lives." *New York Times,* Feb. 4, 2009.

Mayfield, Harold F. "Commentary: The Amateur in Ornithology." *Auk* 96, no. 1 (1979): 168–171.

Maynard, Charles Johnson. *Manual of Taxidermy: A Complete Guide in Collecting and Preserving Birds and Mammals*. Boston: S. E. Cassino, 1883.

Mayr, Ernst. "Ridgway, Robert." *Complete Dictionary of Scientific Biography*, vol. 11. New York: Charles Scribner's Sons, 1981.

McKitrick, Mary C., and Robert M. Zink. "Species Concepts in Ornithology." *Condor* 90, no. 1 (1988): 1–14.

McOuat, Gordon. "Cataloging Power: Delineating 'Competent Naturalists' and the Meaning of Species in the British Museum." *British Journal for the History of Science* 34, no. 1 (2001): 1–28.

Mearns, Barbara, and Richard Mearns. *The Bird Collectors*. San Diego, CA: Academic, 1998.

Meinertzhagen, Richard. "A Preliminary Study of the Relation between Geographical Distribution and Migration with Special Reference to the Palaearctic Region." *Ibis* 61 (1919): 382.

Menand, Louis. *The Metaphysical Club*. New York: Farrar, Straus and Giroux, 2001.

Merriam, Augustus C. "The Coues Lexicon of North American Birds." *Auk* 1, no. 1 (1884): 36–49.

Merriam, Clinton Hart. "Bats of Queen Charlotte Islands, British Columbia." *American Naturalist* 29, no. 345 (1895): 860–861.

Meyers, George S. "Foreword." In *Genera of Fishes and Classification of Fishes*, David Starr Jordan. Palo Alto, CA: Stanford University Press, 1968.

Miller, Alden H. "Review." *Condor* 44, no. 1 (1942): 42.

Morton, William E., and Gordon A. Caron. "Encephalopathy: An Uncommon Manifestation of Workplace Arsenic Poisoning?" *American Journal of Industrial Medicine* 15, no. 1 (1989): 1–5.

Mott, Frank L. *A History of American Magazines*. 5 vols. Cambridge, MA: Harvard University Press, 1930–1968.

"Notes." *American Naturalist* 7, no. 12 (1873): 760–761.

Oberholser, Harry C. "Review—Color Standards and Color Nomenclature." *Condor* 15, no. 3 (1913): 132.

Oesler, Paul H. *The Smithsonian Institution*. New York: Praeger, 1970.

"On the Descent of Man." *British Medical Journal*, April 15, 1871, 401–402.

Orcutt, Charles Russell. "A Chart of Standard Colors." *Garden and Forest*, January 8, 1890, 22–23.

———. "Wanted, a Chart of Standard Colors." *Garden and Forest*, December 25, 1889, 622–623.

Packard, Alpheus S. Jr., et al., eds. "Introductory." *American Naturalist* 1, no. 1 (1867): 1–4.

R.C.M. "'Amateur.'" *Ornithologist and Oologist* 11, no. 8 (1886): 128.

Rees, Abraham, ed. *The Cyclopaedia: or, Universal Dictionary of Arts, Sciences and Literature*, vol. 3. London: Longman, Hurst, Rees, Orne and Brown [etc.], 1819–20.

Reese, Kenneth M., ed. *A Century of Chemistry: The Role of Chemists and the American Chemical Society.* Washington, DC: American Chemical Society, 1976.

Reighard, Jacob. "A Code of Colors for Naturalists." *American Naturalist* 42, no. 500 (1908): 566–568.

Reingold, Nathan. "Definitions and Speculations: The Professionalization of Science in America in the Nineteenth Century." In *The Pursuit of Knowledge in the Early American Republic: American Scientific and Learned Societies from Colonial Times to the Civil War,* 33–69. Baltimore: Johns Hopkins University Press, 1976.

Remsen, J. V., Jr. "The Importance of Continued Collecting of Bird Specimens to Ornithology and Bird Conservation." *Bird Conservation International* 5 (1995): 145–180.

Richmond, Charles Wallace. "Notes and News." *Auk* 18, no. 2 (1901): 221–222.

Ricker, P. L. "A New Color Guide." *Mycologia* 2, no. 37 (1910): 37–38.

Ritterbush, Philip C. "Biology and the Smithsonian Institution." *BioScience* 17, no. 1 (1967): 25–35.

Rivinus, E. F., and E. M. Youssef. *Spencer Baird of the Smithsonian.* Washington, DC: Smithsonian Institution, 1992.

Rossiter, Margaret W. *Women Scientists in America: Struggles and Strategies to 1940.* Baltimore: Johns Hopkins University Press, 1982.

Rothenberg, Marc. "Organization and Control: Professionals and Amateurs in American Astronomy, 1899–1918." *Social Studies of Science* 11, no. 3 (1981): 305–325.

Rudolph, Emanuel D. "Women in Nineteenth Century American Botany: A Generally Unrecognized Constituency." *American Journal of Botany* 69 (1982): 1346–1355.

Rydell, Robert W. *All the World's a Fair: Visions of Empire at American International Expositions, 1876–1916.* Chicago: University of Chicago Press, 1984.

Sandweiss, Martha A. *Passing Strange: A Gilded Age Tale of Love and Deception across the Color Line.* New York: Penguin, 2009.

Schantz, Orpheus Moyer. "Robert Ridgway: Ornithologist, Botanist, Gentleman: 1850–1929." *Wilson Bulletin* (December 1929): 253–255.

Schivelbusch, Wolfgang. *The Railway Journey: The Industrialization of Time and Space in the Nineteenth Century.* Berkeley: University of California Press, 1986.

Scudder, Samuel Hubbard. "Canons of Systematic Nomenclature for the Higher Groups." *American Journal of Science and Arts,* 3rd ser., 3 (1872): 345–351.

———. "Historical Sketch of the Generic Names Proposed for Butterflies: A Contribution to Systematic Nomenclature." *Proceedings of the American Academy of Arts and Sciences* 10 (1874–1875): 91–293.

Seebohm, Henry. "Remarks on Certain Points in Ornithology Nomenclature." *Ibis* 21, no. 4 (1879): 428–437.

Smithsonian Institution. *Annual Report for the Year Ending June 30, 1889.* Washington, DC: Government Printing Office, 1890.

———. *Annual Report of the Board of Regents of the Smithsonian Institution, to 1895.* Washington DC: Government Printing Office, 1896.

———. *Annual Report of the Board of Regents of the Smithsonian Institution for the Year 1874.* Washington, DC: Government Printing Office, 1875.

Société Française des Chrysanthémistes. *Répertoire de couleurs pour aider à la détermination des couleurs des fleurs, des feuillages et des fruits.* Paris: Librairie Horticole, 1905.

Spillman, William Jasper. "Review." *Science,* n.s., 37, no. 965 (1913): 985–989.

Stejneger, Leonhard. "Letter to the Editor." *Condor* 7, no. 3 (1905): 65.

Sterling, Keir, ed. *Contributions to the History of American Ornithology.* New York: Arno, 1974.

Sterling, Keir, and Marianne Ainley. Unpublished manuscript history of the American Ornithologists' Union.

Stimson, Dorothy. *Scientists and Amateurs: A History of the Royal Society.* New York: H. Schuman, 1948.

Stone, Witmer. "In Memoriam: Charles Wallace Richmond, 1868–1932." *Auk* 50, no. 1 (1933): 1–22.

———. "In Memoriam: John Hall Sage." *Auk* 43, no. 9 (1926): 1–17.

———. "Review: Ridgway's 'Color Standards and Color Nomenclature.'" *Auk* 30, no. 3 (1913): 440.

Stresemann, Erwin. *Ornithology from Aristotle to the Present.* Cambridge, MA: Harvard University Press, 1975.

T.D.A.C. "Review: North American Finches." *American Naturalist* 36, no. 424 (1902): 333–336.

Thomas, Phillip Drennon. "Nuttall, Thomas." *Complete Dictionary of Scientific Biography.* New York: Charles Scribner's Sons, 1974.

Turner, Granville. Umbrella. US Patent 787,237, filed May 9, 1904, issued April 11, 1905.

US Bureau of Census. *Eighth Population Census of the United States, Illinois, Wabash County.* Washington, DC: Government Printing Office, 1860.

———. *Seventh Population Census of the United States, Illinois, Wabash County.* Washington, DC: Government Printing Office, 1850.

Vanderpoel, Emily N. *Color Problems: A Practical Manual for the Lay Student of Color.* New York: Longmans, Green, 1902.

Vellai, Tibor, et al. "Genome Economization and a New Approach to the Species Concept in Bacteria." *Royal Society Proceedings: Biological Sciences* 266 (1999): 1953–1958.

Wade, Joseph M. "Gastro-oological." *Ornithologist and Oologist* 9, no. 5 (May 1884): 57.

Warhurst, B. W. *A Colour Dictionary: Giving about Two Hundred Names of Colours Used in Printing, &c., Specially Prepared for Stamp Collectors.* London: Stanley Gibbons, 1899.

"A Washingtonian Succumbs to Acute Pneumonia at Chicago." *Washington Times*, Feb. 24, 1901, 7.

Waterton, Charles. *Wanderings in South America*. New York: Macmillan, 1889.

Weiner, Jonathan. *The Beak of the Finch: A Story of Evolution in Our Time*. New York: Knopf, 1994.

Wetmore, Alexander. "Biographical Memoir of Robert Ridgway, 1850–1929." *National Academy of Sciences, Biographical Memoirs* 15 (1931): 57–101.

Wiebe, Robert. *The Search for Order, 1877–1920*. New York: Hill and Wang, 1967.

Wilkins, Thurman. *Clarence King: A Biography*. 2nd ed. New Mexico: University of New Mexico Press, 1988.

Williams, David M. "Otto Kleinschmidt (1870–1954), Biogeography and the 'Origin' of Species: From Formenkreis to Progression Rule." *Biogeografía* 1 (January 2007): 3–9.

Wilson, Robert. *The Explorer King: Adventure, Science and the Great Diamond Hoax: Clarence King in the Old West*. New York: Scribner, 2006.

Winkler, Kevin. "Obtaining, Preserving, and Preparing Bird Specimens." *Journal of Field Ornithology* 71, no. 2 (2000): 250–297.

Wright, Chauncey. *North American Review* 113, no. 232 (1871): 63–103.

Yoshihisa, Fujino. "Mental Health Burden amongst Inhabitants of an Arsenic-Affected Area in Inner Mongolia, China." *Social Science Medicine* 59, no. 9 (2004): 1969–1973.

Yoon, Carol Kaesuk. *Naming Nature: The Clash between Instinct and Science*. New York: W. W. Norton, 2009.

Zimmer, John Todd. *Catalogue of the Edward E. Ayer Ornithological Library*. Chicago: Field Museum Press, 1926.

Index

Note: Italic page numbers refer to illustrations. RR refers to Robert Ridgway.